Life Cycle Assessment of Wastewater Treatment

T0199237

Life Cycle Assessment and Green Chemistry Series

Series Editor: Vera M. Kolb
Department of Chemistry,
University of Wisconsin-Parkside, USA

Life Cycle Assessment of Wastewater Treatment

Edited by Mu. Naushad

Life Cycle Assessment of Wastewater Treatment

Edited by
Mu. Naushad

CRC Press
Taylor & Francis Group
Boca Raton London New York

CRC Press is an imprint of the
Taylor & Francis Group, an **informa** business

CRC Press
Taylor & Francis Group
6000 Broken Sound Parkway NW, Suite 300
Boca Raton, FL 33487-2742

First issued in paperback 2022

ISBN-13: 978-1-138-05517-9 (hbk)
ISBN-13: 978-1-03-233920-7 (pbk)
DOI: 10.1201/9781315165820

**Visit the Taylor & Francis Web site at
http://www.taylorandfrancis.com**

**and the CRC Press Web site at
http://www.crcpress.com**

Contents

Preface

Water pollution has become a serious issue in the present scenario, and the whole world is facing a water quality crisis resulting from continuous population growth, industrialization, urbanization, food production practices, increased living standards, unsustainable water use practices, and wastewater management strategies. The United Nations predicts that in the coming decade, half of the countries worldwide will be confronted with water stress if not outright shortages. *Life Cycle Assessment of Wastewater Treatment* is a worthwhile addition to the field of wastewater treatment. This book is intended to explore the recent developments in the field of wastewater treatment using a life cycle assessment approach. Life cycle assessment can be used as a tool to evaluate the environmental impacts associated with wastewater treatment and improvement options. Wastewater treatment is the process of converting wastewater—water that is no longer needed or suitable for its most recent use—into an effluent that can be returned to the water cycle with minimal environmental issues. This book is the result of remarkable contributions from experts in interdisciplinary fields of science with comprehensive, in-depth, and up-to-date research and reviews.

Series Editor

Vera M. Kolb received a Chemical Engineering degree and an MS in Chemistry from Belgrade University. She pursued her chemistry studies as a Fulbright scholar, and has received her PhD from Southern Illinois University at Carbondale. She is currently Professor of Chemistry and Director of the Center for Environmental Studies at the University of Wisconsin-Parkside. She had two sabbatical leaves, at the Salk Institute for Biological Studies, and at Northwestern University. As of 2016, she has over 150 publications, including three books and three patents, in the fields such as organic reactions mechanisms, chemistry of opiates and estrogens, organic reactions in water and in the solid state, chemistry under prebiotic conditions, astrobiology, green chemistry, and chemical education. In 2002 she was inducted in the Southeastern Wisconsin Educators' Hall of Fame, and in 2013 she has received Phi Delta Kappa Outstanding Educator Award.

Editor

Dr. Mu. Naushad is currently working as an associate professor in the Department of Chemistry, College of Science, King Saud University, Riyadh, Kingdom of Saudi Arabia. He obtained his PhD in analytical chemistry from Aligarh Muslim University, Aligarh, India in 2007. He has vast research experience in the multidisciplinary fields of analytical chemistry, materials chemistry, and more specifically, environmental science. He holds 5 US patents and has more than 200 publications in international journals of repute, 15 book chapters, and 4 books published by renowned international publishers. He has successfully run several research projects funded by the National Plan for Science and Technology and King Abdulaziz City for Science and Technology, Kingdom of Saudi Arabia. He is the editor and editorial member of several reputed journals, such as *Scientific Report* (Nature); *Process Safety & Environmental Protection* (Elsevier); *Journal of Water Process Engineering* (Elsevier); and *International Journal of Environmental Research & Public Health* (MDPI). He is also associate editor for *Environmental Chemistry Letters* (Springer); *Desalination and Water Treatment* (Taylor & Francis Group); and *Nanotechnology for Environmental Engineering Journal* (Springer). He has been awarded the Scientist of the Year Award (2015) by the National Environmental Science Academy, Delhi, India and the Almarai Award (2017), Saudi Arabia.

Acknowledgements

First and foremost, I would like to thank God for giving me the skills, knowledge, and abilities to achieve anything that I set my mind to do.

This book is the outcome of a remarkable contribution of experts from various interdisciplinary fields of science and covers the most comprehensive, in-depth, and up-to-date research and reviews. I am thankful to all the contributing authors and their co-authors for their esteemed work. I would like to express my gratitude to all the authors, publishers, and other researchers for granting us the copyright permissions to use their illustrations.

I would like to thank the CRC Press and Taylor & Francis Group for giving me the opportunity to edit this book. Special thanks go to Hilary LaFoe and Natasha Hallard, who provided me with editorial assistance and support. Without their patience, this book would not have been possible.

I express my deep gratitude to Prof. Zeid A. ALOthman (Chairman, Department of Chemistry, College of Science, King Saud University, Saudi Arabia) for his valuable suggestions, guidance, and constant inspiration. I am feeling short of words to express my thanks to all my friends and colleagues for their timely help, good wishes, encouragement, and affection.

Above all, I want to thank my parents and all my family members, who supported and encouraged me despite all the time it took me away from them. It was a long and difficult journey for them. Last but not least, I beg forgiveness from all those who have been with me over the course of the years and whose names I have failed to mention.

Finally, I would like to extend my appreciation to the Deanship of Scientific Research at King Saud University for the support.

Contributors

Min Addy
Center for Biorefining, Department
 of Bioproducts and Biosystems
 Engineering
University of Minnesota Twin Cities
Saint Paul, Minnesota

Mümtaz Ak
Department of Civil Engineering
Middle East Technical University
Ankara, Turkey

Esteban Alonso
Department of Analytical Chemistry,
 Escuela Politécnica Superior
University of Seville
Seville, Spain

Gary Anderson
Agricultural and Biosystems
 Engineering Department
South Dakota State University
Brookings, South Dakota

Irene Aparicio
Department of Analytical Chemistry,
 Escuela Politécnica Superior
University of Seville
Seville, Spain

İrfan Ar
Chemical Engineering Department
Gazi University
Ankara, Turkey

Andrea Arias
Department of Chemical Engineering,
 School of Engineering
University of Santiago de Compostela
Santiago de Compostela, Spain

Bruce Bleakley
Department of Biology and
 Microbiology
South Dakota State University
Brookings, South Dakota

Hosmanny Mauro Goulart Coelho
Department of Sanitary and
 Environmental Engineering, School
 of Engineering
Federal University of Minas Gerais
Belo Horizonte, Brazil

Lineker Max Goulart Coelho
Department of Sanitary and
 Environmental Engineering, School
 of Engineering
Federal University of Minas Gerais
Belo Horizonte, Brazil

Juan José Espada
Universidad Rey Juan Carlos
Escuela Superior de Ciencias
 Experimentales y Technología
Madrid, Spain

Gumersindo Feijoo
Department of Chemical Engineering,
 School of Engineering
University of Santiago de Compostela
Santiago de Compostela, Spain

Maria Gamallo
Department of Chemical Engineering,
 School of Engineering
University of Santiago de Compostela
Santiago de Compostela, Spain

Asiye Günal
Chemical Engineering Department
Gazi University
Ankara, Turkey

Arash Jahandideh
Agricultural and Biosystems
  Engineering Department
South Dakota State University
Brookings, South Dakota

Dipika Jaspal
Symbiosis Institute of Technology
Symbiosis International University
Pune, India

G. Janet Joshiba
Department of Chemical Engineering
SSN College of Engineering
Chennai, India

Shraddha Khamparia
Symbiosis Centre for Research and
  Innovation
Symbiosis International University
Pune, India

Serhat Kucukali
Civil Engineering Department
Çankaya University
Ankara, Turkey

Ponnusamy Senthil Kumar
Department of Chemical Engineering
SSN College of Engineering
Chennai, India

Liséte Celina Lange
Department of Sanitary and
  Environmental Engineering, School
  of Engineering
Federal University of Minas Gerais
Belo Horizonte, Brazil

Angela Daniela La Rosa
Department of Civil Engineering and
  Architecture
University of Catania
Catania, Italy

Juan Manuel Lema
Department of Chemical Engineering,
  School of Engineering
University of Santiago de Compostela
Santiago de Compostela, Spain

Sarah Mack
School of Environmental and
  Sustainability Science
Kean University
Union, New Jersey

Arti Malviya
Department of Engineering Chemistry
Lakshmi Narain College of Technology
Bhopal, India

Sara Mardani
Agricultural and Biosystems
  Engineering Department
South Dakota State University
Brookings, South Dakota State

Julia Martín
Department of Analytical Chemistry,
  Escuela Politécnica Superior
University of Seville
Seville, Spain

Rachel McDaniel
Agricultural and Biosystems
  Engineering Department
South Dakota State University
Brookings, South Dakota

Yolanda Moldes-Diz
Department of Chemical Engineering,
  School of Engineering
University of Santiago de Compostela
Santiago de Compostela, Spain

Raúl Molina
Universidad Rey Juan Carlos
Escuela Superior de Ciencias
  Experimentales y Technología
Madrid, Spain

Maria Teresa Moreira
Department of Chemical Engineering,
    School of Engineering
University of Santiago de Compostela
Santiago de Compostela, Spain

Dongyan Mu
School of Environmental and
    Sustainability Science,
Kean University
Union, New Jersey

Daniel Puyol
Universidad Rey Juan Carlos
Escuela Superior de Ciencias
    Experimentales y Technología
Madrid, Spain

Rosalía Rodríguez
Universidad Rey Juan Carlos
Escuela Superior de Ciencias
    Experimentales y Technología
Madrid, Spain

Roger Ruan
Center for Biorefining, Department
    of Bioproducts and Biosystems
    Engineering
University of Minnesota Twin Cities
Saint Paul, Minnesota

Juan Luis Santos
Department of Analytical Chemistry,
    Escuela Politécnica Superior
University of Seville
Seville, Spain

Anbalagan Saravanan
Department of Chemical Engineering
SSN College of Engineering
Chennai, India

Roberto Taboada-Puig
School of Biological Sciences and
    Engineering
Yachay Tech University
San Miguel de Urcuquí, Ecuador

1 Environmental and Health Effects Due to the Usage of Wastewater

Ponnusamy Senthil Kumar and G. Janet Joshiba

CONTENTS

1.1   INTRODUCTION

Water is one of the greatest gifts of nature, and it is the main source from which energy is derived. All living beings in this world depend on water for their survival. Life on earth would be impossible without water. Around 71% of the earth's surface is covered by water; however, around 98% of this water is useless because of its salinity and cannot be used for drinking purposes. The rest of the earth's water is fresh but is not easily accessible, because out of this 2% of fresh water, around 1.6% is trapped in polar ice caps and glaciers. In addition, 0.36% of fresh water resides in aquifers and wells. Thus, only 0.036% of water is available for drinking purposes from rivers and lakes. According to a study conducted by the World Health Organization/United Nations Children's Fund (WHO/UNICEF), the collection of fresh water from nearby water resources for drinking purposes is the highest-priority household work of many women and children in nearly 45 developing countries. People in these countries spend most of their time collecting fresh water instead of attending school, going to work, or looking after their family. From this, we can clearly understand that water has become an unaffordable commodity in our lives today (Reddy and Lee, 2012).

    Water scarcity and the quality of drinking water are among the most significant issues that humankind is confronting in the twenty-first century (Naushad et al., 2016; Alqadami et al., 2017). This water scarcity issue will be exacerbated by changing climatic conditions, which increase the temperature of water, leading to the melting of polar icecaps. As a result, an escalation of the hydrological cycle occurs, which causes flash floods and dry spells. Furthermore, many people globally are seriously affected through contaminated drinking water due to improper sanitary management. In addition to this, other sources such as manufacturing industries, farmlands, and municipal sources play a vital role in increasing water pollution. Water becomes contaminated for various reasons, but the pollution due to chemicals is a major threat globally because of its high toxicity; in many developing countries, nearly 1.1 billion individuals are in need of safe drinking water, and 2.6 billion individuals do not have appropriate sanitation. The unhygienic environment and consumption of contaminated drinking water becomes the primary reason for the 1.6 million deaths taking place every year throughout the world. The proper understanding of public hygiene and the quality of safe potable water varies between people living in agricultural and metropolitan territories, and the possibility of drinking clean and safe water becomes difficult for people living in provincial areas. According to the 2009 Gallup poll, the reports clearly proved that the contamination of potable water was seen to be a major concern in the United States (Schwarzenbach et al., 2010).

Nowadays, a major part of the economy has moved from agriculture to the industrial sector, and this tremendous development and advancement in the industrial sector has increased the origination of many new toxic and hazardous substances in industrial waste that is discharged during manufacturing and processing (Koolivand et al., 2017). As well as industry, increased population and urbanization also play a vital role in the inexorable exploitation of natural resources, the release of some new harmful toxic gases, changing climatic conditions, and an increased amount of waste generation (Vitorino de Souza Melarea et al., 2017). New innovations and technological advancement in industry, urbanization, and transportation have increased the discharge of large amounts of industrial wastewater into the fresh water zones, which remain the most precious source for agriculture. Nitrogen and phosphorus are a noteworthy mixture that contaminates fresh water sources. In addition to this, several harmful substances, such as excessive metals, synthetic toxins, pesticides, and washing liquids from industry released into fresh water sources, are the most frequent causes of increased eutrophication, which leads to the death of aquatic lives. This wastewater does not only affect the crop yield in agriculture; it also affects the livelihood of the poor farmers who depend on these water bodies, such as rivers and lakes, for agriculture (Rajasulochana and Preethy, 2016).

The rapid pace of industrialization, impromptu urbanization, and the population explosion are causing extreme damage to the surrounding water and land resources. The fundamental sources of fresh water contamination can be ascribed to the release of untreated waste from machinery, both clean and toxic, the dumping of modern effluents, and water discharge from agricultural land. Wastewater also leads to the contamination of ground water and causes a deficit of fresh water used for drinking purposes. People in many countries suffer due to this water scarcity. About 70%–80% of fatal diseases are due to contaminated water, and in many countries it affects women and children specifically (Bhatnagar and Sillanpaa, 2010).

The primary destination for wastewater disposal by enterprises is surface water. Most of the rivers in our country are polluted due to effluent discharge. Surface water contamination has increased 20 times due to the blending of industrial effluents in 22 regions of our country. The level of wastewater contamination varies from industry to industry depending on the type of procedure and the size of the enterprise. Rapid industrialization and globalization have prompted the acknowledgment and expanding comprehension of the interrelationship between contamination and general wellbeing. Almost 14,000 deaths take place per day due to this untreated industrial wastewater across developing countries. Because of this mixing of industrial waste into fresh water, a sudden decrease in dissolved oxygen takes place, and this leads to the death of the flora and fauna present in the fresh water. As a result of rapid industrialization, every year, 3.4 million individuals on the planet bite the dust due to waterborne sicknesses (Owa, 2013; Rana et al., 2017).

1.2   SOURCES OF WATER POLLUTION

The majority of water pollution is caused by the generation of contaminants from industrial, commercial, and agricultural sources, transport, dumpsites, and other anthropogenic activities (). Industries such as electroplating, tanning, hydrometal-lurgical, textiles, fertilizers, pesticides, dyeing, electrochemical and motor indus-tries, metallurgical, metal finishing, mine drainage, and battery manufacturing play a major role in the discharge of toxic chemical substances into fresh water (Olanipekun, 2015; Chen et al., 2015). Some of the anthropogenic activities that pre-vail as the cause of water pollution are population growth, increased numbers of dumpsites, leakage of sewage, deforestation, combustion, and littering (Owa, 2013). Thus, we human beings are one of the major reasons for water pollution, and the contaminants that are released from these industrial and commercial sources are explained in detail in the following section.

1.3   CONTAMINANTS

Contaminants are undesirable substances that alter the quality of the water resources. Contaminants originate from natural, industrial, and anthropogenic sources. Among these sources, industrial contaminants are highly toxic and are not easily degradable (Alqadami et al., 2016). The different types of contaminants present in wastewater are heavy metals, dyes, endocrine disrupting compounds, nutrients, and microbes.

1.3.1   Toxic Heavy Metals

Heavy metal contamination is one of the major environmental issues across the globe, because these heavy metals are highly toxic, and it also affects human health and other flora and fauna living on the earth (Sharma et al., 2017). The main sources of the production of heavy metals in the environment are the industrial and domestic sectors. These are not easily degradable, and heavy metals are capable of bioaccumu-lating in living organisms. They can also cause some fatal disorders, such as cancer and mutagenic disorders. The major heavy metals that affect the environment are lead, mercury, nickel, cadmium, arsenic, chromium, zinc, copper, cobalt, and anti-mony (Alqadami et al., 2017; Assadian and Beirami, 2014). Stringent regulations are constituted and applied across many developing countries to remove these toxic heavy metals from industrial effluents for the safety of the people (Ru-shan et al., 2011).

1.3.1.1   Lead

Lead is a unique, well-known industrial heavy metal. It is soft, ductile, and mal-leable in nature. It possesses poor conductivity and high resistance against corro-sion. Lead is not easily biodegradable and is a hazardous heavy metal to human health (Abdelwahab, 2015). The wastewater from industries such as electroplating, metallurgy, paint, electronics, petroleum refining, and storage battery contains high concentrations of lead, and these industrial effluents, when released into nearby fresh water sources, contaminate the water. The admissible levels of lead prescribed by the United States Environmental Protection Agency (USEPA) and the Bureau of Indian Standards in water and wastewater are 0.015 and 0.1 mg L^{-1}, respectively (Pandey et al., 2015).

1.3.1.2 Mercury

Mercury is one of the most life-threatening heavy metal ions. USEPA has indexed mercury as the most hazardous pollutant. It is highly lethal, and it is bioaccumulated easily when it enters the food chain of human beings. Mercury is obtained in various physical and chemical forms, among which mercuric ion is highly noxious. Mercury ions are emitted into the environment from sources such as the mining industry, the paper industry, the battery industry, the cement industry, power stations, and so on. In addition, some natural activities and human activities also increase the concentration of mercury ions in the environment. These mercury ions, even at trace concentrations, can cause serious health hazards (Cunha et al., 2016; Rahman and Singh, 2016; and Du et al., 2015). The desirable limit of mercury in drinking water is 0.001 mg L^{-1} (IS 10500:1991).

1.3.1.3 Nickel

Nickel is an influential heavy metal in our present contemporary society. It serves as the prime raw material in the production of steels, alloys, and battery industries, which are all major industries contributing greatly to our country's economy and growth (Fukuzawa, 2012). Nickel is silvery white in color, and it exhibits sound resistance even in highly alkaline conditions. It is ductile and malleable and possesses good corrosion resistance. Nickel is emitted into the atmosphere by both natural and human activities. Natural sources such as meteor showers, release from soil due to decay, the combustion of woodlands, and so on are the major reasons for the emission of nickel into our environment, and humans also play a major role in increasing nickel pollution by the burning of fossil fuels, the industrial production of nickel, the incineration of community waste, emissions from transport, and so on (Kim et al., 2014). The acceptable limit of nickel in drinking water according to the Bureau of Indian Standards is 0.02 mg L^{-1} (BIS IS 10500:2012).

1.3.1.4 Cadmium

Cadmium is one of the most noxious heavy metals and is also a very big threat to the personal wellbeing of human beings (Balaz et al., 2015). Cadmium has been recorded as an extremely risky heavy metal by many organizations around the world due to its high organic half-life of 10–30 years. Cadmium is a well-known mutagen, which particularly affects the excretory system of human beings, and it is lethal to other organs as well (Bilal, 2016). The manufacturing units of cement, batteries, plastics, pesticides, pigments, dyes, and so on that release untreated effluents into nearby water bodies are largely responsible for the increased cadmium concentration in wastewater (Iqbala et al., 2016). According to the regulations of USEPA and the European Union (EU), the allowable limit of cadmium in drinking water was set as 0.5 mg L^{-1} (Moinfar and Khayatian, 2017).

1.3.1.5   Arsenic

Arsenic is an unhealthy heavy metal, which is found extensively in the earth's crust and also in the deeper regions of the earth (El-Moselhy et al., 2017). Basically, this toxic heavy metal is released into our ecosystem through natural activities, such as volcanic eruption and leakage of arsenic from deposits in the earth's crust, and also some anthropogenic activities, such as pesticide production, glass production, mining of metals, the metallurgical industry, and increasing amounts of municipal waste (Jia et al., 2016; Heppa et al., 2017). As arsenic is abundantly found in nature, it has a higher possibility of affecting living beings by entering into their food chain through air, water, and soil (Kampouroglou and Economou-Eliopoulos, 2017). According to the directions of the WHO, the acceptable limit of arsenic in drinking water is 10 µg L^{-1}. Arsenic is one of the heavy metals that even in minute concentrations can cause serious damage to living beings (Ali et al., 2016).

1.3.1.6   Chromium

Chromium is an important heavy metal, used across the board in several industries, and hence, the concentration of chromium in the environment is increasing due to the untreated release of chromium into nearby water bodies and inadvisable storage of chromium waste by the industries. Of the different forms of chromium, it has been found that hexavalent chromium (+VI) is the most toxic to living beings, and it is easily transportable (Gheju et al., 2017). Chromium stands out among the most infamous excess metals discharged by different businesses: for example, the tanning and calfskin businesses, the electroplating industry, fabricating enterprises, fungicides, earthenware production, creates, glass, photography, impetus, and colors (Dehghani et al., 2016). USEPA has strictly established the allowable limit of Cr(VI) ions in shallow water as 0.1 mg L^{-1} and in drinking water as 0.05 mg L^{-1} (Rangabhashiyam and Selvaraju, 2014).

1.3.1.7   Zinc

Zinc is a prudential heavy metal, often used in industrial practices among other heavy metals such as iron, aluminum, and copper. It is generally used as a part of several applications, for example, in metallurgy, development, electrical and chemical divisions, and oil businesses, because of its resistance to erosion and its trademark mechanical properties. Due to the unique molecular nature of zinc, it is widely used in the dye industries in pigments to reduce corrosion, and it is also used to increase the reaction speed of chemical reactions (Ebin et al., 2016). According to the WHO regulations, the desirable limit of zinc in drinking water is 5 mg L^{-1} (Kumar and Puri, 2012).

1.3.1.8   Copper

Copper is a significant heavy metal that is dangerous to human wellbeing and its aura. It is used generally in industry and is accumulating in nearby water bodies (Al-Harahsheha et al., 2015). Copper ions are discharged into nature by means of distinctive industrial activities such as mining, electroplating, oil refining, manure production, wood mash, breaking down of rocks, and metal purifying and composite producing units. On the one hand, copper ions are fundamental to the good health of human beings, but on the other hand, the consumption of these heavy metal ions in a high concentration causes serious health effects to human beings. As per the Bureau of Indian Standards, the ultimate permissible limit of copper ions in drinking water is 0.05 mg L^{-1} (Kiruba et al., 2014; Kumar et al., 2015; Prabu et al., 2016).

1.3.1.9   Cobalt

Cobalt is a characteristic component present in the outer layer of the earth. Cobalt is said to possess a rocky, sparkly, steely, dim, unscented appearance and has distinctive forms. The concentration of cobalt is increased in the environment through both natural and anthropogenic activities. Soil particles, water bodies, and dust in the air are the indigenous sources of cobalt in our ecosystem, and a specific amount of cobalt is present in these natural sources (Rengaraj and Moon, 2002). The anthropogenic activities that cause a significant appearance of cobalt in water bodies are effluent discharges from the production units of industries such as processing of minerals, purification of metals, electroplating, and paints and pigments. Cobalt is a component of vitamin B12 and a necessary component for human wellbeing. Cobalt from industrial sources enters the food web and is bioaccumulated in all living organisms in our ecosystem (Abbas et al., 2014).

1.3.1.10   Antimony

Antimony is a growing threat universally due to its highly toxic nature. According to the reports of USEPA and EPA, Antimony is perceived as a major noxious heavy metals. Antimony is generally used for an assortment of purposes, including combustion of fossil fuels, incineration, batteries, fire retardants, polyethylene glycol terephthalate, brake linings, coal mines, and extraction of metals from their ores. Antimony is eliminated as a scum during the extraction of metals such as gold, silver, and copper from their diverse mineral feedstocks. Through the gas passages and effluent discharge of the extraction zone, this antimony reaches the atmosphere and pollutes the habitat (Multani et al., 2017; Hu et al., 2017).

1.3.2   Dyes

The world's first financially effective engineered color, Mauvein, was discovered by Henry Perkin in 1856. Since that point, nearly 10,000 various colors and shades have been produced and used as part of different enterprises. On average, nearly 7×10^5 tons of dyes are produced every year globally for diverse applications in the industrial sector. During the synthesis and manufacture of the products, these dyes are released into the environment through the effluent discharge of the industries. Dyes are produced as an integral part of many anthropogenic and industrial activities, such as textile industry, paper manufacturing, food industry, tanning, in rural explore and in light gathering exhibits. The mixing of the dye-contaminated wastewater into the nearby water bodies results in nutrient excess, leading to the death of aquatic life, and it also causes hazardous repercussions through chemical reactions such as oxidation, hydrolysis, or other compound responses occurring in the wastewater. The chemical classes of dyes most often used in industry are azo, anthraquinone, sulfur, indigoid, triphenylmethyl (trityl), and phthalocyanine compounds (Zangeneh et al.,

2015; Forgacs et al., 2004). Dyes are classified into three types: Anionic, cationic, and non-ionic. The anthraquinone dyes are complex to synthesize, resulting in a high cost of production. The decomposition of the anthraquinone dyes is difficult, and it depends on the type of microbial community used. The anthraquinone dyes are the second biggest class of dyes, possessing a complicated aromatic structure, which provides strong resistance against degradation (Rizi et al., 2017).

1.3.3 Microbial Contamination

Microbial contamination is also one of the major sources of water pollution. It is caused by harmful microorganisms in drinking water. Improper sanitation is the reason for the spreading of noxious pathogens in drinking water. The waste materials excreted by human beings and animals contain *Escherichia coli* (*E. coli*), which is a basic organism habitually present in drinking water. Moreover, *E. coli* is a marker for the presence of other pathogens, such as *Vibrio cholerae, Salmonella typhi*, and S. *paratyphi*, in drinking water. According to WHO reports, on average, nearly 1.1 billion people around the world consume perilous drinking water for their survival. Lack of cleanliness, contaminated unsafe drinking water, and improper sanitation are responsible for 88% of diarrheal illness on the planet and, moreover, are the major reason for 3.1% of deaths annually (i.e., 1.7 million) and 3.7% of illnesses annually (i.e., 54.2 million) of human beings globally. This microbial contamination causes many harmful diseases in various developing nations, and it is necessary to monitor these noxious microorganisms and remove them from potable water to prevent the transmission of toxic disease-causing strains in drinking water, even in many developed nations (Ashbolt, 2004; Fawell and Nieuwenhuijsen, 2003).

1.3.4 Agricultural Runoff

The pesticides and chemicals released from agricultural land during crop cultivation are one of the major sources of water pollution. Agriculture is the practice through which a civilized society cultivates and manufactures food, and it consequently incorporates ranger service, trim culture, and biomass creation for the cultivation of fuel and livestock. The relationship between agriculture and fresh water is unpredictable, and they are interrelated in various dimensions. Moreover, agricultural practices contribute to some of the water-related environmental problems, such as a deficit of fresh water for irrigation purposes, leakage of pesticides into fresh water sources, suspended loads from soil erosion, alteration in the water cycle, destruction of aquatic life, and alteration in the food cycle. The most important contaminants originating from agricultural land are nitrogen and phosphorus compounds. The release of nitrogen compounds such as nitrate and nitrite is found to be toxic to human health, and they can cause many dreadful disorders, such as blue baby syndrome and methemoglobinemia, which seriously affects children. Knowing the harmful effects of nitrate, WHO has restricted the allowable limit of nitrate to 50 mg L^{-1} in drinking water, but nowadays, the levels of nitrate appear to be elevated from 50 to 100 mg L^{-1}. The usage of many modern pesticides and fertilizers for enhancing crop production has also created a major concern. The runoff from agricultural land with nutrient overdose and harmful pesticides and chemicals contaminates fresh water and causes toxic ill effects to the human habitat (Fawell and Nieuwenhuijsen, 2003; Moss, 2008).

1.3.5 ENDOCRINE DISRUPTING COMPOUNDS

Endocrine disrupting compounds (EDC) are a group of harmful chemical compounds present in industrial wastewater that are capable of disturbing the human endocrine system. The EDC are troublesome to assimilate and interpret because of their complex nature, and they incorporate organic and instrumental strategies. These compounds have diverse structures and workings. The WHO, the EU, and USEPA have catalogued the various types of endocrine disrupting compounds present in the environment and have also set certain allowable limits for EDC. As understanding about the wastewater and the contaminants increases, it is clearly seen that many chemical compounds possess these noxious endocrine disrupting properties, and this creates a need for the removal of EDC from wastewater (Bolong et al., 2009).

1.4 WASTEWATER CHARACTERISTICS

The understanding and study of the characteristics of wastewater is essential for building the administration of ecological quality. Wastewater treatment involves skilled people from various domains. Scientific experts have worked hard to improve the testing strategies for studying substantial and synthetic changes (Muttamara, 1996). The physical, chemical, and biological characteristics of wastewater are explained in detail in this section.

1.4.1 PHYSICAL CHARACTERISTICS

The vital physical qualities of wastewater are pH, temperature, color, and total solid content of the water. These physical characteristics are explained in detail in this sub-section.

1.4.1.1 Color

Color is one of the important parameters that are used to evaluate the nature of wastewater. The color of the wastewater usually varies depending on the source from which it is discharged. The important sources of wastewater are the municipal sector, the industrial sector, and the decomposition of biological compounds. Table 1.1 clearly explains the various colors of wastewater and their characteristics.

1.4.1.2 Odor

Fresh wastewater released from industry is usually odorless. Scents have been evaluated as the principal worry of the general population with respect to the running of wastewater treatment systems. In numerous areas, many wastewater treatment projects have been rejected as a result of the dread of potential smells. Fresh wastewater does not smell obnoxious, but due to the discharge of an assortment of musty mixes, a bad odor arises. The odors of industrial wastewater are no worse than those of

TABLE 1.1

Various Colors of Wastewater and their Characteristics

Color	Type	Nature of Wastewater
Brown	Industries	Fresh condition
Grey	Domestic wastewater	Wastewater has experienced some level of decomposition
Black	Septic systems	Broad bacterial deterioration under anaerobic conditions has taken place

septic systems. Wastewater in septic systems is highly noxious due to the hydrogen sulfide produced during the decomposition of sulfate to sulfide by anaerobic micro-organisms. Chemical compounds such as cadaverin, mercaptan, indol, and skatol are the offensive compounds that produce a rotten smell during decomposition in industrial wastewater.

1.4.1.3 Temperature

Temperature is one of the most important parameters to monitor in wastewater, because the treatment of wastewater using microbes is commonly temperature limited. The temperature of the wastewater depends on the geographic location. The temperature will range from 13°C to 24°C in the warmer provinces and from 7°C to 18°C in the colder parts of the world. The higher temperature of wastewater affects the aquatic organisms living in the water bodies by creating a deficiency of oxygen, which reduces the reaction rate of the mixture and biological response rates. The wastewater must be maintained at an optimal temperature, because a high temperature of wastewater results in the growth of undesirable organisms, whereas when the temperature drifts down, the treatment of wastewater becomes tedious and time consuming.

1.4.1.4 Solids

During the dissipation of wastewater at 103°C–105°C, some aggregate solid substances are deposited as residual debris. Wastewater contains suspended, volatile suspended, and dissolved solids. Suspended solids increase the turbidity of the wastewater, but they can be easily removed by subjecting the wastewater to a filtration process. The other form of solids present in wastewater is volatile suspended solids, which are made up of organic compounds such as carbohydrates, proteins, and fats. They can easily be ignited at high temperature and affect the oxygen concentration of the water resources. Settleable solids are another component of the wastewater, which can be removed only by a sedimentation process. Most of the suspended solids are settleable in nature, and a certain concentration of dissolved solids is also present in the wastewater, which alters the characteristics of the wastewater (Muttamara, 1996).

1.4.2 Chemical Characteristics of Wastewater

Wastewater is basically composed of harmful gases, organic chemical compounds, inorganic chemicals, and excessive nutrients. Gases such as methane, nitrogen, carbon dioxide, hydrogen sulfide, and ammonia are usually present in fresh wastewater sources. Wastewater is composed of inorganic chemicals, organic chemicals, harmful gases, and dissolved oxygen. It consists of natural organic compounds such as carbon, oxygen, hydrogen, and nitrogen. The primary components present in wastewater are carbohydrates (25%–50%), proteins (40%–60%), and oils (10%). It also contains iron, sulfur, and phosphorus. These organic compounds present in wastewater are easily biodegradable, whereas some of the synthetic organic compounds present in wastewater cannot be composted easily, which complicates the wastewater treatment. Inorganic compounds such as chloride, nitrogen, sulfur, phosphorus, hydrogen ions, and heavy metals are present in wastewater. Plant forms such as algae and other organisms use these inorganic substances, such as carbon, phosphorus, and ammonia-nitrogen, as a source of food. These compounds are capable of affecting the organisms at even minute concentrations. Nutrients such as phosphorus and nitrogen in higher concentrations affect the water resources by causing algal bloom, which causes the destruction of aquatic organisms. Undesirable levels of nitrogen and phosphorus in the water stream cause excess algal growth, which creates an oxygen deficient environment and destroys the aquatic organisms. The monitoring of nutrient levels in wastewater is essential to inhibit the undesirable growth of phytoplankton in water.

1.4.3 BIOLOGICAL CHARACTERISTICS OF WASTEWATER

The effluents discharged from industry contain harmful pathogens: bacteria, protozoa, and viruses. Bacteria are notable microorganisms that help in the decomposition of organic compounds in wastewater using the oxygen in the wastewater. Measuring the quantity of microorganisms is one of the important parameters in checking the effectiveness of wastewater treatment. The acceptable range of microbial load in the wastewater is 10^5–10^8 mL^{-1}. Harmful pathogens such as intestinal viruses can easily be transmitted through water and can cause harmful diseases. The harmful pathogens should be removed from the wastewater during treatment to avoid the transmission of dreadful diseases (Muttamara, 1996). The biological system is an important part of the wastewater, as it helps in the breakdown of complex substances in the wastewater. The biological characteristics of the wastewater should be clearly understood to decide the type of wastewater treatment. Thus, these are the physical, chemical, and biological characteristics of wastewater, which help to design the wastewater treatment system. The environmental effects of wastewater are discussed in detail in the following section.

1.5 IMPACTS OF WASTEWATER ON ENVIRONMENT

Wastewater is well known for both its beneficial activities and its drawbacks. As wastewater contains many essential nutrients for the growth of plants, it is used as a growth enhancer in agriculture. On the other hand, wastewater poses some serious negative impacts to the environment (Akpor and Muchie, 2011). Effluent discharge releases different contaminants into the fresh water resources. The bulk of fresh water is profoundly devoured by humans for drinking and other domestic purposes. The blending of industrial contaminants into the water resources destroys the aquatic organisms living in the water, makes the water unfit for irrigation purposes, and diminishes the crop yield. The harmful impacts of contaminated wastewater on the quality of accepting water bodies are complex, and they completely depend on the combination of effluents, the volume of the industrial release, the type of source, chemical reactions, and microbiological focus. Wastewater creates a strong impact on the environment, and it causes some deleterious effects, such as depletion of dissolved oxygen, discharge of toxins, bioaccumulation, substantial changes in water bodies, eutrophication, transmission of water borne disease, and destruction of aquatic organisms (Okereke et al., 2016). Wastewater affects not only the water bodies but also other components such as air, soil, plants, and amphibians. Through the contamination of the receiving water bodies, the nature of the environment is altered and mutated; the damage caused to the environment by wastewater is irreversible; and the damage caused by wastewater to the environment is of concern to all the living organisms on our planet.

1.5.1 WATER SCARCITY

Water is an abundant source of nature to be conserved. Rapid industrialization is one of the biggest reasons behind the contamination caused to the fresh water sources. When compared with lack of sanitation, the impact of industrial pollution is notably toxic in the fresh water sources, because of the discharge of harmful non-biodegradable contaminants from industry. In China, more than 25 billion tons of unfiltered toxins were dumped into the conduits in a solitary year. The incompetent monitoring of contaminants in water and the lack of stringent rules and regulations for industrial effluents increase the extent of contamination of potable water. As result of negligence, many underdeveloped countries are poorly prepared to outline and execute sufficient directions for controlling wastewater and contamination (Curry, 2010). The far-reaching debasement of water quality around the world is the most genuine water issue, undermining human wellbeing and the trustworthiness of biological systems. Additionally, water scarcity has become a noteworthy worry for people living all over the world, and this has become a great challenge to the development of humans and their financial condition (UNESCO IHP Bureau).

1.5.2 CLIMATIC CHANGE

Climatic change is one of the major contributors to water pollution. The addition of harmful contaminants into the receiving water bodies changes the quality of the fresh water and makes it unfit for drinking and other domestic processes. Climatic change results in the increase of global temperature and causes global warming. As a result of increased heat in the atmosphere, the melting of polar ice caps takes place, and this causes sudden floods and dry spells. The sudden imbalance in the natural environment affects the hydrological cycle of our planet and causes sudden climatic change (Schwarzenbach et al., 2010). Climatic change also has a negative impact on water quality. It increases the temperature of the water, which gradually increases the content of contamination and alters the quality of the water. The uneven distribution of rainfall fails to satisfy the water demand of people globally (Curry, 2010).

1.5.3 BIOACCUMULATION

Bioaccumulation is the process by which certain substances such as heavy metals, pesticides, and toxic chemicals that are in low concentrations in water can be found in high concentrations in the tissues of plants and animals. Substances that bioaccumulate are stable, possess a long half-life period, and are not processed by the human digestive system. Bioaccumulated substances are further magnified by entering the food chain (Okereke et al., 2016). Discharge of effluents and domestic waste is the main source of deposits of bioaccumulatives in wastewater. This alters the quality of the fresh water bodies and transfers the toxic substance into the receiving water bodies (Akpor and Muchie, 2011). Bioaccumulation with the support of supplements and vitality takes place within the sight of living cells. There are two phases in the expulsion of substantial metals from the condition. The first is the maintenance of the metal particles on the cell surface, and the second is the transportation of the particles of metal into the cell (Alqadami et al., 2017).

1.5.4 EUTROPHICATION

Nitrogen and phosphorus are the nutrients responsible for the growth of plants, and when these nutrients are released from agricultural runoff into the nearby water bodies, it causes eutrophication. These nutrients also boost the development of aquatic vegetation, but when they exceed an optimum level, they may cause the development of unnecessary weeds and algal bloom. This gives a bad odor and taste to the water, which make it unfit for drinking (Owa, 2013). Eutrophication causes a deficiency of dissolved oxygen. Eutrophication of water sources may likewise create natural conditions that support the development of toxin-producing cyanobacteria. Eutrophication leads to the depletion of dissolved oxygen in water bodies, which kills fish and other aquatic organisms (Okereke et al., 2016). The pernicious effects caused by the eutrophication of wastewater are

- Reduced quality of receiving water bodies, with bad odor and formation of algal blooms
- Increased depletion of dissolved oxygen
- Undesirable growth of phytoplankton
- Inhibition of aeration of aquatic plants, leading to the death of the plants
- Decreased wastewater treatment efficiency

Effects may also include the expansion of growth of new varieties of plants and also the extinction of rare species of other plants and animals in our habitat (Akpor and Muchie, 2011).

1.5.5 DECREMENT OF DISSOLVED OXYGEN

During the degradation of organic compounds and complex chemical compounds by bacteria, dissolved oxygen is consumed as a substrate for the decomposition reaction. This further leads to deficiency of dissolved oxygen in the receiving fresh water bodies. The depletion of dissolved oxygen affects the aquatic organisms by decreasing the immunity of fish, resulting in growth inhibition and early death, and it affects the population of the living organisms (Okereke et al., 2016). The optimum level of dissolved oxygen in the wastewater is around 8–10 g m^{-3}. When the biological oxygen demand (BOD) levels of wastewater are below 4 g m^{-3}, the stream is said to possess healthy levels of dissolved oxygen, and the stream framework can manage the amount of waste without influencing the fish. At a dissolved oxygen level of 5 g/m^3, the fish become noticeably stressed, and at 2 g m^{-3}, the fish will die from lack of oxygen unless they can move to more oxygenated waters (Ministry for the Environment New Zealand).

1.5.6 PERSISTENT ORGANIC POLLUTANTS

A group of chemicals which are discharged in waste from certain industries have high stability, and these are a growing threat to living organisms globally. These pollutants are generally released as a final waste product during manufacture and can also be by-products. Highly chlorinated compounds, such as polychlorinated biphenyls (PCB), dichlorodiphenyltrichloroethane (DDT), polychlorinated dioxins (PCD), dibenzofurans and polycyclic aromatic hydrocarbons (PAH), are highly toxic and bioaccumulative. They seriously affect the health of human beings by causing cancer, thyroid imbalance, psychological issues, diabetes, and many fatal diseases. They are highly harmful and act as EDC (Schwarzenbach, 2010).

1.6 IMPACTS OF WASTEWATER ON PUBLIC HEALTH

Basically, wastewater contains contaminants such as heavy metals, dyes, pesticides, EDC, persistent organic pollutants, and nutrients. Each contaminant has a unique negative impact on the human body and poses a serious threat to human health.

1.6.1 HEAVY METAL POISONING

Heavy metal poisoning is one of the major issues caused by industrial wastewater, and it results from the aggregation of heavy metals in lethal concentrations in the delicate tissues of the human body (Naushad et al., 2017). The extent of toxicity varies according to the type of metal accumulated. In general, low concentrations of heavy metals are required for the proper functioning of the human body, and when the bio-recommendation for heavy metals is exceeded in the tissues, heavy metal poisoning occurs. The most toxic heavy metals, which cause harmful diseases even in trace concentrations, are mercury, lead, arsenic, and cadmium. Industrial activities and other anthropogenic activities are the main sources of heavy metal poisoning. Table 1.2 clearly explains the different ill effects caused by heavy metals.

Lead is the most noteworthy of the metals that cause heavy metal poisoning, and it is mainly taken in through ingestion in food, water and mostly through air. Lead poisoning affects hemoglobin production, the kidneys, the joints, the regenerative systems, and the cardiovascular framework and causes major and irreversible harm to the brain. Its different impacts include harm to the gastrointestinal tract (GIT), the urinary tract, the central nervous system (CNS), and the peripheral nervous system (PNS). Lead affects children by interfering with the development of gray matter, resulting in low IQ. Intense and continual exposure to lead results in psychosis.

Cadmium is harmful at extremely low levels. It is destructive to both human well-being and aquatic biological systems. It is a cancer-causing, embryotoxic, and mutagenic chemical compound. Cadmium in the body has been shown to bring about damage to vital organs such as kidney, liver, and bone structures. The inhalation of cadmium at high concentrations causes obstructive lung sickness and cadmium pneumonitis. It may cause hyperglycemia and iron deficiency anemia.

Mercury is a toxic heavy metal, and it is not an essential heavy metal for human physiology and function. Inorganic forms of mercury cause premature birth, congenital deformity, and gastrointestinal issues such as destructive esophagitis and hematochezia. Due to its complex some portion of the PNS, coming about in nerve irritation that causes muscle shortcoming (Duruibe et al., 2007).

1.6.2 Impacts of Microbes

Contaminated wastewater contains many harmful microbes originating from the fecal waste of animals and humans. The major disease-causing microbes present in wastewater are bacteria, viruses, fungi, protozoa, and parasites. Mixing with this microbe-contaminated wastewater results in transfer of harmful microorganisms into the fresh water bodies. Consumption of this contaminated wastewater causes diseases transmitted specifically through water. Ailments caused by microscopic organisms, infections, and protozoa are the most widely recognized dangers to health related to untreated drinking and recreational waters. Contaminated water is a vehicle for waterborne infections, for example, giardiasis, cholera, campylobacteriosis, typhoid fever, hepatitis A, shigellosis, salmonellosis, and cryptosporidiosis (Akpor, 2011).

1.6.2.1 Diseases Caused by Bacteria

Bacterial species are the primary organisms found in wastewater. The most common bacterial organisms present in wastewater are *E. coli*, *Campylobacter*, *Listeria*, *Leptospira*, *Salmonella*, and *Vibrio*. Diverse bacterial species are present in the wastewater; mostly, these are innocuous to human beings, but there are some pathogenic species present in wastewater that cause ailments such as typhoid, diarrhea, and other intestinal problems. Bacterial species such as *E. coli* and *Pseudomonas* mainly affect the digestive system, causing gastroenteritis, and they specifically affect newborns.

1.6.2.2 Diseases Caused by Viruses

Viruses are among the most dangerous contaminants in wastewater. They are capable of causing some dreadful and untreatable health hazards to humans. As it is difficult to diagnose the presence of viruses, bacterial viruses such as bacteriophages are used to analyze the fecal contamination and also to improve the viability of treatment procedures to get rid of enteric infections. The most harmful enteric viruses present in wastewater are retroviruses, enteroviruses, caliciviruses, Norwalk viruses, rotaviruses, adenoviruses, echoviruses, and hepatitis A virus. These viruses most commonly cause infections in the liver and eye, causing diseases such as meningitis and liver damage.

1.6.2.3   Diseases Caused by Parasites

Among the protozoa living in wastewater, *Cryptosporidium parvum*, *Cyclospora*, and *Giardia lamblia* are notable for causing disturbance in the intestine, resulting in serious diarrhea, eye infections, lactose intolerance, and hypothyroidism, and even leading to death. Parasitic worms such as nematodes, flatworms, and leeches cause infection in humans and are transmitted through wastewater.

1.6.2.4   Effects of Agricultural Chemicals

Agricultural wastes consist of phosphorus and nitrogen, and the presence of excess nutrients in the wastewater causes eutrophication, which results in the growth of unwanted toxin-synthesizing algal species. The consumption of such toxins causes skin diseases, gastroenteritis, nervous system damage, and liver damage. The toxins produced by the cyanobacteria seem to be carcinogenic. The intake of nitrites, even at low concentration, results in a deadly disease called methemoglobinemia. Methemoglobinemia is related to nitrates in drinking water above the most extreme contaminant level ($10 \, \text{mg L}^{-1}$) as set by the US Environmental Assurance Agency.

1.6.2.5   Effects of Endocrine Disrupters

EDC are highly toxic to human and animal health. They disturb the endocrine system by emulating, blocking, or upsetting the secretion of hormones that are most essential for the health of humans and other species. EDC specifically affect the reproductive system of human beings by diminishing sperm count and damaging the reproductive system. EDC also cause tumors in specific regions such as the prostate, the ovary, and the testicles. They cause alterations in the function of the reproductive system, change sexual behavior, and disturb the immune system (Bolong et al., 2009; Akpor and Muchie, 2011).

1.6.3   Effects of Dyes

Azo colors are among the first natural compounds to be related to human tumors. Many are not cancer-causing agents. An azo dye such as Congo dye affects the respiratory, gastrointestinal, and reproductive systems. It causes eye infection and irritates the eye (Chawla et al., 2017). Malachite green is a frequently used dye, and it is a major carcinogen, causing cancer in the vital organs (Pathania et al., 2016). Malachite green suppresses the immune system and the reproductive system of human beings. It is highly toxic to fish at trace concentrations and affects the chromosomes, as it is a mutagen (Srivastava et al., 2004).

1.7   CASE STUDY ON WASTEWATER EFFECTS ON HUMANS

Minamata disease is one of the most dreadful genuine instances of illness caused by the release of wastewater from a modern plant into the environment. Methylmercury from the industrial waste discharge contaminated the marine life in the encompassing waters, and as a result, the individuals who consumed the mercury-contaminated fish and seafood were severely affected by Minamata disease. The methylmercury-affected Shiranui Sea is an island ocean in the southwest locale of Japan's Kyushu Island. It is around 60 km long and 20 km wide, covering a range of 1200 km². At the time of methylmercury contamination, around 200,000 individuals living on the shoreline of the ocean who engaged in the fishing business were affected by Minamata disease. In Minamata Bay, 71,000 m² of slope with high mercury content (over 100 ppm) was secured with soil (recovery region), and 1,539,000 m² of the water region containing mercury at 25 ppm or higher on the ocean depths was dug (digging region). Furthermore, there are no effective remedies for the sickness; only palliative treatment, restoration treatment, physiotherapy, needle therapy, and moxibustion are being offered to patients. As of March 1992, 2252 patients had been authoritatively diagnosed as having Minamata sickness, of whom 1043 have died. A total of 12,127 people claiming to have the sickness have not been recognized as Minamata disease patients, and what is more, 1968 people are still awaiting the result of examination by the Examining Board for the Acknowledgment of Minamata Disease Patients. Around 2000 people not authoritatively recognized as patients are presently in court. Minamata disease has not stopped, and these individuals require help and assistance.

1.8   FUTURE PERSPECTIVES

As we have discussed in this chapter, wastewater poses a serious threat to the environment. Fresh water is the only promising source of drinking water and water for irrigation. The undesirable addition of harmful substances to the environment damages the fresh water sources. Contaminants such as heavy metals, dyes, pesticides, and other harmful chemicals are not easily degradable, and they are capable of being bioaccumulated into the tissues of living organisms. These contaminants enter the food chain and are thus accumulated into the organism. Due to this water pollution, fresh water sources are severely damaged, and it also causes water scarcity. Primarily, governments should introduce stringent rules to control the discharge of harmful pollutants into water bodies. Adequate monitoring of the wastewater should be done before it is released into the water bodies. The only promising remedy to reduce the toxic effects of wastewater is wastewater treatment. Through wastewater treatment, the harmful contaminants can be removed from the water, and the toxicity of the wastewater can be reduced before it is released into the receiving waters. To accomplish unpolluted wastewater release into receiving water bodies, there is a need for cautious planning, satisfactory and appropriate treatment, and the observation and proper enactment of standards. This involves both science and human support, and proper information about the ill effects of the wastewater must be given to the public. These are some of the measures through which the severe effects of the wastewater can be minimized.

1.9   CONCLUSION

Water is a fundamental source of energy for all the living beings in the world, and it is the basic requirement for the survival of human beings. The quantity and quality of the water are being progressively degraded due to industrial and anthropogenic activities. The rapid growth of industrialization and globalization has increased the emission of harmful pollutants into the environment. Wastewater discharged from industry contains harmful substances such as heavy metals, dyes, pesticides, and other toxic complex chemical compounds. When these harmful pollutants reach the receiving water bodies, they completely alter the quality of the water and make it unfit for drinking and irrigation. The contaminants in the water sources are bioaccumulated into living organisms and progressively biomagnified. When contaminants in the water exceed the permissible limit, they cause fatal disorders and other genetic disorders. Wastewater shows negative impacts on the environment, such as eutrophication, depletion of dissolved oxygen, and bioaccumulation. These negative impacts are irreversible, and the effects on human beings and other living organisms are severe. Many serious health effects and fatal disorders are caused by wastewater. In addition, due to the increasing water pollution, there occurs a scarcity of potable water. To avoid the ill effects of this wastewater, stringent regulation must be accomplished with sustainable water treatment technologies. The clear understanding and monitoring of wastewater is a must. Implementing biological technologies for treating wastewater will help in controlling secondary waste production.

REFERENCES

Abbas, M., Kaddour, S., and M. Trari. 2014. Kinetic and equilibrium studies of cobalt adsorption on apricot stone activated carbon. *Journal of Industrial and Engineering Chemistry.* 20: 745–751.

Abdelwahab, N. A., Ammar, N. S., and H. S. Ibrahim. 2015. Graft copolymerization of cellulose acetate for removal and recovery of lead ions from wastewater. *International Journal of Biological Macromolecules.* 79: 913–922.

Akpor, O. B. 2011. Wastewater Effluent Discharge: Effects and Treatment Processes. http://en.unesco.org/themes/water-security/hydrology/water-scarcity-and-quality.

Akpor, O. B. and M. Muchie. 2011. Environmental and public health implications of wastewater quality. *African Journal of Biotechnology.* 10: 2379–2387.

Al-Harahsheha, M. S., Zboon, K. A., Al-Makhadmeh, L. et al. 2015. Fly ash based geopolymer for heavy metal removal: A case study on copper removal. *Journal of Environmental Chemical Engineering.* 3: 1669–1677.

Ali, J., Tuzen, M., Kazi, T. G. et al. 2016. Inorganic arsenic speciation in water samples by miniaturized solid phase micro extraction using a new polystyrene polydimethylsiloxane polymer in micro pipette tip of syringe system. *Talanta.* 161: 450–458.

Alqadami, A. A., Naushad, M., Abdalla, M. A., Khan, M. R., and Z. A. ALOthman. 2016. Adsorptive removal of toxic dye using Fe3O4–TSC nanocomposite: Equilibrium, kinetic, and thermodynamic studies. *Journal of Chemical* and *Engineering Data.* 61: 3806–3813.

Alqadami, A. A., Naushad, M., Abdalla, M. A., Ahmad, T., ALOthman, Z. A., ALShehri, S. M., and A. A. Ghfar. 2017. Efficient removal of toxic metal ions from wastewater using a recyclable nanocomposite: A study of adsorption parameters and interaction mechanism. *Journal of Cleaner Production.* 156: 426–436.

Alqadami, A. A., Naushad, M., Alothman, Z. A., and Ghfar, A. A. 2017. Novel Metal–Organic Framework (MOF) Based Composite Material for the Sequestration of U(VI) and Th(IV) Metal Ions from Aqueous Environment. *ACS Appl Mater Interfaces* 9: 36026–36037.

Ashbolt, N. A. 2004. Microbial contamination of drinking water and disease outcomes in developing regions. *Toxicology.* 198: 229–238.

Assadian, F. and P. Beirami. 2012. An optimization model for removal of zinc from industrial wastewater. *Paper Presented on Industrial Engineering and Engineering Management* (IEEM), 2012 IEEE International Conference, Hong Kong, China, 19–21 Nov. 2015

Balaz, M., Bujnakova, Z., Balaz, P. et al. 2015. Adsorption of cadmium(II) on waste biomaterial. *Journal of Colloid and Interface Science.* 454: 121–133.

Bhatnagar, A. and M. Sillanpaa. 2010. Utilization of agro-industrial and municipal waste materials as potential adsorbents for water treatment—A review. *Chemical Engineering Journal.* 157: 277–296.

Bilal, M., Kazi, T. G., Afridi, H. I. et al. 2016. Application of conventional and modified cloud point extraction for simultaneous enrichment of cadmium, lead and copper in lake water and fish muscles. *Journal of Industrial and Engineering Chemistry.* 40: 137–144.

Bolong, N., Ismaila, A. F., Salim, M. R. et al. 2009. A review of the effects of emerging contaminants in wastewater and options for their removal. *Desalination.* 239: 229–246.

Chawla, S., Uppal, H., Yadav, M. et al. 2017. Zinc peroxide nanomaterial as an adsorbent for removal of Congo red dye from wastewater. *Ecotoxicology and Environmental Safety.* 135: 68–74.

Chen, C. S., Shih, Y. J., and Y. H. Huang. 2015. Remediation of lead (Pb(II)) wastewater through recovery of lead carbonate in a fluidized-bed homogeneous crystallization (FBHC) system. *Chemical Engineering Journal.* 279: 120–128.

Cunha, R. C., Patrício, P. R., Vargas, S. J. R. et al. 2016. Green recovery of mercury from domestic and industrial waste. *Journal of Hazardous Materials.* 304: 417–424.

Curry, E. 2010. Water scarcity and the recognition of the human right to safe freshwater. *Northwestern Journal of International Human Rights.* 9: 104–121.

Dehghani, M. H., Sanaei, D., and I. Ali. 2016. Removal of chromium(VI) from aqueous solution using treated waste newspaper as a low-cost adsorbent: Kinetic modeling and isotherm studies. *Journal of Molecular Liquids.* 215: 671–679.

Du, M., Wei, D., Tan, Z., Lin, A., and Y. Du. 2015. Predicted no-effect concentrations for mercury species and ecological risk assessment for mercury pollution in aquatic environment. *Journal of Environmental Sciences.* 28: 74–80.

Duruibe, J. O., Ogwuegbu, M. O. C., and J. N. Egwurugwu. 2007. Heavy metal pollution and human biotoxic effects. *International Journal of Physical Sciences.* 2: 112–118.

Ebin, B., Petranikova, M., and B. M. Steenari. 2016. Effects of gas flow rate on zinc recovery rate and particle properties by pyrolysis of alkaline and zinc-carbon battery waste. *Journal of Analytical and Applied Pyrolysis.* 121: 333–341.

El-Moselhy, M. M., Ates, A., and A. Çelebi. 2017. Synthesis and characterization of hybrid iron oxide silicates for selective removal of arsenic oxyanions from contaminated water. *Journal of Colloid and Interface Science.* 488: 335–347.

Fawell, J. and M. J. Nieuwenhuijsen. 2003. Contaminants in drinking water: Environmental pollution and health. *British Medical Bulletin.* 68: 199–208.

Forgacs, E., Cserháti, T., and G. Oros. 2004. Removal of synthetic dyes from wastewaters: A review. *Environment International.* 30: 953–971.

Fukuzawa, R. 2012. Climate change policy to foster pollution prevention and sustainable industrial practices—A case study of the global nickel industry. *Minerals Engineering.* 39: 196–205.

Gheju, M., Balcu, I., Enache, A. et al. 2017. A kinetic approach on hexavalent chromium removal with metallic iron. *Journal of Environmental Management.* 203: 937–941.

Heppa, L. U., Pratas, J. A. M. S., and M. A. S. Graça. 2017. Arsenic in stream waters is bioaccumulated but neither biomagnified through food webs nor biodispersed to land. *Ecotoxicology and Environmental Safety.* 139: 132–138.

Hu, X., He, M., S. Li et al. 2017. The leaching characteristics and changes in the leached layer of antimony-bearing ores from China. *Journal of Geochemical Exploration.* 176: 76–84.

Iqbala, M., Iqbalb, N., Bhattib, I. A. et al. 2016. Response surface methodology application in optimization of cadmium adsorption by shoe waste: A good option of waste mitigation by waste. *Ecological Engineering*. 88: 265–275.

Jia, X., Gong, D., and J. Wang. 2016. Arsenic speciation in environmental waters by a new specific phosphine modified polymer microsphere preconcentration and HPLC-ICPMS determination. *Talanta*. 160: 437–443.

Kampouroglou, E. E. and M. Economou-Eliopoulos. 2017. Assessment of arsenic and associated metals in the soil-plant-water system in Neogene basins of Attica, Greece. *Catena*. 150: 206–222.

Kim, K. H., Shon, Z. H., and P. T. Mauulida. 2014. Long-term monitoring of airborne nickel (Ni) pollution in association with some potential source processes in the urban environment. *Chemosphere*. 111: 312–319.

Kiruba, U. P., Kumar, P. S., Gayatri, K. S. et al. 2014. Study of adsorption kinetic, mechanism, isotherm, thermodynamic, and design models for Cu(II) ions on sulfuric acid modified Eucalyptus seeds: Temperature effect. *Desalination and Waste Water Treatment*. 56: 2948–2965.

Koolivand, A., Mazandaranizadeh, H., Binavapoor, M. et al. 2017. Hazardous and industrial waste composition and associated management activities in Caspian industrial park, Iran. *Environmental Nanotechnology, Monitoring and Management*. 7: 9–14.

Kumar, M. and A. Puri. 2012. A review of permissible limits of drinking water. Indian Journal of Occupational *and* Environmental Medicine 16: 40–44.

Kumar, P. S., Deepthi, A., Bharani, R. et al. 2015. Study of adsorption of Cu(II) ions from aqueous solution by surface-modified Eucalyptus globulus seeds in a fixed-bed column: Experimental optimization and mathematical modelling. *Research on Chemical Intermediates*. 41: 8681.

Moinfar, S. and G. Khayatian. 2017. Continuous sample drop flow-based microextraction combined with graphite furnace atomic absorption spectrometry for determination of cadmium. *Microchemical Journal*. 132: 293–298.

Moss, B. 2008. Water pollution by agriculture. *Philosophical Transactions of the Royal Society B*. 363: 659–666.

Multani, R. S., Feldmann, T., and Demopoulos, G. P. 2017. Removal of antimony from concentrated solutions with focus on tripuhyite (FeSbO4) synthesis, characterization and stability. *Hydrometallurgy*. 169: 263–274.

Muttamara, S. 1996. Wastewater characteristics. *Resources, Conservation and Recycling*. 16: 145–159.

Naushad, M., Ahamad, T., Al-Maswari, B. M., Alqadami, A. A., and Alshehri, S. M. 2017. Nickel ferrite bearing nitrogen-doped mesoporous carbon as efficient adsorbent for the removal of highly toxic metal ion from aqueous medium. *Chemical Engineering Journal*. 330: 1351–1360.

Naushad, M., Ahamad, T., Sharma, G. et al. 2016. Synthesis and characterization of a new starch/SnO2 nanocomposite for efficient adsorption of toxic Hg2+ metal ion. *Chemical Engineering Journal*. 300: 306–316.

Okereke, J. N., Ogidi, O. I., and K. O. Obasi. 2016. Environmental and health impact of industrial wastewater effluents in Nigeria—A review. *International Journal of Advanced Research in Biological Sciences*. 3: 55–67.

Olanipekun, O., Oyefusi, A., Neelgund, G. M. et al. 2014. Adsorption of lead over graphite oxide. *Spectrochimica Acta Part A: Molecular and Biomolecular Spectroscopy*. 118: 857–860.

Owa, F. D. 2013. Water pollution: Sources, effects, control and management. *Mediterranean Journal of Social Sciences*. 4: 8.

Pandey, P. K., Sharma, S. K., and S. S. Sambi. 2015. Removal of lead(II) from waste water on zeolite-NaX. *Journal of Environmental Chemical Engineering*. 3: 2604–2610.

Pathania, D., Katwal, R., Sharma G. et al. 2016. Novel guar gum/Al2O3 nanocomposite as an effective photocatalyst for the degradation of malachite green dye. *International Journal of Biological Macromolecules* 87: 366–374.

Prabu, D., Parthiban, R., Kumar, P. S. et al. 2016. Adsorption of copper ions onto nano-scale zero-valent iron impregnated cashew nut shell. *Desalination and Water Treatment.* 57: 6487–6502.

Rahman, Z. and V. P. Singh. 2016. Assessment of heavy metal contamination and Hg-resistant bacteria in surface water from different regions of Delhi, India. *Saudi Journal of Biological Sciences.* DOI:10.1016/j.sjbs.2016.09.018.

Rajasulochana, P. and V. Preethy. 2016. Comparison on efficiency of various techniques in treatment of waste and sewage water—A comprehensive review. *Resource-Efficient Technologies.* 2: 175–184.

Rana, R. S., Singh, P., and V. Kandari. 2017. A review on characterization and bioremediation of pharmaceutical industries' wastewater: An Indian perspective. *Applied Water Science.* 7: 1–12.

Rangabhashiyam, S. and N. Selvaraju. 2014. Evaluation of the biosorption potential of a novel Caryota urens inflorescence waste biomass for the removal of hexavalent chromium from aqueous solutions. *Journal of the Taiwan Institute of Chemical Engineers.* 47: 59–70.

Reddy, H. K. and S. M. Lee. 2012. Water pollution and treatment technologies. *Environmental and Analytical Toxicology.* 2: 5.

Rengaraj, S. and S. Moon. 2002. Kinetics of adsorption of Co(II) removal from water and wastewater by ion exchange resins. *Water Research* 36: 1783–1793.

Rizzi, V., D'Agostino, F., Fini, P. et al. 2017. An interesting environmental friendly cleanup: The excellent potential of olive pomace for disperse blue adsorption/desorption from wastewater. *Dyes and Pigments* 140: 480–490.

Ru-shan, R., Fa-en, S., Da-hua, J. et al. 2011. Notice of Retraction Sorption Equilibrium and Kinetic Studies of Cu(II) from Wastewater with Modified Sepiolites, *2011 5th International Conference Bioinformatics and Biomedical Engineering* (iCBBE), Wuhan, China. 10–12 May, 2011.

Schwarzenbach, R. P., Egli, T., Hofstetter, T. B. et al. 2010. Global water pollution and human health. *Annual Review of Environment and Resources.* 35: 109–136.

Sharma, G., Naushad, M., AlMuhtaseb, A. H. et al. 2017. Fabrication and characterization of Chitosan-crosslinked-poly(alginic acid) nanohydrogel for adsorptional removal of Chromium metal ions from aqueous medium. *International Journal of Biological Macromolecules.* 95: 484–493.

Srivastava, S, Sinha. R., and Roya, D. 2004. Toxicological effects of malachite green. *Aquatic Toxicology.* 66: 319–329.

Vitorino de Souza Melarea, A., Gonzalez, M., and K. Faceli. 2017. Technologies and decision support systems to aid solid-waste management: A systematic review. *Waste Management.* 59: 567–584.

Water Quality Standards. Indian Standard for Drinking Water Specification IS 10500:1991. www.indiawaterportal.org/../indian-standard-drinking-water—bis-specifications-10500.

Zangeneh, H., Zinatizadeh, A. A. L., Habibi, M. et al. 2015. Photocatalytic oxidation of organic dyes and pollutants in wastewater using different modified titanium dioxides: A comparative review. *Journal of Industrial and Engineering Chemistry* 26: 1–36.

2 Ecological Assessment
Use of Hydropower and Biogas Energy in Waste Water Treatment Plants

Serhat Kucukali, İrfan Ar,
Mümtaz Ak, and Asiye Günal

CONTENTS

2.1 INTRODUCTION

After the enactment of Urban Waste Water Treatment Directive, The Council Directive 91/271/EEC concerning urban waste water treatment, the installation of waste water treatment plants (WWTPs) in Europe has been accelerated. Hence, the amount of sludge production increased by over 62% between 1992 and 2005 (Méndez, 2005; Appels et al., 2008). Sludge production is increasing year by year parallel to the increasing number of WWTPs, and moreover, the use of sludge in agriculture is completely forbidden in some countries, such as the Netherlands, South Korea, and Switzerland (Bachmann, 2009). Therefore, the lack of useful applications for sludge has become a very serious problem, and this has driven research to focus on finding a reasonable and environmentally friendly solution to this environmental problem. Regarding this environmental concern, various technologies have been developed. The most widely used ones are (Patricia Sinicropi, 2012)

Energy source: WWTP sludge

1. Anaerobic digestion → Energy products: biogas, compressed natural gas, pipeline gas
2. Thermal conversion methods
   i. Incineration   Heat → Process heat
   ii. Co-combustion   Power generation
                              Syngas

iii. Gasification →   Compressed natural gas
                              Pipeline quality gas
iv. Pyrolysis → fuel product → process heat

3. Potential energy from falling wastewater or effluent (hydropower)
4. Biofuels and microbial fuel cells

In this chapter, the use of biogas and hydro energy will be addressed using the real-time operational data of a WWTP in Turkey.

2.2 ANAEROBIC DIGESTION: USE OF BIOGAS ENERGY IN WWTP

In all communities, two vitally important services are the supply of clean drinking water and efficient wastewater treatment (EPRI, 2013). Wastewater treatment is an energy-intensive process, since pumping, treating, disinfecting, aerating, and processing biosolids consumes a large amount of energy. For example, WWTPs in the United States consumed approximately 30.2 billion kWh of electricity annually as of 2013 (EPRI, 2013). Public awareness of environmental issues, especially greenhouse gas (GHG) emissions, decarbonization of energy sources, and energy, efficiency directed WWTPs toward producing their own energy and hence, net neutral energy consumption (Bachmann, 2015). Fortunately, the organic matter in raw wastewater contains 10 times more energy than is necessary to treat it. Also, WWTPs can produce 100% of the energy that they need to operate (Patricia Sinicropi, 2012). Anaerobic digestion (AD) is a proven technology used for stabilization of sewage sludge, which allows the generation of biogas from the same process (Bonnier, 2008; Lebuhn et al., 2015). In the AD process, the organic contents of sewage sludge are decomposed by bacteria in the absence of free oxygen to yield the biogas, which consists mainly of methane and carbon dioxide (Table 2.1).

In fact, AD is a natural process that can occur in lakes, dams, and swamps, and the *ignis fatuus* arises due to the presence of biogas (therefore, biogas is also called *marsh gas*). Although AD is given as a two-step reaction in some texts (Mountein, 2011), it is a complex process, as follows (Mes et al., 2003):

1. *Hydrolysis*: Conversion of non-soluble biopolymers (saccharides, proteins, lipids) to soluble organic compounds
2. *Acidogenesis*: Conversion of soluble organic compounds to volatile fatty acids (VFA) and CO_2
3. *Acetogenesis*: Conversion of VFA to acetate and H_2
4. *Methanogenesis*: Conversion of acetate and CO_2 plus H_2 to methane gas.

These processes also consist of several elementary reactions, but the detailed kinetic analysis of these reactions is outside the scope of this chapter; detailed knowledge can be found in Al Seadi et al. (2008). Biogas contains 45%–70% methane and has a calorific value close to lignite (biogas containing 65% methane has a lower heating value [NCV] of 23.3 MJ m^{-3} [European Environment Agency, 2005]) and can be

TABLE 2.1

Biogas Composition

Gas Component	Concentration Range	Mean Value
Methane (CH_4)	45%–70%	60%
Carbon dioxide (CO_2)	25%–55%	35%
Water vapor	0%–10%	3%–10%
Nitrogen (N_2)	0.01%–5%	1%
Oxygen (O_2)	0.01%–2%	0.3%
Hydrogen (H_2)	0%–1% <1%	—
Ammonia (NH_3)	0.01–2.5 mg m^{-3}	0.7%
Hydrogen sulfide (H_2S)	10–30,000 mg m^{-3}	<500 mg m^{-3}

Source: Rakican, K. and Sobota, M., Biogas for farming, energy conversion and environment protection. Department of Agroenergetics Lithuanian University of Agriculture Studentu 11, LT-53361, Akademija, Kaunas distr. Lithuania. 9, 2007.

1. Used directly in engines for electricity generation
2. Used in combined heat and power (CHP) engines, producing both electricity and heat
3. Burned to produce heat (which could also produce steam)
4. Cleaned/scrubbed to become bio-methane and used in the same way as natural gas for cooking, lighting, or as a vehicle fuel
5. Used as fuel for a high-temperature fuel cell (Turco et al., 2016)

2.3   USE OF HYDROPOWER IN WWTPS

There are two possible ways to generate hydroelectricity from wastewaters. The first is before the WWTP, and the second is after the WWTP. In the second case, the treated wastewater is diverted to a hydro turbine before being discharged to a river, lake, or sea (ESHA, 2010). The benefits and simplifications of these facilities compared with river-type hydropower plants could be summarized as follows: (1) all infrastructure access is present, which will reduce the investment cost and risk considerably, (2) the facility has no significant environmental impacts, and it has a guaranteed discharge through the year, (3) the generated electricity will be used at the infrastructure system, and the excess electricity can be sold to the government, (4) there is no land acquisition (Figure 2.1) and significant operating costs, and (5) since the hydropower potential is a function of the available discharge and head, the required data (i.e., discharge, head, flow duration curve) are usually recorded (Kucukali, 2010; Cottin et al., 2012). From a technological point of view, various solutions and configurations exist to adapt to different network set-ups. As an example, Switzerland has significant experience in the development of such multipurpose schemes, where energy production by the hydropower plant is the secondary purpose of the infrastructure (Table 2.2). For instance, Samra wastewater treatment plant in Jordan is an example of electricity production from wastewaters before and after the

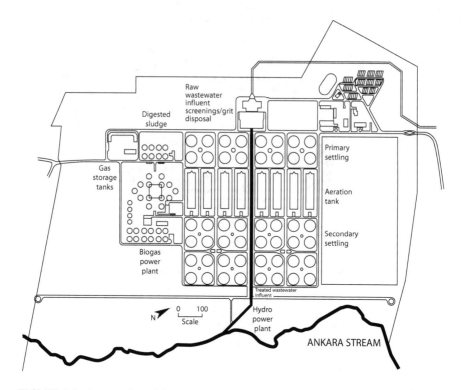

FIGURE 2.1 Layout plan of the Tatlar WWTP.

WWTP. The four hydro turbines installed at Samra WWTP will generate 21 GWh per year electrical energy, which has important environmental benefits, as Jordan generates its electricity mainly from oil. This production is estimated to avoid the emission of more than 17,000 tons of CO_2 per year. Also, the facility covers 98% of Amman city's electricity consumption by using the biogas and hydropower (at both inlet and outlet) energies together (ESHA, 2010).

In the case of hydro energy, electric generation, E, is proportional to the discharge, Q, and the head, H, as follows:

$$E = \rho g Q H \eta t \tag{2.1}$$

where:

    ρ is the density of fluid ($\rho_{wastewater} = 1080$ kg m^{-3})
    g is acceleration due to gravity
    η is the efficiency of the hydro turbine for the corresponding discharge
    t is the working time of the turbine for the relevant time period

Assessing reliable and sufficient discharge data during the planning phase of a hydropower development is necessary to accurately estimate the hydro energy yield. Accordingly, long-term (at least 1 year) on-site discharge measurements

TABLE 2.2

Multipurpose Schemes in Switzerland: Operating and Remaining Potential

Water Network Type	Potential Type	Number of Sites	Output (MW)	Production (GWh per year)
Drinking water	Operating	90	17.8	80
	Remaining	380	38.9	175
Treated waste water	Operating	6	0.7	2.9
	Remaining	44	4.2	19
Raw waste water	Operating	1	0.38	0.85
	Remaining	86	7.1	32

Source: ESHA (European Small Hydropower Association). Energy Recovery in Existing Infrastructures with Small Hydropower Plants: Multipurpose Schemes—Overview and Examples, 2010.

and the analysis of the measured data by an expert are key steps in reducing the uncertainty related to the energy yield assessments. Ak et al. (2016) used a fuzzy logic decision analysis tool to select the most appropriate low-head hydro turbine alternative at the outlet of WWTPs, and they found the Archimedean screw to be the most suitable technology for hydropower development at the outlet of a WWTP due to its superior performance according to economic and environmental criteria. In this context, this study aims to show the possible benefits of the installation of a hydro turbine at the outlet of a WWTP. This could be an alternative clean energy solution to reduce the consumption of energy supplied by the national electric grid, mostly fed by fossil fuels, and to induce the minimization of CO_2 emissions to the atmosphere. Energy recovery efficiency, ER, is calculated from

$$ER = \frac{E}{E_c} \qquad (2.2)$$

where:

E is the generated electricity

E_c is the electricity consumed by the facility

2.4 CASE STUDY: TATLAR WWTP

Tatlar WWTP, which is the one of the biggest WWTPs in Europe, has been operated by the Municipality of Ankara since 1993. The combined sewer system carries domestic and industrial wastewater and rainwater by gravity, and there is no need for pumping. Because of its high population and industrial activities, Ankara produces an enormous amount of wastewater. The wastewater of Ankara metropolitan area is discharged into Ankara Creek. The plant was built as an active sludge project with 765,000 m³ per day wastewater treatment capacity (Ak et al., 2016). The treatment process is anaerobic sludge stabilization, active sludge technique, and removal of sludge water using the strip filter technique. By the year 2025, it is planned to increase the capacity of the plant to ensure the treatment of 6.3 million people's wastewater which will be about 1,377,000 m³ per day (Ak et al., 2016). The function of the WWTP is to enhance the water quality of the receiving water body from Class IV to Class II by applying physical treatment (screenings/grit disposal), primary settling, activated sludge aeration, and secondary settling (Figure 2.1). Then, the treated wastewater effluent is discharged into Ankara Creek. The discharge is measured at the outlet of the plant at the Parshall flume by an ultrasonic displacement meter (Figure 2.2).

Assessing reliable and sufficient discharge data during the planning phase of a hydropower development is necessary to accurately estimate the energy yield. Accordingly, long-term (at least 1 year) on-site discharge measurements and the analysis of the measured data by an expert are key steps in reducing the uncertainty related to the energy yield assessments. Uncertainty analysis is becoming an increasingly important tool in measurements. The uncertainty in the discharge measurements in the Parshall flume is expected to be less than 3%, while the error in the level measurements at the project site is expected to be below 0.1%. In this study, to calculate the annual hydroelectricity generation, the daily flow duration curve of Tatlar WWTP for the year 2013 (Figure 2.3) is used. From these data, monthly hydroelectricity generation is computed.

The construction duration for the Archimedean screw is significantly shorter (i.e., 9 months) than that of the conventional hydro turbine, which creates an

FIGURE 2.2 Discharge measurement at the outlet of the Tatlar WWTP by an ultrasonic displacement meter.

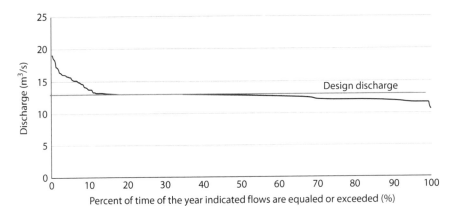

FIGURE 2.3   Discharge flow duration curve of Tatlar WWTP for the year 2013.

opportunity cost due to the early electric generation of the screw. Also, it has lower operation and maintenance costs compared with other conventional low-head hydro turbines (Ak et al., 2016). The supply and installation of the Archimedean screw require less than 10 months, due to the fact that the screw is a compact technology and does not require the assembly of many parts (Figure 2.4). This characteristic of the screw also reduces the supply risk (Steel and Lilleland, 2015). Also, the Archimedean screw is the only low-head hydropower technology that causes aeration by creating turbulence and mixing at the free surface of the flow. Hence, the screw enhances the dissolved oxygen content of the passing water. This is important, because Çakir (2014) reported that the artificial aeration process alone is responsible for half of the total energy cost in a wastewater plant. Also, a 3 MW biogas power plant has been operated in Tatlar WWTP. The monthly electric generation from the biogas power plant is shown in Figure 2.5. Biogas power plants avoid GHG emissions considerably relative to other power plants. In Table 2.3, the GHG emission reduction potential of different types of renewable energy projects in Turkey is presented. As can be clearly seen from Table 2.3, the biogas power plants have nearly four times higher GHG emission reduction potential compared with other renewable power plants in terms of CO_2e tons MWh^{-1}. This is due to the fact that in biogas power plants, methane is tapped and converted into useful energy. Methane is a GHG with a global warming potential nearly 23 times greater than CO_2; hence, 1 tonne of methane equals about 23 tons CO_2e (Bayon et al., 2009). Monthly hydroelectricity and biogas generation based on the measured values and their comparison with the monthly electric consumption of the Tatlar WWTP for the year 2013 are shown in Figure 2.5. The facility recovers over 100% of its electricity consumption by harnessing the biogas and hydro energies. Moreover, this combined electricity production is estimated to avoid the emission of more than 60,000 tons CO_2 per year.

FIGURE 2.4   Photos of the Archimedean screws at the outlet of Tatlar WWTP: (a) top view, (b) side view. Project data: $Q=15$ m^3 s^{-1}, $H_g=9.25$ m, $H_n=8.76$ m, $P=4\times250$ kW $=1000$ kW.

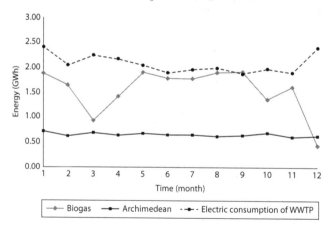

FIGURE 2.5   Monthly hydroelectricity and biogas generation based on the measured values and their comparison with the monthly electricity consumption of the Tatlar WWTP for the year 2013.

TABLE 2.3
Performance of the Different Types of Renewable Power Plants in Turkey

Type of Renewable	Name of Power Plant	P (MW)	E (GWh per year)	Capacity Factor (%)	Annual CO₂e Reduction (tons per year)	CO₂e Emission Reduction Factor (ton MWh⁻¹)
Landfill[a]	Mamak	25.4	203.2	91	500,000	2.46
Landfill	Sincan	5.6	44.8	91	125,000	2.79
Wind[b]	Zeytineli	50	149	34	89,000	0.60
Diversion type hydro	Sayan	14.9	48.3	37	26,550	0.55
Storage type hydro	Karasu	5.9	38.9	75	17,918	0.46

[a] ITC, http://www.itcturkiye.com/, 2012.
[b] Futurecamp, http://fc-tk.futurecamp.de/, 2012.

2.5 CONCLUSIONS

This study focuses on the energy aspect of WWTPs. The use of the existing biogas and hydropower energy potential in WWTPs has been analyzed. The new energy laws and the market conditions in most countries around the world create an opportunity to harness this potential. As a case study, Tatlar WWTP in Ankara has been investigated in a detailed manner. The annual electricity consumption of Tatlar WWTP was 26.5 GWh in 2013. The installation of an Archimedean screw at the outlet of the WWTP has a payback period of less than 3 years, and hydroelectricity generation exceeds the electricity consumption by about 30%, producing clean and feasible energy. The findings of the case study demonstrated that by using biogas and hydro energy together, the energy recovery efficiency of a WWTP can exceed 100%, which is consistent with the operational data of Samra WWTP in Jordan. These energy-related technological solutions can make considerable contributions to sustainable WWTP development.

REFERENCES

Ak, M., Kentel, E., and Kucukali, S., 2016 A fuzzy logic tool to evaluate low-head hydropower technologies at the outlet of wastewater treatment plants, *Renewable and Sustainable Energy Reviews*, 68, 727–737.

Al Seadi, T., Rutz, D., Prassl, H., Köttner, M., Finsterwalder, T., Volk, S., and Janssen, R., 2008. *Biogas Handbook*, Published by University of Southern Denmark Esbjerg, Niels Bohrs Vej, 9–10, ISBN 978-87-992962-0-0, pp. 21–29.

Appels, L., Baeyens, J., Degrève, J., and Dewil, R., 2008. Principles and potential of the anaerobic digestion of waste-activated sludge. *Progress in Energy and Combustion Science*, 34, 755–781.

Bachmann, N., 2009. Vorteile und Grenzen der Vergarung von leicht abbaubaren Industrie- und Lebensmittelabfallen in Abwasserrinigungsanlagen. Federal Office of Energy, Switzerland, No.290101/102963.

Bachmann, N., 2015. Sustainable Biogas Production in Municipal Waste Water Treatment Plants, Technical Brochure, IEA Energy Technology Network.

Bayon, R., Hawn, A., and Hamilton, K., 2009. *Voluntary Carbon Markets: An International Business Guide to What They Are and How they Work*, Sterling, Earthscan.

Bonnier, S., 2008. Technical Synthese, the State of the Promotion of Biogas from Waste Water Plants in France and Europe. https://www.agroparistech.fr/IMG/pdf/syn08-eng-Bonnier.pdf.

Çakır, C., 2014. Atıksu Arıtma Tesislerinde Hava Dağılımı Optimizasyonu ile Enerji Tasarrufu. 4. Turkish-Ge. Water Partnership-Day, Antalya, Turkey (in Turkish).

Cottin, C., Choulot, A., and Denis, V., 2012. Multipurpose schemes: Feedback on electro-mechanical equipments set within drinking water networks. *Proceeding of the 2012 HidroEnergia Conference*, Wroclaw, Poland.

EPRI (Electric Power Research Institute), 2013. Electricity Use and Management in the Municipal Water Supply and Wastewater Industries 3002001433 Final Report.

ESHA (European Small Hydropower Association), 2010. Energy Recovery in Existing Infrastructures with Small Hydropower Plants: Multipurpose Schemes—Overview and Examples.

European Environment Agency, 2005. Climate change and a European low-carbon energy system. Report 1/2005. EEA.

Futurecamp, 2012. http://future-camp.de/?locale=en_US (accessed March 2016).

ITC, 2012. www.itcturkiye.com/ (accessed April 2016).

Kucukali, S. 2010. Hydropower potential of municipal water supply dams in Turkey: A case study in Ulutan Dam. *Energy Policy*, 38(11), 6534–6539.

Lebuhn, M., Weiß, S., Munk, B., and Guebitz, G.M., 2015. Microbiology and molecular biology tools for biogas process analysis, diagnosis and control, in *Biogas Science and Technology*, Editors: G. Gübitz, A. Bauer, G. Bochmann, A. Gronauer, Springer Applied Science, Stefan Weiss, pp. 1–40.

Méndez A., 2005. Obtención de biocombustibles por tratamiento térmico de lodos de depuradora. In: *Congreso Europeo de Energías Renovables y Eficiencia Energética*. Ávila, Spain.

Mes, T.Z.D., Stams, A.J.M., Reith, J.H., and Zeeman, G., 2003. Methane production by anaerobic digestion of wastewater and solid wastes. *Bio-Methane & Bio-Hydrogen: Status and Perspectives of Biological Methane and Hydrogen Production*. Dutch Biological Hydrogen Foundation, The Hague, The Netherlands.

Mountein, H. 2011. Energy Recovery in Wastewater Treatment - More than Biogas. Energy Efficient Wastewater Treatment and Solid Waste Management Seminar, Edmonton, AB.

Patricia Sinicropi, J.D., 2012. Biogas Production at Wastewater Treatment Facilities. Congressional Briefing-May 16, National Association of Clean Water Agencies.

Rakican, K. and Sobota, M., 2007. Biogas for farming, energy conversion and environment protection. Department of Agroenergetics Lithuanian University of Agriculture Studentu 11, LT-53361, Akademija, Kaunas distr. Lithuania. 9,

Steel, E. and Lilleland, O., 2015. Using supplier performance data to improve project performance. *Proceeding of Hydro 2015 Conference*, Bordeaux, France, 10.02, USB.

Turco, M., Ausiello, A., and Micoli, L., 2016. *Treatment of Biogas for Feeding High Temperature Fuel Cells Removal of Harmful Compounds by Adsorption Processes*, Springer International Publishing Switzerland.

3 Life Cycle Assessment of Municipal Wastewater and Sewage Sludge Treatment

Maria Teresa Moreira, Andrea Arias, and Gumersindo Feijoo

CONTENTS

3.1 INTRODUCTION

The increasing demand for clean water by citizens and environmental organizations led the European Commission to consider water protection as one of its priorities as outlined in the Water Framework Directive (WFD) adopted in 2000. Water consumption is expected to increase by 16% in 2030, which is especially worrying in the existing scenario of water scarcity that affects 11% of the European population and 17% of EU territory. The EU policy paper on climate change published in 2009 highlights the need for further measures to enhance the efficiency of water use and to increase resilience to climate change (European Commission, 2009). Concerns are associated not only with the decreasing availability of freshwater but also with the deterioration of water quality due to eutrophication or toxicity (Loubet et al., 2014).

Wastewater treatment plants (WWTPs) are the key stakeholders responsible for effluent discharges into the aquatic environment. From the development of the activated sludge process in 1914, the main focus in wastewater treatment has been devoted to improving effluent quality, which has evolved in parallel with increasingly strict discharge limits. In the European context, the conventional configuration of WWTPs is designed to remove mainly organic matter and nitrogen. However, the implementation of new discharge limits in accordance with the WFD requires a search for environmentally efficient treatment schemes. In particular, the contribution of sewage treatment to climate change is considered as a major indirect environmental impact. It has been estimated that around 1% of the average daily electricity consumption in Western Europe is due to municipal and industrial wastewater treatment. If hydrological management and agricultural demand are included, the consumption could reach levels of 4%–5%, as reported in COST Action Water2020. Therefore, the possibility of energy saving in WWTPs should acquire increasing importance in the water treatment sector.

Beyond the target of reduced energy consumption, the idea of sustainable WWTPs during conception, design, upgrading, and operation should be based on those technologies capable of maintaining a balance between desirable quality of emissions (effluent, sludge, and gases) and recovered resources (reclaimed water, fertilizers, and energy) in a multi-disciplinary approach. In this framework, WWTP performance needs to be evaluated under technical, environmental, energetic, social, and economic aspects adapted to the specific needs of each scenario (size, location, point of discharge, etc.).

3.2   LIFE CYCLE ASSESSMENT METHODOLOGY IN THE FRAMEWORK OF WASTEWATER TREATMENT

Among the different environmental management tools, life cycle assessment (LCA) is a method that allows the environmental impact of a product or a service to be assessed in relation to its function over its whole life cycle, from raw material acquisition to production, use, and disposal. The standard procedure for LCA developed by the International Organization for Standardization (ISO, 1997) established the stages of the methodology when addressing an LCA study: goal and scope definition, the life cycle inventory (LCI) analysis phase, life cycle impact assessment (LCIA), and the interpretation of results (Figure 3.1).

3.2.1   GOAL AND SCOPE

This is the first stage of the methodology and probably the most important, as it includes essential elements of the study such as the purpose, scope, and main hypotheses (ISO, 2006a,b).

To correctly set up the study, the functional unit (FU) has to be selected, that is, the reference value to which all flows and emissions are referred. The adequate

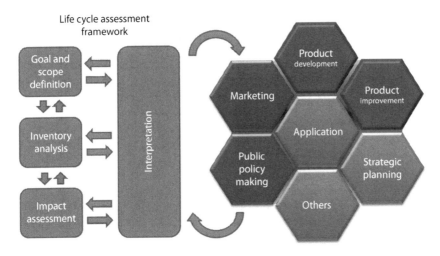

FIGURE 3.1   Framework and applications of LCA methodology. (Adapted from ISO 14040. Environmental Management and Life Cycle Assessment—Principles and Framework. International Organization for Standardization, 2006a.)

selection of the FU is of major importance, because different FUs could lead to different results for the same product systems (Panesar et al., 2017).

The presentation of the product/process to be assessed will provide the framework for the system boundaries as well as the limitations of the study. Accordingly, the specificities for each case study must be clearly stated from the very beginning. Regarding the selection of the system boundaries, physical, geographical, and temporal approaches will determine the stages of the life cycle to be assessed (research and development, material extraction, component manufacturing, use, and management and disposal at the end of life).

In the specific case of wastewater treatment, the system boundaries will typically include primary and secondary treatment (tertiary processes can also be present in some facilities) as well as the production of energy and chemicals and the emissions associated with the process (Figure 3.2). When evaluating lab-scale technologies, scale-up has great importance, especially with regard to infrastructure, operational yields, and energy consumption. Over- or underestimation of these items can totally

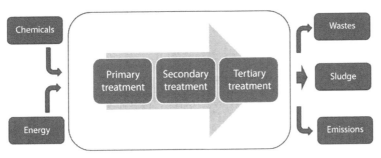

FIGURE 3.2   The framework of wastewater treatment.

mask the real impacts. Other major elements that may or may not be included in the system boundaries are impacts that it is possible to avoid. An example of this is within the particular situation of agricultural application of treated sludge, which allows the positive effects of the nutrient value of the sludge to be taken into account and the system to be expanded to include the avoidance of production of synthetic fertilizers as well as the negative consequences related to their application to soil in terms of eutrophication and toxicity impacts.

3.2.2 LIFE CYCLE INVENTORY ANALYSIS

The LCI involves data collection and calculation procedures to estimate the consumption of resources and the qualities of waste flows and emissions attributable to the life cycle of a product/process (ISO, 2006a). It usually starts by constructing a flowchart of the process, representative of the goal and scope considered, and continues with data collection. Within this phase, the major concern is often related to data availability and quality. When it comes to assessing the potential benefits of innovative solutions, it is foreseen that data collection will come from laboratory experiments and in the best cases, from pilot facilities. Only when well-established technologies have been assessed can field measurements and data from real plants be handled.

The option of modeling and simulation of a new facility, the use of a detailed design project, and relevant literature may be useful alternatives for primary data collection, but they are considered complementary. Background information (e.g., electricity generation systems, concrete and chemicals production processes) is normally provided by LCI databases, for example the Ecoinvent (www.ecoinvent.ch). It is also an iterative process, because as data collection proceeds, more information becomes available, and the goal and scope of the study might change accordingly.

3.2.3 LIFE CYCLE IMPACT ASSESSMENT

The evaluations of the potential environmental impacts associated with inventory data are performed in the LCIA. This stage converts the potential contributions of the resource extractions and wastes/emissions of the inventory into a number of potential impacts. For this purpose, the ISO standards define three mandatory stages (1 to 3) plus two optional ones (4 and 5) depicted in Figure 3.3 and are as follows:

1. Selection of impact categories, which can be classified as midpoint impact categories (climate change, ozone layer depletion, eutrophication, ecotoxicity, human toxicity, resource depletion, land use, acidification, radiation, ozone layer formation, and respiratory inorganics) or end-point categories (human health, ecological health, and resource depletion).
2. Classification of the different emissions according to the impact category they affect. For example, CO_2, CH_4, and N_2O emissions are classified in the category of climate change. We need to be aware that some substances can affect several impact categories.

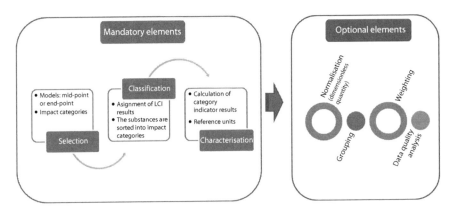

FIGURE 3.3 Elements of Life Cycle Impact Assessment. (Adapted from ISO 14040. Environmental Management and Life Cycle Assessment—Requirements and Guidelines. International Organization for Standardization, 2006a.)

3. Characterization, the quantification of the impact associated with each emission, which provides a way to directly compare the LCI results within each category. Characterization factors are considered as equivalency factors; for example, CO_2 for climate change.
4. Normalization, the stage when the different impact potentials and consumption of resources are expressed on a common scale.
5. Weighting, which aggregates all impact categories into one single score using subjective criteria such as expert opinion, monetization, or policy goals.

Among the range of available methods, the choice of impact assessment models should be tailored according to their relevance to urban wastewater systems. It is recommended to use the EU recommended midpoint impact categories as described in the International Life Cycle Data (ILCD)/Product Environmental Footprint (PEF) guide (European Commission, 2010). Furthermore, the ILCD approach should be combined with more complete models such as ReCiPe (Goedkoop et al., 2013). Other LCIA methods, such as CML (Guinée et al., 2002), EDIP20036 (Hauschild and Potting, 2003), or Impact 2002+ (Jolliet et al., 2003), present a set of impact categories such as eutrophication, global warming, and ecotoxicity. Thus, most LCA practitioners would only need to choose the methodology that fits best the objective pursued and the impact categories that are relevant for each scenario.

3.2.4 INTERPRETATION OF RESULTS

The interpretation phase assesses the outcomes of the study in a context of alternative scenarios as well as through sensibility and uncertainty analysis (Bauman and Tillman, 2004). This stage should identify the hotspots of the process as well as communicating conclusions, limitations, and recommendations. One of the widely used ways of presenting the results is to take a reference scenario for which the impacts are calculated and relate the impacts of the other scenarios to the one considered as reference.

3.3 EVALUATION OF CONVENTIONAL WWTPS: METHODOLOGICAL ASSUMPTIONS AND ENVIRONMENTAL INDICATORS

The evaluation of conventional WWTPs by LCA started in the 1990s (Emmerson et al., 1995; Roeleveld et al., 1997), and two approaches were followed: the identification of improvement alternatives for a single plant (Hospido et al., 2004) and the comparison of different technologies toward an equivalent target (Coats et al., 2011; Gallego et al., 2008; Meneses et al., 2010; and Rodriguez-Garcia et al., 2011).

The major differences found among the different LCA reports on wastewater treatment systems are the selection of the FU, the system boundaries, and the environmental indicators. The most common FUs used in LCA studies of WWTPs are the flow rate of treated water (the simplest approach) and the quantification of the environmental load associated with a one-person equivalent (Table 3.1). The former has the advantage of being based on physical data, while the latter tends to be used for comparative purposes, since the design project of wastewater facilities includes this parameter as a design value that integrates the composition and flow of the influent. However, neither alternative reflects the function of the system and operational efficiency in terms of eutrophication reduction. One interesting alternative is the Net Environmental Benefit (NEB) approach developed by Godin et al. (2012), who calculated the difference between avoided potential impacts, such as those associated with influent discharge and those associated with WWTP operation.

The quality of LCI data is another issue of special interest, considering that inventory data may be subject to substantial variability in terms of flow and composition of the influent (Yoshida et al., 2014a). When it comes to analyzing data representativeness, it is evident that the operation of a WWTP is not a satisfactory example of

TABLE 3.1
Functional Units Used in Some LCA Studies of Wastewater Treatment Plants

Functional Unit	
Value	**References**
Cubic meters of treated water	Brix (1999), Hospido et al. (2004), Ortiz et al. (2007), Høibye et al. (2008), Renou et al. (2008), Pasqualino (2011), Venkatesh and Brattebø (2011), Hospido et al. (2012), and Li et al. (2013)
Cubic meters and kilograms of PO_4^{3-} removed	Rodríguez-García et al. (2011)
Population equivalent	Emmerson et al. (1995), Tillman et al. (1998), Mels et al. (1999), Lundin et al. (2000), Kärrman and Jönsson (2001), Dixon et al. (2003), Machado et al. (2007), Gallego et al. (2008), and Hospido et al. (2008)

steady-state conditions. Hospido et al. (2004) evaluated the influent variability of a WWTP during two different seasons (humid and dry). However, the outcomes from the environmental analysis demonstrated that variations within each period turned out to be more significant than those variations attributed to seasonality. Water discharge and sludge application to agricultural soil were found to be the main contributors to the environmental performance of a WWTP.

It is evident that the configuration of a WWTP will definitely affect the treatment performance in terms of removal yields of target pollutants, but it will also have indirect related impacts. It is important to compare the impacts generated from a WWTP scenario against a non-treatment option.

In general, the most significant environmental impacts in the different LCA studies on wastewater treatment can be classified within three impact categories: energy consumption associated with aeration in the secondary treatment, accounting for global warming; the presence of heavy metals in the sewage sludge, which influences the toxicity-impact categories; and the remaining content of chemical oxygen demand (COD), N, and P in the treated effluent, linked to eutrophication. Other relevant impact categories, but with comparatively less importance, are ozone layer depletion (especially when including N_2O emissions) and abiotic depletion (associated with fossil energy and material depletion).

3.4   FOCUS ON SLUDGE MANAGEMENT

Sewage sludge is an unavoidable waste product from the treatment of wastewater, and its final disposal plays a relevant role in the global impact of the WWTP (Gallego et al., 2008). According to the figures provided to the European Commission, sludge production associated with sewage treatment has steadily increased between 1995 and 2008 (6.7–10.13 million tons dry matter [DM]). Moreover, it is estimated that sludge production will reach about 13 million tons DM in 2020 in the EU27 (European Commission, 2008). This can be attributed mainly to the implementation of the Urban Waste Water Treatment Directive (91/271/EEC), which means a higher amount of wastewater treated and, in many cases, an increase in the production of sludge.

There is an increasing tendency to consider sludge as a resource rather than a waste material; it can be used for nutrients (phosphorus and nitrogen), and carbon can be recovered and reused in agriculture (Hospido et al., 2005; Johansson et al., 2008; Lederer and Rechberger, 2010; and Linderholm et al., 2012). In a life cycle perspective, agricultural sludge application substitutes for the production and use of mineral fertilizers, reducing the depletion of virgin resources such as mineral phosphorus extracted from phosphate rock (Cordell et al., 2011). However, the difference in the environmental performance on applying sludge instead of mineral fertilizer to soil depends on the product quality, especially in terms of heavy metals content.

As an alternative to agricultural sludge application, energy can be recovered from sewage sludge via anaerobic digestion or incineration (Rulkens, 2008). The energy generated from biogas may be used internally in the WWTPs, thus reducing their energy demand from external sources, for example from the electricity grid (Appels et al., 2008). The advantage of incineration is the production of energy, whereas its main drawback is the air emission of toxic substances into the environment.

Focusing on energy and greenhouse gas (GHG) emissions, Houillon and Jolliet (2005) compared a range of scenarios for wastewater sludge treatment: agricultural spreading, fluidized bed incineration, wet oxidation, pyrolysis, incineration in cement kilns, and landfill. From an energy perspective, the first two obtained the best results. In another work, anaerobic digestion, pyrolysis, and incineration of sludge were compared (Hospido et al., 2005). Regarding eutrophication, the benefits due to avoidance of fertilizers if digested sludge is applied to the land are not present in the other two options, since no nutrients are recovered. However, land application also has its drawbacks, since the toxic impacts associated with heavy metal release are strongly reduced when thermal processes are considered.

The trade-off between anaerobic digestion, aerobic stabilization, and incineration of sewage sludge indicated that for eutrophication- and toxicity-related categories (except for terrestrial ecotoxicity), facilities performing anaerobic digestion followed by sludge incineration show higher environmental impacts compared with plants treating the sludge by aerobic stabilization prior to agricultural sludge application (Niero et al., 2014).

In the context of sustainable wastewater treatment and sludge management, future research is needed for the quantification of long-term toxic impacts on humans and ecosystems and the potential for carbon sequestration, as well as regarding the assessment of the economic and social repercussions of different management options.

3.5   NEW GENERATION OF WWTPS UNDER A LIFE CYCLE PERSPECTIVE

Although the main objective of conventional WWTPs has traditionally been the removal of organic matter, nowadays, key words in wastewater treatment facilities include terms such as micropollutants, GHG emissions, nutrient removal, resource recovery, reclaimed water, and high-quality effluents. Consequently, LCA studies should comprise these new targets in the analysis of wastewater treatment technologies.

3.5.1   QUANTIFYING HIDDEN IMPACTS: EMERGING POLLUTANTS AND DIRECT GHG EMISSIONS

Micropollutants are considered as a potential hazard for aquatic organisms and human health (Carballa et al., 2004). However, their inclusion in life cycle environmental assessments of WWTPs was not possible until their characterization factors (CFs) for toxicity were available. The potential impacts on ecotoxicity and human toxicity of wastewaters containing priority and emerging pollutants were quantified according to the methodology developed by Muñoz et al. (2008), which was later updated by Alfonsín et al. (2014) by means of the USEtox (Rosenbaum et al., 2008) and USES-LCA 2.0 (van Zelm et al., 2009) methodologies.

Different tertiary treatments based on advanced wastewater treatment technologies (sand filtration, ozonation, and membrane bioreactors [MBRs]) have been assessed for their influence on the removal of a range of micropollutants, specifically organic substances, heavy metals, estrogens, and pathogens (Høibye et al., 2008; Wenzel et al., 2008; and Muñoz et al., 2009).

More recently, Rodriguez-Garcia et al. (2014b) reported a more extensive analysis including seven WWTPs using MBR and four novel MBR configurations at pilot-plant scale. The former (i.e., the seven full-scale plants) showed the importance of removing eutrophying substances, whereas the latter (the four novel MBRs) focused on the removal of pharmaceuticals and personal care products (PPCPs), of which only hormones were found to be significant for the toxicity impact categories.

Fugitive emissions of CH_4 and N_2O in WWTPs are an environmental issue with increasing importance within the scientific and technical literature (Nair et al., 2014; Pijuán et al., 2014; Venkatesh et al., 2014; and Yoshida et al., 2014b). CH_4 is principally formed in the WWTP units, where anaerobic conditions prevail (Daelman et al., 2012), while N_2O is mainly generated in anoxic areas of activated sludge reactors (Ahn et al., 2010). However, Guisasola et al. (2008) reported that dissolved CH_4 present in the sewer influent is emitted via stripping in the aerated units. Detailed emission inventories from WWTPs are being conducted in several countries (Kampschreur et al., 2009; Ahn et al., 2010), and models have been developed to estimate these emissions (Foley et al., 2010b; Rodríguez-García et al., 2012), demonstrating the importance of considering the environmental impacts of wastewater treatment under a holistic perspective. A long-term study performed in the Kralingseveer WWTP (The Netherlands) reported that the quantification of direct emissions of N_2O and CH_4 exceeded the plant's indirect CO_2 emissions related to electricity use (Daelman et al., 2013). A more recent study performed in two different Spanish WWTPs included the results from GHG sampling campaigns in the environmental assessment of the facilities (Figure 3.4); the authors concluded that ruling out this type of data will lead to an underestimation of between 15% and 35% of the final global warming potential (GWP) impact (Lorenzo-Toja et al., 2015).

3.5.2   THE PARADIGM OF NUTRIENT REMOVAL OR NUTRIENT RECOVERY FROM WASTEWATERS

Gallego et al. (2008) quantified the environmental impact of several configurations to remove N and P in 13 WWTPs of less than 20,000 persons equivalent. The content of P, N, and organic matter in the treated effluent and the content of heavy metals in the sludge were identified as the most significant impacts for all WWTPs. Electricity use plays an important role in five of seven impact categories and presents the highest importance in four of them. When technologies were compared, BioDenipho® and aerobic-anoxic treatments were found to be the less damaging options for secondary treatment, as they attained higher removal efficiencies of N and P than extended aeration. The potential for P removal was assessed in different

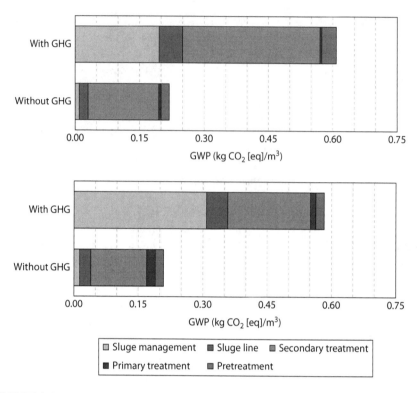

FIGURE 3.4 Contribution to GWP in scenarios with GHG measurements.

technologies based on biological and/or chemical processes from an environmental perspective by Coats et al. (2011), who highlighted the benefits of biological treatments over chemical processes.

Rodriguez-Garcia et al. (2011) performed a broader analysis of 24 WWTPs, which were classified into six types according to their quality requirements for their final discharge or reuse. This report introduced the novelty of using two different FUs: one based on volume (cubic meters) and the other on eutrophication reduction (kilograms PO_4^{3-} eq. removed). The FU of kilograms PO_4^{3-} eq removed reflects the real objective of a WWTP and the sustainability among the different options. Those WWTPs designed for organic matter removal showed better performance in both environmental and economic terms if the volume was used as the FU, whereas more complex types such as reuse plants showed variable results: lower eutrophication but higher cost and global warming. The use of the second FU showed the superior performance of those configurations that allowed water reuse.

A stream with high characteristic values of N and P is the centrate that comes from the anaerobic digester. Nitritation, Anammox, and nitrite shortcut technologies presented lower environmental impacts than struvite crystallization (Rodriguez-Garcia et al., 2014a). However, when the three options are presented in their context and implemented in a full-scale WWTP, their contribution to the global impact can be considered negligible.

3.5.3   TOWARD WATER REUSE: HIGH QUALITY EFFLUENTS

Membrane-based technologies are usually considered a recommendable option for tertiary treatment due to the possible reuse of the effluent. Tangsubkul et al. (2005) compared several treatment technologies for the potential agricultural reuse of wastewater, including MBR and continuous microfiltration (CMF). In a specific work on MBR for water reuse, tertiary treatments were found to provide high-quality effluents without increasing the environmental impact associated with the WWTP (Ortiz et al., 2007). Technologies specifically designed for high-quality effluents (chlorination plus ultraviolet treatment, ozonation, or ozonation with hydrogen peroxide) were compared by Meneses et al. (2010) and Pasqualino et al. (2011), who concluded that the options evaluated presented similar environmental profiles and that the environmental impact of the plant was only slightly increased when compared with traditional ones.

3.6   INTEGRATING LCA IN THE DESIGN AND SIMULATION OF NEW FACILITIES

The use of simulation and modelization to compare technologies can be of remarkable interest in the decision-making process. Foley et al. (2010a) compared six different technologies for nutrient removal. The normalized results of the potential impacts showed that to justify stringent nutrient removal, it would be necessary to weight the eutrophication impact category three times more than other categories, such as global warming or toxicity. A very similar approach was followed by Wang et al. (2012) for three discharge levels, each one more demanding than the previous one, finding that more stringent requirements for water quality led to a higher environmental burden.

LCA and modeling have also been combined to select the most appropriate treatment scenario for a given situation (wastewater to be treated, location, etc.) by means of the implementation of the LCA methodology within the knowledge basis of a decision support system (Garrido-Baserba et al., 2012).

Along these lines, the PIONEER_STP project will assess the impact of innovative units (nowadays at laboratory or pilot scale or in their early stages of industrial implementation) on the global plant efficiency and sustainability, taking into account nutrients, energy, emerging pollutants (EPs), GHGs, and cost/benefit balances (Figure 3.5). As organic matter (OM) and nutrients will be targeted separately, two of them will be focused on energy recovery from valuable organic matter and two on nutrient removal/recovery. The processes can be combined into different plant layouts (using a superstructure-based optimization framework), which will be further optimized by modeling and simulation based on technical, environmental, energetic, and economic aspects.

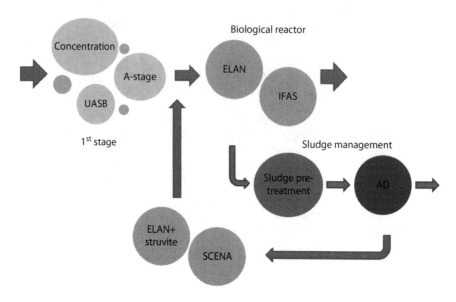

FIGURE 3.5 PIONEER_STP layout.

3.7 CONCLUSIONS

This chapter has delved into the main environmental findings that have been reported in the literature in the analysis of wastewater treatment technologies by life cycle perspective methodologies. LCA has been shown to be a key environmental methodology to identify key hotspots in the analysis of wastewater treatment technologies and provide extensive information to define criteria and establish priorities in the selection of technologies for wastewater treatment.

ACKNOWLEDGEMENTS

This research has been supported by the EU project (PIONEER_STP). The authors belong to the Galician Competitive Research Group GRC 2013-032, program co-funded by FEDER: Fondo Europeo de Desarrollo Regional, as well as to CRETUS (AGRUP2015/02)

REFERENCES

Ahn, J., Kim, S., Park, H., Rahm, B., Pagilla, K., and Chandran, K., 2010. N$_2$O emissions from activated sludge processes, 2008–2009: Results of a National Monitoring Survey in the United States. *Environmental Science and Technology* 44, 4505–4511.

Alfonsín, C., Hospido, A., Omil, F., Moreira, M.T., and Feijoo, G., 2014. PPCPs in wastewater—Update and calculation of characterization factors for their inclusion in LCA studies. *Journal of Cleaner Production* 83, 245–255.

Bauman, H. and Tillman, A.-M., 2004. *The Hitch Hiker's Guide to LCA*. Studentlitteratur AB, Lund (Sweden).

Brix, H., 1999. How 'green' are aquaculture, constructed wetlands and conventional wastewater treatment systems? *Water Science and Technology* 40(3), 45–50.

Carballa, M., Omil, F., Lema, J.M., Llompart, M., García, C., Rodríguez, I., Gómez, M., and Ternes, T., 2004. Behavior of pharmaceuticals, cosmetics and hormones in a sewage treatment plant. *Water Research* 38(12), 2918–2926.

Coats, E.R., Watkins, D.L., and Kranenburg, D., 2011. A comparative environmental life-cycle analysis for removing phosphorus from wastewater: Biological versus physical/chemical processes. *Water Environment Research* 83(8), 750–760.

Cordell, D., Rosemarin, A., Schröder, J.J., and Smit, A.L., 2011. Towards global phosphorus security: A systems framework for phosphorus recovery and reuse options. *Chemosphere* 84(6), 747–758.

Daelman, M.R.J., van Voorthuizen, E.M., van Dongen, U.G.J.M., Volcke, E.I.P., and van Loosdrecht, M.C.M., 2012. Methane emission during municipal wastewater treatment. *Water Research* 46(11), 3657–3670. doi:10.1016/j.watres.2012.04.024.

Daelman, M.R.J., van Voorthuizen, E.M., van Dongen, L.G.J.M., Volcke, E.I.P., and van Loosdrecht, M.C.M., 2013. Methane and nitrous oxide emissions from municipal wastewater treatment—results from a long-term study. *Water Science and Technology* 67(10), 2350–2355. doi:10.2166/wst.2013.109.

Dixon, A., Simon, M., and Burkitt, T., 2003. Assessing the environmental impact of two options for small-scale wastewater treatment: Comparing a reedbed and an aerated biological filter using a life cycle approach. *Ecological Engineering* 20(4), 297–308.

Emmerson, R.H.C., Morse, G.K., Lester, J.N., and Edge, D.R., 1995. The life-cycle analysis of small-scale sewage-treatment processes. *Water and Environment Journal* 9, 317–325.

European Commission, 2008. Environmental, economic and social impacts of the use of sewage sludge on land. Final Report. Part I: Overview Report.

European Commission, 2009. White Paper. Adapting to climate change: Towards a European framework for action. Commission of the European Communities, of Brussels on 1st of April of 2009 (COM (2009) 147 final).

European Commission, 2010. European Commission—Joint Research Centre—Institute for Environment and Sustainability: International Reference Life Cycle Data System (ILCD) Handbook—General guide for Life Cycle Assessment—Detailed guideline. First edition March 2010. EUR 24708 EN. Luxembourg. Publications Office of the European Union.

European Cost Action ES1202 "Water 2020" (www.water2020.eu).

Foley, J., De Haas, D., Hartley, K., and Lant, P., 2010a. Comprehensive life cycle inventories of alternative wastewater treatment systems. *Water Research* 44, 1654–1666.

Foley, J., De Haas, D., Yuan, Z., and Lant, P., 2010b. Nitrous oxide generation in full-scale biological nutrient removal wastewater treatment plants. *Water Research* 44, 831–844.

Gallego, A., Hospido, A., Moreira, M.T., and Feijoo, G., 2008. Environmental performance of wastewater treatment plants for small populations. *Resources, Conservation and Recycling* 52(6), 931–940.

Garrido-Baserba, M., Reif, R., Hernández, F., and Poch, M., 2012. Implementation of a knowledge-based methodology in a decision support system for the design of suitable wastewater treatment process flow diagrams. *Journal of Environmental Management* 112, 384–391.

Godin, D., Bouchard, C., and Vanrolleghem, P.A., 2012. Net environmental benefit: Introducing a new LCA approach on wastewater treatment systems. *Water Science and Technology* 65, 1624.

Goedkoop, M., Heijungs, R., Huijbregts, M., De Schryver, A., Struijs, J., and Van Zelm, R., 2013. ReCiPe 2008. A life cycle impact assessment method which comprises harmonised category indicators at the midpoint and the endpoint level. Report I: Characterisation, first ed.; May 2013 (www.lcia-recipe.net).

Guinée, J.B., Gorrée, M., Heijungs, R., Huppes, G., Kleijn, R., de Koning, A. et al., 2002. *Life Cycle Assessment. An Operational Guide to the ISO Standards.* Centre of Environmental Science, Leiden 2002.

Guisasola, A., de Haas, D., Keller, J., and Yuan, Z., 2008. Methane formation in sewer systems. *Water Research* 42, 1421–1430.

Hauschild, M. and Potting, J., 2003. Spatial differentiation in life cycle impact assessment— The EDIP2003 methodology. Institute for Product Development Technical University of Denmark.

Hospido, A., Moreira, M.T., Fernández-Couto, M., and Feijoo, G., 2004. Environmental performance of a municipal wastewater treatment plant. *International Journal of Life Cycle Assessment* 9, 261–271.

Hospido, A., Moreira, T., Martín, M., Rigola, M., and Feijoo, G., 2005. Environmental evaluation of different treatment processes for sludge from urban wastewater treatments: Anaerobic digestion versus thermal processes. *International Journal of Life Cycle Assessment* 10, 336–345.

Hospido, A., Moreira, M.T., and Feijoo, G., 2008. A comparison of municipal wastewater treatment plants for big centres of population in Galicia (Spain). *International Journal of Life Cycle Assessment* 13(1), 57–64.

Hospido, A., Sanchez, I., Rodriguez-Garcia, G., Iglesias, A., Buntner, D., Reif, R., Moreira, M.T., and Feijoo, G., 2012. Are all membrane reactors equal from an environmental point of view? *Desalination* 285, 263–270.

Houillon, G. and Jolliet, O., 2005. Life cycle assessment of processes for the treatment of wastewater urban sludge: Energy and global warming analysis. *Journal of Cleaner Production* 13, 287–299.

Høibye, L., Clauson-Kaas, J., Wenzel, H., Larsen, H.F., Jacobsen, B.N., and Dalgaard, O., 2008. Sustainability assessment of advanced wastewater treatment technologies. *Water Science and Technology* 58(5), 963–968.

ISO, 1997. ISO 14040. Environmental Management—Life Cycle Assessment—Principles and Framework. International Organization for Standardization. Switzerland.

ISO, 2006a. ISO 14040. Environmental Management and Life Cycle Assessment—Principles and Framework. International Organization for Standardization. Switzerland.

ISO, 2006b. ISO 14044. Environmental Management and Life Cycle Assessment—Requirements and Guidelines. International Organization for Standardization. Switzerland.

Jolliet, O., Margni, M., Charles, R., Humbert, S., Payet, J., Rebitzer, G., and Rosenbaum, R., 2003. IMPACT 2002+: A new life cycle impact assessment methodology. *The International Journal of Life Cycle Assessment* 8(6), 324–330.

Kampschreur, M.J., Temmink, H., Kleerebezem, R., Jetten, M.S.M., and Van Loosdrecht, M.C.M., 2009. Nitrous oxide emission during wastewater treatment. *Water Research* 43, 4093–4410.

Kärrman, E. and Jönsson, H., 2001. Normalising impacts in an environmental systems analysis of wastewater systems. *Water Science and Technology* 43, 293–300.

Lederer, J. and Rechberger, H., 2010. Comparative goal-oriented assessment of conventional and alternative sewage sludge treatment options. *Waste Management* 30(6), 1043–1056.

Li, Y., Luo, X., Huang, X., Wang, D., and Zhang, W., 2013. Life cycle assessment of a municipal wastewater treatment plant: A case study in Suzhou, China. *Journal of Cleaner Production* 57, 221–227.

Lorenzo-Toja, Y., Vázquez-Rowe, I., Chenel, S., Marín-Navarro, D., Moreira, M.T., and Feijoo, G., 2015. Eco-efficiency analysis of Spanish WWTPs using the LCA+ DEA method. *Water Research*, 68, 651–666.

Loubet, P., Roux, P., Loiseau, E., and Bellon-Maurel, V., 2014. Life cycle assessments of urban water systems: A comparative analysis of selected peer-reviewed literature. *Water Research*, 67, 187–202.

Lundin, M., Bengtsson, M., and Molander, S., 2000. Life cycle assessment of wastewater systems: Influence of system boundaries and scale on calculated environmental loads. *Environmental Science and Technology* 34(1), 180–186.

Machado, A.P., Urbano, L., Brito, A.G., Janknecht, P., Salas, J.J., and Nogueira, R., 2007. Life cycle assessment of wastewater treatment options for small and decentralized communities. *Water Science and Technology* 56(3), 15.

Mels, A.R., Van Nieuwenhuijzen, A.F., Van der Graaf, J.H.J.M., Klapwijk, B., De Koning, J., and Rulkens, W.H., 1999. Sustainability criteria as a tool in the development of new sewage treatment methods. *Water Science and Technology* 39(5), 243–250.

Meneses, M., Pasqualino, J.C., and Castells, F., 2010. Environmental assessment of urban wastewater reuse: Treatment alternatives and applications. *Chemosphere* 81, 266–272.

Muñoz, I., Gómez, J.M., Molina-Díaz, A., Huijbregts, M.A.J., Fernández-Alba, A.R., and García-Calvo, E., 2008. Ranking potential impacts of priority and emerging pollutants in urban wastewater through life cycle impact assessment. *Chemosphere* 74, 37–44.

Muñoz, I., Rodríguez, A., Rosal, R., and Fernández-Alba, A.R., 2009. Life Cycle Assessment of urban wastewater reuse with ozonation as tertiary treatment: A focus on toxicity-related impacts. *Science of the Total Environment* 407, 1245–1256.

Nair, S., George, B., Malano, H.M., Arora, M., and Nawarathna, B., 2014. Water–energy–greenhouse gas nexus of urban water systems: Review of concepts, state-of-art and methods. *Resources, Conservation and Recycling* 89, 1–10.

Niero, M., Pizzol, M., Gundorp, H., and Thomsen, M., 2014. Comparative life cycle assessment of wastewater treatment in Denmark including sensitivity and uncertainty analysis. *Journal of Cleaner Production* 68, 25–35.

Ortiz, M., Raluy, R.G., and Serra, L., 2007. Life cycle assessment of water treatment technologies: Wastewater and water-reuse in a small town. *Desalination* 204(1–3), 121–131.

Panesar, D.K., Seto, K.E., and Churchill, C.J. 2017. Impact of the selection of functional unit on the life cycle assessment of green concrete. *International Journal of Life Cycle Assessment*. doi:10.1007/s11367-017-1284-0.

Pasqualino, J.C., Meneses, M., and Castells, F., 2011. Life cycle assessment of urban wastewater reclamation and reuse alternatives. *Journal of Industrial Ecology* 15(1), 49–63.

Pijuán, M., Torà, J., Rodríguez-Caballero, A., César, E., Carrera, J., and Pérez, J., 2014. Effect of process parameters and operational mode on nitrous oxide emissions from a nitritation reactor treating reject wastewater. *Water Research* 49, 23–33.

Renou, S., Thomas, J.S., Aoustin, E., and Pons, M.N. 2008. Influence of impact assessment methods in wastewater treatment LCA. *Journal of Cleaner Production* 16(10), 1098–1105.

Rodriguez-Garcia, G., Molinos-Senante, M., Hospido, A., Hernandez-Sancho, F., Moreira, M.T., and Feijoo, G., 2011. Environmental and economic profile of six typologies of wastewater treatment plants. *Water Research* 45, 5997–6010.

Rodriguez-Garcia, G., Hospido, A., Bagley, D.M., Moreira, M.T., and Feijoo, G., 2012. A methodology to estimate greenhouse gases emissions in Life Cycle Inventories of wastewater treatment plants. *Environmental Impact Assessment Review* 37, 37–46.

Rodriguez-Garcia, G., Frison, N., Vázquez-Padín, J.R., Hospido, A., Garrido, J.M., Fatone, F., Moreira, M.T., and Feijoo, G., 2014a. Life cycle assessment of nutrient removal technologies for the treatment of anaerobic digestion supernatant and its integration in a wastewater treatment plant. *Science of the Total Environment* 490, 871–879. doi:10.1016/j.scitotenv.2014.05.077.

Rodriguez-Garcia, G., Molinos-Senante, M., Gabarron, S., Alfonsin, C., Hospido, A., Corominas L., Hernandez-Sancho, F., et al., 2014b. Cost benefit and life cycle assessment. In *Membrane Biological Reactors: Theory, Modelling, Design, Management and Applications to Wastewater Reuse*. F.I. Hai, K. Yamamoto, and C.H. Lee (ed.), IWA Publishing, London.

Roeleveld, P., Klapwijk, A., Eggels, P., Rulkens, W., and van Starkenburg, W., 1997. Sustainability of municipal wastewater treatment. *Water Science and Technology* 35, 221–228.

Rosenbaum, R.K., Bachmann, T.M., Gold, L.S., Huijbregts, M.A.J., Jolliet, O., Juraske, R., Koehler, A., et al., 2008. USEtox – The UNEP-SETAC toxicity model: Recommended characterisation factors for human toxicity and freshwater ecotoxicity in life cycle impact assessment. *International Journal of Life Cycle Assessment* 13(7), 532–546.

Rulkens, W., 2008. Sewage sludge as a biomass resource for the production of energy: Overview and assessment of the various options. *Energy Fuels* 22(1), 9–15.

Tangsubkul, N., Beavis, P., Moore, S.J., Lundie, S., and Waite, T.D., 2005. Life cycle assessment of water recycling technology. *Water Resource Management* 19, 521–537.

Tillman, A.M., Svingby, M., and Lundström, H., 1998. Life cycle assessment of municipal waste water systems. *International Journal of Life Cycle Assessment* 3, 145–157.

Venkatesh, G. and Brattebø, H., 2011. Environmental impact analysis of chemicals and energy consumption in wastewater treatment plants: Case study of Oslo, Norway. *Water Science and Technology* 63(5), 1018–1031.

Venkatesh, G., Chan, A., and Brattebø, H., 2014. Understanding the water-energy-carbon nexus in urban water utilities: Comparison of four city case studies and the relevant influencing factors. *Energy* 75, 153–166.

Wang, X., Liu, J., Ren, N.-Q., Yu, H.-Q., Lee, D.-J., and Guo, X., 2012. Assessment of multiple sustainability demands for wastewater treatment alternatives: A refined evaluation scheme and case study. *Environmental Science and Technology* 46, 5542–5549.

Wenzel, H., Larsen, H.F., Clauson-Kaas, J., Høibye, L., and Jacobsen, B.N., 2008. Weighing environmental advantages and disadvantages of advanced wastewater treatment of micro-pollutants using environmental life cycle assessment. *Water Science and Technology* 57, 27.

Yoshida, H., Clavreul, J., Scheutz, C., and Christensen, T.H., 2014a. Influence of data collection schemes on the life cycle assessment of a municipal wastewater treatment plant. *Water Research* 56, 292–303.

Yoshida, H., Mønster, J., and Scheutz, C., 2014b. Plant-integrated measurement of greenhouse gas emissions from a municipal wastewater treatment plant. *Water Research* 61, 108–118.

4 Life Cycle Assessment of Beneficial Reuse of Waste Streams for Energy in Municipal Wastewater Treatment Plants

Dongyan Mu, Sarah Mack,
Roger Ruan, and Min Addy

CONTENTS

4.1   INTRODUCTION

As sustainable energy gains momentum in today's society, municipal wastewater shows promise as the next source of clean energy. In 2012, the Environmental Protection Agency (EPA) found that over 14,000 publicly owned wastewater treatment plants were serving 238.2 million Americans (United States Environmental Protection Agency, 2016). In the United States, 76% of the population send their wastewater to public facilities, while the remaining population use private septic systems (United States Environmental Protection Agency, 2016).

When municipal wastewater is collected and sent to a centralized municipal wastewater treatment facility, it goes through a series of treatment processes to be cleaned and purified. The treated water is then discharged to a local body of water. However, in addition to the cleaned water, treatment plants also produce a series of waste streams, which are likely to facilitate ecological harm when discharged into the environment. A commonly known waste stream produced by municipal wastewater treatment plants is biosolids, which are mainly derived from the excess activated sludge in aeration tanks and are usually deposited to landfills, which has raised significant concerns in waste management. Other wastes include centrate, which is the wastewater collected from the sludge dewatering; this is not typically discharged out of the plant directly, but recycled back into the activated sludge. As it contains high chemical oxygen demand (COD) and nutrient levels, centrate can raise the treatment loading of aeration tanks.

To reduce the waste generated, a variety of technologies have been developed and applied in wastewater treatment facilities. One of these technologies is sludge digestion, a process that has been used for many years to stabilize raw sludge and recover energy in sludge. In addition, the digested sludge has been proposed to improve soil nutrients under federal and state standards. Besides conventional sludge digestion, which produces bioelectricity, other technologies have been developed in the recent past to reduce and reuse wastes generated in treatment facilities. In particular, due to the rising oil prices since 2005, interest has increased in the creation of technologies that can produce transportation biofuels using waste streams. Using nutrient-rich centrate, for example, algae can be cultivated and then converted into transportation fuels for vehicles. Pyrolysis or hydrothermal liquefaction has also been tested for converting sludge to biofuels. A specific technology has been developed to convert the lipids in scum to biodiesel using acid hydrolysis and glycerolysis. Rather than being perceived as waste, biosolids have come to be viewed as a valuable resource for water, energy, and plant nutrient conservation (Asano et al., 2007).

A detailed description of each of the waste streams has been introduced, along with several technologies that convert the various waste streams within the municipal wastewater treatment plant to biofuels or bioenergy. More importantly, this chapter discusses the environmental impacts and benefits of these technologies in comparison to conventional petroleum and electricity. A life cycle assessment (LCA) is included in this chapter as a case study, which took place at a wastewater treatment plant located in Saint Paul, MN USA that integrated various technologies of biofuel/bioenergy production to recycle and reuse waste streams.

4.2   WASTE STREAMS GENERATED IN MUNICIPAL WASTEWATER FACILITIES

Before introducing the reuse technologies, it is necessary to first understand the major waste streams created within municipal wastewater treatment facilities. This section will describe each of the waste streams, how they are treated, their characteristics, and their ecological impacts.

4.2.1   BIOSOLIDS

Biosolids are all solid streams discharged out of wastewater treatment facilities. A major source of biosolids is sludge collected from the bottom of primary and secondary clarifiers in the wastewater treatment plants. As the water content can be as high as 99.5%, the sludge collected is first sent to the sludge thickening/storage tank to remove water and increase the solid content. It is then stabilized through chemical stabilization or anaerobic digestion. Lastly, the sludge is dewatered mechanically until the water content reaches 70% and then sent out of the plants for final disposal. Anaerobic digestion can also serve as a facility to recycle energy from biosolids. When biosolids are sent outside of the plant, they are typically incinerated, landfilled, or applied to land as fertilizer (Davis and Slaughter, 2006).

When biosolids are incinerated, they emit energy, heat, and harmful contaminants. Burning biosolids leaches polychlorinated biphenyls (PCBs), a group of priority pollutants controlled by the US EPA, into the environment (Rice et al., 2003). While trace exposure to PCBs is more commonly found, extreme exposure to PCBs has led to the death of fish, wildlife, flora, and fauna. Some common ailments reported are cancerous lesions, immune suppression, change in enzyme activity, and liver and reproductive failure (United States Fish and Wildlife Service, 2015).

When biosolids are sent to a landfill, biogas is created and sometimes collected for generating energy. Ecological harm may arise via leaching of harmful chemicals into water/soil or via toxic chemicals released into the atmosphere from decomposition of the compounds in the biosolids. Leachate from waste streams contains nitrate, metals, organics, and/or pathogens, which potentially enter groundwater if the landfill is incorrectly placed geographically or has damage to the lining (Kajitvichyanukul et al., 2008). Potential ecologically harmful impacts from landfills also include degradation to air quality, public health, wildlife, and habitats of endangered species (US EPA, 2003).

Applying biosolids to land as fertilizer also raises many concerns. The pathogen content of sludge is of high concern, as there have been instances in the past when poor management has led to harmful effects on municipalities. A tragic instance of this was seen at Walkerton, Ontario when *Escherichia coli* 0157:H7 and other pathogens contaminated the drinking water after sludge was sprayed on rural land. In addition to the concern about pathogens, contamination by heavy metals and persistent organic pollutants (POPs) is also of high concern. Heavy metals such as zinc and copper are micronutrients that facilitate plant growth but can be damaging to the plants in excessive amounts. Heavy metals are not very mobile, tend to accumulate on the surface of soils, and cannot be easily removed once applied to land. Of all the heavy metals, cadmium and mercury are of greatest concern, as they tend to enter the food chain and be biomagnified (Reilly, 2001). Lastly, POPs found in sludge, such as polychlorinated biphenyls (PCBs) and organochlorine pesticides (OCPs), have the potential to accumulate in soil, thus negatively impacting flora and fauna (Youcai, 2016).

Given the environmental impacts from conventional biosolid disposal methods, many researchers have focused on new technologies for beneficial reuse. Sludge contains potential energy of 8000 Btu per dry pound (McCarty et al., 2011), suggesting that it is a good source of alternative energy. The idea of converting biosolids into green transportation fuels, that is, biodiesel, has increasingly become a topic of interest. New technologies such as pyrolysis, gasification, and liquefaction have been tested to convert biosolids into various products such as biodiesel and gasoline.

4.2.2 SCUM

Scum is a layer skimmed from the top of the primary settling tank during the wastewater treatment process. Scum contains cooking oils, animal fats, food wastes, soaps, waxes, and other impurities produced by restaurants, households, and various other entities. For many wastewater treatment facilities, scum is combined and treated with activated sludge in the same anaerobic digestion process. When sludge digestion is applied for energy recycling, the scum will always float on the top of the digester and form a thick layer, which will impede the digester's performance. Therefore, other wastewater treatment facilities will treat scum as a waste product, either incinerating it or sending it to a landfill, which raises similar environmental problems to those of biosolids.

Studies have shown that dried scum contains approximately 62% fats and oils and almost 14% organic biosolids (Anderson et al., 2016), which make it attractive as an alternative energy source. Different attempts have been made to convert scum to biodiesel (Anderson et al., 2016). A novel process for extracting oil from scum and successfully converting it to fuel grade biodiesel has been achieved at the University of Minnesota (Bi et al., 2015). Oil extraction from wastewater scum via acid hydrolysis and glycerolysis was studied as the pretreatment method for biodiesel production. The results showed that free fatty acids (FFA) in scum can be reduced effectively through glycerolysis kinetics, and zinc-based catalysts increased the reaction rate significantly (Anderson et al., 2016).

4.2.3 CENTRATE AND OTHER LIQUID WASTE STREAMS

Centrate is the liquid portion that is formed from the sludge dewatering process in the wastewater treatment plant. Once extracted, the centrate is typically sent back to the activated sludge process and combined with untreated wastewater for further treatment. Centrate usually contains high levels of pollutants, which can result in COD levels up to 5000 mg L^{-1}. The nitrogen (N) and phosphorus (P) levels can be 30–40 times higher than those of untreated municipal wastewater; see Table 4.1. If centrate is sent back to the secondary treatment process, it will increase the pollutant levels of inflow to the aeration tank, consequently reducing the treatment effect, and will raise the energy use in the treatment. The energy use will be higher, since the elevated quantities of pollutants present require aeration for a longer time to remove them.

TABLE 4.1
Characteristics of Wastewater Streams in MWTP

	COD (mg L^{-1})	TSS (g L^{-1})	VSS (g L^{-1})	NH3-N (mg L^{-1})	TN (mg L^{-1})	TP (mg L^{-1})
Centrate	3027 ± 779	0.46 ± 0.28	0.35 ± 0.26	113 ± 18	150	250
Municipal wastewater	224.0 ± 4.2	–	–	–	38.95 ± 1.91	6.86 ± 0.05

COD: Chemical Oxygen Demand; TSS: Total Suspended Solid; TN: Total Nitrogen; TP: Total Phosphorous.

Fortunately, the carbon, nitrogen, and phosphorus composition ratios in centrate are favorable for growing algae. This makes centrate attractive as a nutrient source for the production of microalgae, which can be converted into biofuels for vehicle use (Wang et al., 2010). Studies have shown that algae grown with centrate usually have higher yields, higher algae growing density, and fewer inputs than other wastewater streams such as raw municipal wastewater and manure (Mu et al., 2014). One study shows that centrate can be directly added into algae growing reactors without dilution (Min et al., 2011). The pollutants in the effluent of algae cultivation can be reduced significantly, which will reduce the load in the aeration tank.

Besides centrate, other liquid waste is created, such as supernatant of sludge thickening and sludge digestion. The liquid waste contains high levels of COD, N, and P as well. Instead of being discharged from the plant, these other liquid streams are usually recycled to the activated sludge process, which raises the treating load of the aeration tank. Injecting those streams into algae cultivation can reduce the pollutants in those steams and co-produce biofuels for the plant's use.

4.3 TECHNOLOGIES FOR RENEWABLE ENERGY FROM WASTE CREATED IN WASTEWATER TREATMENT FACILITIES

4.3.1 ANAEROBIC DIGESTION OF BIOSOLIDS

Anaerobic digestion is a commonly used method to convert biosolids into biogas, which generates electricity and heat. It is a natural biochemical process in which microorganisms break down and digest the solid biomass in anaerobic conditions. As the microorganisms are breaking down the biosolids, they create biogas—usually comprised of methane (CH_4) and carbon dioxide (CO_2)—and solid and liquid residues called *digestate*. A liquid layer, supernatant, is also created on the top of the digestate and discharged out of the digester. In a municipal wastewater treatment plant, digestion is usually done in mesophilic conditions, where the biosolids are in tanks at ~30–38°C to increase yield and eliminate harmful pathogens, because more methane can be generated.

Biogas is typically made up of approximately 50%–80% methane and approximately 20%–50% carbon dioxide. In the Greenhouse Gases, Regulated Emissions, and Energy Use in Transportation (GREET) database, methane generation rate is around 0.3 L CH_4 per kg total suspended solids (TSS) of sludge digested. In some cases, there may be trace amounts of nitrogen gas (N_2), hydrogen gas (H_2), ammonia (NH_3), and/or hydrogen sulfide (H_2S). Biogas outputs from the digestion can be burned to power internal combustion engines or upgraded into biomethane and generate electricity to be returned to the electrical grid (USA, 2010). When biogas is being used for energy production, it offsets the energy necessity for fossil fuels, which could reduce greenhouse gas (GHG) emissions from burning fossil fuels and slow down global climate change. The digestion process itself consumes heat and electricity for process mixing and heating, biogas clean up, and digestate dewatering. The energy created by biogas combustion can cover the energy use for the process operation.

The digestate from the sludge digestion process has to be further dewatered for final disposal or other reuse. This stream is rich in nitrogen and phosphorus and can be applied to fertilize land and fields after dewatering, thus bringing environmental benefits by replacing synthetic fertilizer use, improving soil quality, and sequestrating carbon in soil. The typical digested sludge contains 1.6–3.0 wt.% of N, 1.5–4.0 wt.% of P_2O_5, and 0–3.0 wt.% of K_2O. When applied to land, the nutrient use efficiency depends on the compositions of certain nutrients in the digestate. For example, nitrogen use efficiency is different for NO_3, NH_4, and organic nitrogen, at 100%, 75%, and 40%, respectively (Metcalf and Eddy, 2003). In addition to the environmental benefits from applying biosolids to land, environmental concerns, which will be described in Section 4.2.1, have been raised. The digestate can also be combusted or directly landfilled.

The liquid/supernatant collected from the top of the digester contains high levels of nutrients. The COD level may be up to 1000 mg L^{-1}. When recycled back to the activated sludge process, the supernatant will increase loading in the aeration tank. A recent study by Min et al. (2011) proposed using this stream to grow algae, which are then converted to algal biofuels for vehicles. As the supernatant is created by a moderate temperature digestion, the nutrients are easier for the algae to absorb and can promote high yields in algae cultivation. The supernatant recycling offers environmental benefits by reducing energy use in the activated sludge process, saving nutrient use in algae cultivation, and replacing fossil fuel use in transportation. The total nutrients in supernatant depend on the amount of nutrients contained in biogas and digestate. The nutrient balance calculation can help to estimate the nutrient content in supernatant.

Many LCA studies have been conducted for biosolid digestion to evaluate the environmental impacts of this process (Yoshida et al., 2013; Pradel et al., 2016). A majority of those studies concluded that compared with incineration, composting, and landfilling, the scenario of anaerobic digestion combined with agricultural land application was the most environmentally friendly in terms of emissions and consumption of energy, but heavy metals released from sludge contributed significantly to human toxicity and ecotoxicity. The study by Gourdet et al. (2017) identified four key factors that are related to environmental impacts in anaerobic digestion. They are: (1) the biodegradation rate of volatile solids, (2) the nitrogen mineralization rate during anaerobic digestion, (3) phosphorus and nitrogen capture rates during the thickening and dewatering processes, and (4) the consumption of chemicals such as $FeCl_3$ in the dewatering process.

4.3.2 PYROLYSIS AND OTHER THERMAL CHEMICAL PROCESSES TO TREAT SLUDGE

Besides incineration, several thermal chemical processes, including pyrolysis, gasification, and liquefaction, have been proposed as an energy source using dewatered or digested biosolids in wastewater treatment plants. Although they have still only been developed at pilot scale, these technologies have attracted increasing attention, as they can produce liquid fuels such as petroleum and diesel, which can be directly used in vehicles and replace fossil fuels.

Pyrolysis is a process in which the decomposition of biosolids is facilitated by high temperatures in anaerobic conditions. For pyrolysis to have maximum yields and efficiency, the biosolids must first be pretreated to the desired moisture content (<10%) and particle size. If the feedstock has an overabundance of moisture, it must be dried first. If the particle size is too large, it will need to be reduced in size before continuing the process. The pretreated biosolids are added to the pyrolysis reactor, where the conditions are anaerobic. In general, the pyrolysis of organic substances produces gas and solid residue, char, which is high in carbon content. Next, the raw gases and char are separated. The gases are then quenched with cold water. In this step, the cold water will quickly cool the gases, and oil vapor will be condensed into bio-oil. The non-condensable gases are collected and recycled back into the pyrolysis reactor (Speight, 2008). The operation temperature of pyrolysis is around 300–600°C, and the pressure is 0.1–0.5 MPa.

The yield of bio-oil, gas, and char depends on many factors, including operating temperature, pressure, time, and heating speed as well as the composition of feedstocks. Generally, bio-oil yields are high in conditions of fast heating, high temperature (500–1300°C), and low pressure (50–150 bars). For example, using a fluidized bed, Fonts et al. (2008) conducted that pyrolysis of sewage sludge at the temperature of 540°C could obtain the maximum liquid yield of about 33 wt.%. By using a pyrolysis centrifugal reactor (PCR), Trinh et al. (2013) obtained the maximum liquid yield of about 41 wt.% on a dry ash free feedstock basis (daf) and a sludge oil energy recovery of 50% at the temperature of 575°C. Chang et al. (2016) completed a pilot-scale pyrolysis experiment on municipal sludge and proceeded to operational effectiveness evaluation, and the result showed that the optimal operating conditions were a pyrolysis temperature of 450–500°C and a pyrolysis time of 30–40 min (Chang et al., 2016).

The bio-oil collected in pyrolysis has to go through the thermal upgrading processes to be converted to commercialized transportation fuels. The upgrading usually includes two steps. First, the hydro treating removes impurities that could affect downstream equipment. H_2 is imported and added at a rate of 5 wt.% of bio-oil. The hydrocracking process then breaks down heavy molecules into shorter-chain fuels, diesel and gasoline. The yields of biofuels depend on the composition of the bio-oil. The total fuel yield varies from 40 to 80 wt.% of raw bio-oil sent to upgrading (NREL, 2010). As sludge-derived bio-oil upgrading has not been well studied, the exact biofuel yields and quality are still unclear. Many studies proposed to send bio-oil to a local oil refinery where bio-oil can be combined with regular petroleum raw oil for upgrading. In this situation, the impact of bio-oil on the final products will be mitigated.

Char is co-produced in the pyrolysis process. Elements such as phosphorus and magnesium are concentrated in the pyrolysis process; the char containing these elements can be used as a soil amendment or fertilizer (Smith et al., 2009). The pyrolysis and sludge drying processes consume energy during operation. Char has to be burned to generate the amount of heat required for the drying and pyrolysis processes. The energy required for drying depends on the water content of the sludge. If the sludge has been dewatered or digested, the water content is around 70%–80%. That sludge requires at least 2600 MJ of heat per cubic meter of wet sludge to reduce the water content to 10%. In some cases, the heat from char is not enough for sludge drying, so some bio-oil will need to be burned for heat as well. Although the oil yield is reduced, the process can keep energy self-contained, and therefore, no fossil fuel will be consumed.

Another thermal chemical conversion technology proposed is sludge gasification, in which the sludge breaks into gaseous fuels called *syngas* in an oxygen-deficient condition. The gasification has to operate at a higher temperature (800–1300°C) and a higher pressure (>0.1 MPa) than pyrolysis. The syngas contains CO (15%–20%), CO_2 (10%–12%), H_2 (20%–24%), CH_4 (0%–4%), and N_2 (48%–52%). As the gas exits the gasifier at 300–400°C, it contains tar and particulates; therefore, it must be cooled and cleaned. The syngas can be used directly for gas engines, or it can be converted into biofuels. A very famous process that converts clean syngas into transportation fuel is called the Fischer-Tropsch (F-T) process, in which H_2 and CO react to form liquid fuels (diesel/gasoline) with a catalyst. The gasification process requires more operation energy than pyrolysis; however, the diesel produced through gasification and F-T has higher quality; it is free of sulfur and nitrogen and low in aromatics, which makes the process attractive.

Both pyrolysis and gasification need dry biosolids before the processes can begin, which consumes a significant amount of energy. In the past 2 years, hydrothermal liquefaction technology (HTL) has been introduced to treat sludge due to its ability to treat biomass with moisture content up to 80–95 wt.% (PNNL, 2016). The energy-intensive thermal drying process, therefore, could be eliminated in the production pathway. In HTL, sludge is directly converted to liquid oil at a reaction temperature of less than 400°C. Char, gas, and aqueous phases are co-produced in HTL. The product yields depend on multiple factors, including operation temperature, pressure, water content of sludge, and the catalyst. A study by the PNNL (2016) showed that when primary sludge was treated, the product yields from daf sludge were 40 wt.% for oil, 35 wt.% for aqueous, 22 wt.% for gas, and 3 wt.% for char. The bio-oil produced needed to be further updated with the hydrotreating and hydrocracking processes. The energy in biochar and biogas from HTL was assumed to be recycled into the heat and power system (HPS) as heat for in-plant use. The aqueous portion of material from the HTL process containing C, N, and P was assumed to be recycled to grow algae in the open pond.

Many LCAs have been conducted for the thermal technologies that produce biofuels and bioenergy. Hospido et al. (2005) compared the environmental performance of sludge treatment with anaerobic digestion (AD) followed by land application versus pyrolysis. The results showed that the AD process was better for eutrophication, global warming, and acidification, whereas pyrolysis was better for human and terrestrial toxicity. Cao and Pawlowski (2013) conducted an LCA of pyrolysis to treat digested sludge versus nondigested sludge, and found that the environmental performance of treating digested sludge was better than that of treating nondigested sludge. Mills et al. (2014) conducted an LCA comparing sludge treatment technologies and concluded that adding pyrolysis to further recycle energy in digested sludge improved the environmental performance of regular AD technology. Buonocore et al. (2016) analyzed the environmental impacts of gasification of digested sludge with the LCA tool. The results showed that adding gasification to treat digested sludge improved the environmental performance on all impact categories examined, particularly on eutrophication and human health potential.

4.3.3 Scum to Biodiesel

Scum has a high oil content, and there is scope for a technology to convert the lipids into bio-oil. Researchers from the University of Minnesota proposed and conducted an LCA of a technology that converts scum to American Society for Testing and Materials-grade diesel. Figure 4.1 shows the dried scum and biodiesel produced from scum. Figure 4.2 shows the pilot refinery machine of Scum-to-Biodiesel. The technology proposes a six-step energy conversion pathway: Filtering, acid washing, acid-catalyzed esterification, base-catalyzed esterification, and finally, refining (distillation). These steps are described in the following list.

1. *Filtering*: In this process, larger particles are filtered out from the raw scum flow. The scum is heated to 60°C and filtered for 1 h. In this way, 95 wt.% of scum oil can be filtered out. The water separated from this process can be used to dilute acid in the following acid wash processes. The solid waste is filtered out and sent to the combustion or digestion process to recover additional energy. Energy is consumed for heating scum.
2. *Acid wash*: The scum flow is then washed by 5% H_2SO_4 solution to convert soap present in scum into FFA. The process aims to maximize the biodiesel

(a)   (b)

FIGURE 4.1 Scum-to-Biodiesel technology. (a) Scum collected from Metropolitan Wastewater Treatment Plant at St Paul, MN. (b) Final product (biodiesel).

FIGURE 4.2   Scum-to-Biodiesel Demo Facility developed at University of Minnesota.

yield, break the emulsion for better water/oil separation, and further remove impurity from the scum oil, thus speeding up the esterification reaction in the following acid-catalyzed esterification process. When the acid solution is introduced into the raw scum oil, it first mixes with the raw oil for 0.5 h at 60°C. The mixture is then allowed to settle for about half an hour to collect the oil in the upper phase and sediment/water in the lower phase. The energy is required to heat up the scum oil, to remove the water in the exit oil, and for mixing.

3. *Acid-catalyzed esterification:* In this step, the FFA is reacted with methanol to form fatty acid methyl ester (FAME) with the catalyst H_2SO_4. Methanol (30% of oil weight) and 98% sulfuric acid (5% of oil weight) are added to the oil and mixed for 1 h at 60°C. Then, the mixture is allowed to settle for 0.5 h to separate FAME from excess methanol and acid. The FAME collected is dried at 105°C to remove all water. The total energy use in this process includes heat for maintaining the reaction at 60°C for 1.5 h, heat to remove water from the raw oil, and power for mixing. After acid-catalyzed esterification, the scum oil would meet the criteria as a feedstock for base-catalyzed transesterification.

4. *Base-catalyzed transesterification:* In this process, the triglyceride in the raw scum oil is converted to FAME with base catalysts, that is, potassium methoxide (CH_3OK). Methanol and CH_3OK are first mixed with raw oil from the acid esterification at 60°C for 0.5 h. The mixture is then sent to a rotary evaporator to remove excess methanol. The energy use in this

process includes heat for keeping the reaction at 60°C for 30 min, power for mixing, and power for the rotary evaporator to remove methanol.

5. *Glycerol wash:* Extra glycerol was added to the mixture oil to separate FAME from glycerol, catalysts, and impurities. Based on the experiment conducted, the best conditions are an added glycerol-to-oil mixture ratio of 1:2 by weight, mixed at 60°C for 15 min. Then, the upper layer is collected for further upgrading. The collected FAME is 92.6 wt.% of the total oil input of the process. The lower layer is sent to the glycerol recycling process. The energy use for this process is heat for evaporating methanol and power for mixing.

6. *Distillation:* The FAME mixture from the glycerol wash is introduced into a customized vacuum rectification column system, and finally, biodiesel is obtained. The biodiesel yield in this process is 88.2 wt.% of FAME input. The heat used in distillation is 2.51 MJ kg^{-1} biodiesel, and the power used for the vacuum is 0.045 kWh kg^{-1} biodiesel, which is based on industrial data.

7. *Methanol and glycerol recycling:* Two additional processes are added to recycle extra methanol and glycerol. Extra methanol applied in acid esterification and base transesterification can be recycled for use in the next batch. A rotary evaporator is used after acid and base esterification to separate excess methanol out. The energy use in the rotary evaporator is 1.1 MJ kg^{-1} methanol, which is based on theoretical heat for methanol evaporation. A distillation process will be used to recover the glycerol to a technical grade. Energy use will be electricity and heat for distilling glycerol (10% of oil weight). The extra glycerol produced in the process will be used as carbon resources for algae production if mounted in the plant.

The LCA and technical economic analysis (TEA) tools were used to evaluate the Scum-to-Biodiesel technology and compare it with incineration and anaerobic digestion (Mu et al., 2016). The energy conversion efficiency of the Scum-to-Biodiesel technology was found to be 60%, which is much higher than for conventional methods of scum treatment (digestion 5% and combustion 33%). The LCA concluded that the Scum-to-Biodiesel technology has lower fossil fuel depletion, GHG emissions, and eutrophication potential compared with combustion and digestion. The TEA showed that the Scum-to-Biodiesel process has the greatest financial potential, while the incineration of scum yielded the largest potential of reclaimed energy. Altogether, the Scum-to-Biodiesel technology can achieve higher revenue and significant environmental benefits when compared with scum digestion and combustion processes (Mu et al., 2016).

4.3.4 Algae Cultivation and Algal Biofuel Production from Liquid Waste Streams

Recently, algal biofuels have gained popularity among the scientific community, showing promise as a sustainable energy alternative (Menetrez, 2012). Microalgae have been proposed as a biomass feedstock for producing transportation fuels due to the variety of their outstanding properties. Algae offer relatively high area oil productivity compared with terrestrial bioenergy feedstocks, and they grow with short cycles that facilitate frequent harvesting (National Algal Biofuels Technology Roadmap, 2010). Algae can also be cultivated in marginal land with harsh conditions and can grow in saline conditions and wastewater (Ruiz et al., 2013; Soh et al., 2014). In addition, algae have the ability to grow in flue gas derived from coal power plants and thus sequestrate CO_2 from the atmosphere (Singh and Olsen, 2013). Much existing literature supports the argument that algal biofuels would be a potential alternative to petroleum-derived fuels in the transportation sector in the United States.

Numerous studies have suggested that using nutrient-laden wastewater to cultivate algae can improve the environmental performance of algal biofuels, because it reuses waste N, P, and water, which reduces the energy use and emissions from acquiring these inputs, as well as reducing overall biofuel production costs. Additionally, algae remove nutrients from the wastewater streams, which reduces loading in treatment facilities and in turn, decreases electricity consumption. Algae cultivation has been examined using various wastewaters, such as municipal wastewater, industrial wastewater, and animal manures. The liquid wastes created in the wastewater treatment facilities contain high levels of nutrients, which are a good source for growing algae. A study by Min et al. (2011) tested algae growing in various wastewaters and found that the centrate from municipal wastewater yielded the highest algae growth.

Algal biofuel production can be divided into three major stages: algae cultivation, algae conversions, and algal bioproduct use.

4.3.4.1 Algae Cultivation Stage

This stage includes the algae cultivation facility and algae–water separation and dewatering processes. There are two types of algae cultivation facility. One is the raceway open pond structure with wide surface area (100–1000 m^2) and shallow tank (~0.3 m). The algae yield ranges from 13.6 to 24.7 g m^{-2} d^{-1} in various locations across the United States. Another type is photobioreactors (PBRs) with various designs and structures. Figure 4.3 shows some PBR designs. These designs facilitate obtaining higher algae yields (up to 50 g m^{-2} d^{-1}), use fewer resources, and offer easier operation conditions. The open pond cultivation has a lower initial investment cost and is fit for large-scale production, but it requires a relatively flat area and more easily becomes contaminated. In contrast, PBRs cost more for installation and operation, but they produce more algae per unit reactor and have lower requirements for the installation site.

The algae species are critical to algae production, because different species have different growth rates and lipid content. Several species, such as *Chlorella* sp. and *Spirulina* sp., have been extensively examined as energy feedstocks. A lot of research has focused on the cultivation of algae species that have high yield and energy content or that are capable of growing in harsh conditions, because the productivity of algal biofuel depends on the biomass yield and lipid/heat content of the algae. However, algae with higher lipid/heat content normally have lower yields, and improving both lipid content and yield tends to require greater inputs of energy, nutrients, or other resources into the system, which increases the environmental burden. Therefore, algae productivity and electricity use are actually often a tradeoff in algae cultivation and need to be better understood to produce more biofuel.

(a)   (b)

FIGURE 4.3   PBR designs at the University of Minnesota. (a) Multi-layer cultivation PBR and (b) tubular PBR.

Based on a report by Argonne National Labs (2011), the carbon and nutrient requirements are 1.9 kg CO_2 kg^{-1} dry algae, 0.0125 kg P kg^{-1} dry algae, and 0.056 kg N kg^{-1} dry algae. This is based on the composition of algal cells (C : N : P = 103:10:1). The actual nutrient requirement may be higher, considering that the nutrient uptake efficiencies are different for different algae species. When wastewater is used to grow algae, not all the nutrients in wastewater are available to algae. For example, wastewater from thermal conversion processes such as pyrolysis and liquefaction contains a high level of bio-unavailable nutrients. Even though the total nutrients are high, the algae yields are often low. In a study by Min et al. (2011), four wastewater streams created in a wastewater treatment plant were examined for growing *Chlorella* sp. The results showed that algae grown in centrate had the highest algae yield (34 g m^{-2} d^{-1}) and lipid yield. The COD, N, and P removal rates reached 70%, 61%, and 61%, respectively. Besides nutrients, wastewater also contains organic carbon, which algae can take up and therefore reduce the inorganic carbon (CO_2) requirement for growing algae. When an injection of CO_2 is required, CO_2 collected from coal fire power plants should be injected to sequestrate CO_2 from the atmosphere. Electricity is used for flue gas capture, purification, and injection, which should be accounted for in a life cycle inventory (LCI).

Flocculation and sedimentation processes are used to separate algae from water in the algae slurry exiting the cultivation systems. After the process, the solid content in algae slurry can reach around 1%–2%. The supernatant collected after separation is recycled back to the cultivator, which recycles nutrients and reduces the need for fresh water. After flocculation, the algae slurry usually goes through a dewatering process such as centrifuging to further remove water and increase solid content to 20–30 wt.%. The algae–water separation stage is an energy-intensive stage. Electricity use causes major environmental impact.

4.3.4.2 Algal Conversion Stage

All technologies that can covert biosolids into energy products can be used for converting algae. These technologies are divided into two groups. One is based on biochemical conversion, including anaerobic digestion, fermentation, and so on. Another group is thermal chemical conversion, including pyrolysis, combustion, gasification, hydrothermal liquefaction, and so on. Many of these technologies will be described in Sections 4.3.1 and 4.3.2. Factors that influence the selection of the conversion process to use include the type and quantity of algae, the desired end products, investment and operation costs, and site conditions.

In addition, a technological pathway can use chemicals such as hexane to extract the lipid content out of algae and then convert the lipids to biodiesel through a trans-esterification process using methanol. Glycerin is co-produced with diesel in this pathway. Pressure homogenization is used as a pretreatment before lipid extraction to break the algal cell walls and facilitate lipid extraction. The technical pathway is also called *wet lipid extraction*, as no intensive drying process is applied. It is a frequently modeled technology for algal biofuel production in current literature. The diesel yield of this technological pathway depends on the lipid content of the algae. The residue of algae from liquid extraction could go through an anaerobic digestion process to recycle energy remaining in the residue. The biogas collected from the digestion is used to generate electricity for plant use. The digestate can be used as soil nutrients.

4.3.4.3 Algal Bioproduct Use Stage

The products from the various conversion technologies include electricity; solids (char, digestate); gas (methane, syngas); and biofuels (diesel, petroleum, and ethanol). They can be used for multiple purposes inside the plant or sold to outside markets. Some products can be used to generate heat or electricity. Some could be used as transportation fuels. Others can be applied to land for soil nutrients. Based on the study by Clarens et al. (2011), the nutrients contained in the digestate are around 40 mg N g^{-1} digestate and 12 mg P g^{-1} digestate. Nutrient use efficiencies are approximately 25% and 7% by weight for N and P, respectively. In the use stage, environmental impacts can be increased.

Lastly, beneficial reuse of waste streams in wastewater treatment facilities still faces many uncertainties and challenges. For example, the composition of wastewater created in a plant varies by source, which increases the difficulty and uncertainty of algal biomass cultivation. Furthermore, the nutrient profile of wastewater from some sources may render it unsuitable for algae cultivation. For example, the presence of inhibitors can result in poor nutrient assimilation and significantly reduce algal biomass productivity. In addition, the characteristics of wastewater-derived algae are quite different from those of freshwater algae, which leads to variation in biofuel yields, resource use, and environmental impacts from algal biomass conversion. The performance of algal biofuels depends on which conversion technology is used, as each technology varies greatly in productivity, efficiency, and operating conditions.

Many studies have conducted LCAs to assess the environmental performance of wastewater-driven algal biofuels. The study by Clarens et al. (2010) compared the ecological harm from biomass production of algae using municipal wastewater versus terrestrial biomass (corn, switchgrass, and canola). The results showed a reduction of environmental impacts with algae biomass cultivation, because conventional fertilizers are replaced by wastewater in these pathways. Another, later study by Clarens et al. (2011) also compared wastewater-derived algal biodiesel and bioelectricity production via lipid extraction, combustion, and digestion technologies. Their results showed that algae combustion outperformed the other two technologies in energy use and GHG emissions, because it required less upstream electricity and heat and fewer chemical inputs. A life cycle environmental impacts analysis for wastewater-derived biofuels was conducted by Mu et al. in 2014. The study found that using centrate as a feedstock produces environmental benefits relative to conventional fuel. No matter what algae conversion technology is chosen, centrate-derived bioenergy/biofuels have better life cycle performance than petroleum-based fuels. Algae cultivation is the most energy-intensive stage in all pathways compared, and it consumes over 50% of total life cycle fossil fuels. The environmental benefits arise because of the replacement of wastewater treatment, the absorption of CO_2, and the removal of nitrogen and phosphorus from wastewater. When growing algae in wastewater, the PBR is better than using an open pond.

4.4   LIFE CYCLE ASSESSMENT (LCA) OF WASTE REUSE IN WASTEWATER TREATMENT FACILITIES

The LCA of waste reuse from the waste streams, like all other LCA projects, should follow the ISO 14,000 standards, which include four standard steps: goal and scope definition, LCI, life cycle impact analysis, and interpretation. This chapter only highlights considerations on LCA and modeling when focusing on the reuse of waste streams.

1. *Scope and system boundary*: Figure 4.4 shows the potential technology pathways of waste reuse. The system boundaries for LCA analysis start with the various wastes collected from the treatment facility, through waste conversion technologies, all the way to the final use and disposal in land/ vehicles/landfills. The reuse of waste streams could also be combined into the LCA of a wastewater treatment plant. In this scenario, the reuse brings environmental benefits to the plant. The LCA could just focus on the waste to products stage, excluding the final use and disposal. This technique could be useful in comparing different waste reuse technologies.

2. *Functional unit*: Functional units within the LCA of waste reuse could be input functional units (wastes) or output functional unit (products). The selection of the functional unit depends on the goal and scope of the

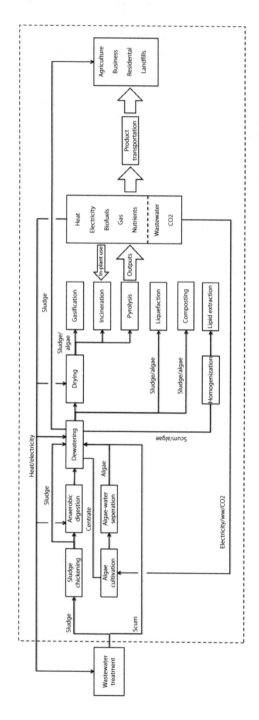

FIGURE 4.4   Waste stream reuse technology pathways.

project. If the project's goal is to evaluate waste reduction, the functional unit should be based on the mass of waste. In contrast, the functional unit should be based on products such as diesel or electricity if the goal is to evaluate the wastewater treatment facility as an energy source. Based on current literature, the majority of LCA projects used input functional units (Pradel et al., 2016).

3. *Waste allocation*: LCA modeling can treat waste streams as "zero burden" and charge them with no environmental burden. Arguments may be raised, because real products are generated from wastes. If the waste is treated as a resource, it should carry an upstream environmental burden. However, allocating an environmental burden between wastewater treatment and waste streams is complicated. The study by Pradel et al. (2016) proposed that allocation should consider the factors between the sludge production and the water treatment for each step that generates waste.

4. *Reuse benefits*: When the LCA aims to assess the waste reduction in treatment facilities, the functional unit will be based on inputs, that is, the waste flows. The products generated could be treated as credits to replace petroleum-based products or fertilizers or energy in markets. The LCA modeling could subtract the impacts of producing these products from the final impacts. Similarly, if the project includes algal biofuel production, it would be better to account for the removal of pollutants in wastewater as credits to wastewater treatment.

5. *Global warming potential accounting*: Wastes for reuse in the treatment facility are mainly biological wastes. During bioenergy production and consumption, CO_2 is emitted to the atmosphere, with other gases such as methane (CH_4), carbon monoxide (CO), and nitrous oxide (N_2O). The CO_2 is usually accounted for as biological CO_2, which is not included in the global warming potential calculation. However, the CH_4, N_2O, and other GHGs emitted from the production and consumption processes should be accounted for, because they have much higher warming potential than CO_2.

6. *Impact analysis*: The LCA could use existing analysis methods, such as TRACI 2 or Impacts 2000+, and so on. The LCA could also focus on several key impacts. Based on the project's goal and scope, the LCA of waste reuse should include impact categories for both waste reduction and energy production. For example, impacts of GHG emissions and fossil fuel use should be included, because they are key to evaluating the energy production. When the biosolids are disposed as land nutrients, human health and ecotoxicity should be included. If algae cultivation includes treating wastes, as well as these impacts, acidification and eutrophication should be included in the impact analysis, because the wastewater will be discharged from the algae cultivation stage.

4.5 AN LCA CASE STUDY OF BENEFICIAL REUSE OF WASTE STREAMS WITHIN A MUNICIPAL WASTEWATER TREATMENT PLANT

4.5.1 A METROPOLITAN WASTEWATER TREATMENT PLANT (MWTP) AND ITS WASTE STREAMS

The Metropolitan Wastewater Treatment Plant (MWTP) is located in St Paul, MN. It treats 250 million gallons of municipal wastewater per day with 98.9% pollutant removal efficiency (Metropolitan Council Environmental Services, 2014). The LCA study was assumed to be conducted within this real-life plant. The major solid waste flow at this plant is the vast amount of sludge, 265 dry tons daily, collected from the bottom of primary and secondary tanks. The currently practiced waste management of the sludge is anaerobic digestion and then landfilling. Besides sludge, the plant creates approximately 1 million gallons of centrate per day in the sludge dewatering. Currently, centrate is sent back to the aeration tank and combined into municipal wastewater. The characteristics of municipal wastewater treated at the plant, centrate, and sludge can be found in Tables 4.1 and 4.2.

As mentioned in Section 4.3.3, the centrate is a good nutrient source for oil-rich microalgae, which can be converted to algal biodiesel to replace petroleum-based fuels. The algal biofuel production can benefit not only the environment but also the wastewater treatment plant by saving energy or compensation by selling biofuels to the market. Therefore, this study proposed to integrate centrate algal biofuel production into the existing sludge digestion to produce more bioenergy for the MWTP. In addition, the study proposed to use digestate as land nutrients instead of dumping it in landfills.

The process integration of the waste treatment system is shown in Figure 4.5. The sludge collected at the bottom of the sludge-thickening tank was assumed to be converted to biogas through conventional anaerobic digestion. The biogas created was cleaned and used to generate electricity and heat in turbines. The centrate collected from the biosolids dewatering was assumed to grow algae in the PBRs. In addition, the supernatant from digestion and sludge thickening was also assumed to be sent to the PBRs to grow algae. Algae harvest from the PBRs was assumed to be dewatered in centrifuges and converted to algal biodiesel and glycerol with the wet lipid extraction technology. The non-lipid algae residual was assumed to be combined with sludge and digested to recover energy. Finally, the digested sludge and algae residual was dewatered in centrifuges and transported out for land application.

TABLE 4.2

Characteristics of Sludge from MWTP in St Paul, MN

Proximate analysis (wt.%)				Elemental analysis (wt.%)				Higher Heating Value (HHV) (MJ kg⁻¹)	Lower Heating Value (LHV) (MJ kg⁻¹)
Moisture	Ash	Volatile	Fix C	C	H	N	O		
4.53	15.01	68.57	16.42	53.24	7.39	6.12	33.25	24.42	21.77

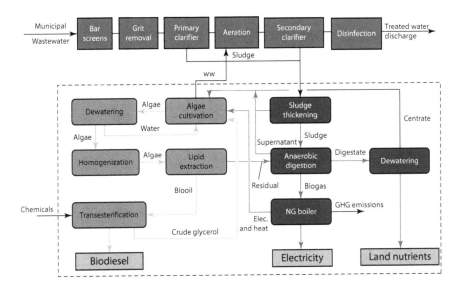

FIGURE 4.5   LCA process scope of the case study.

In the proposed system, algal biofuel conversion and sludge digestion shared some unit processes and outputs/emissions; therefore, when integrated, they produced less emissions and waste. Although LCAs of the two technologies have been conducted separately, as described in Sections 4.3.3 and 4.3.4, the environmental impacts of this integrated system could be different from those of the two single technologies. Therefore, it was still necessary to conduct a full-scale LCA for the integrated system to improve the understanding of technology integration for waste reuse in the wastewater treatment plant.

4.5.2   LCA Description

The purpose of this project was to conduct a well-to-wheel LCA of an integrated waste reuse system using the sludge and centrate within the MWTP to produce energy and land nutrients. The LCA followed the standards developed by ISO 14,041 and 14,044.

As multiple inputs and outputs are used in the system, it was difficult to define any single input or output as a functional unit. The functional unit, therefore, was set as the plant, which treats 250 million gallons of wastewater and generates 1 million gallons of centrate and 265 dry tons of dry sludge daily. The process scope for analysis is shown in Figure 4.5. One of the study's hypotheses is that the integration of waste reuse technologies could bring more environmental benefits to the plants. Therefore, the study also compared the integrated system with the single sludge digestion currently used by the plant.

The LCI was established by process modeling, in which the inputs, outputs, emissions, and energy uses of the processes were calculated and determined. The team at Kean University and University of Minnesota has modeled two technology pathways separately in previous studies. This study used models for unit processes and integrated them into one system model. A summary of the parameters for modeling and sources of data can be found in Table 4.3. The LCI included all direct inputs and emissions and also their upstream impacts. All direct and upstream impacts were either provided by researchers at the University of Minnesota or based on existing literature and LCA databases such as GREET 2015, EcoInvent 3, and the US-EI database. Labor, facilities manufacturing, and other impacts were not included in this analysis.

TABLE 4.3
Process Design Parameters for the Integrated Waste Reuse System in the MWTP

Unit processes	Assumptions	References
	Algal Biofuel Production	
Algae cultivation in PBRs	Growth rate: 36 g m^{-3} per day	Frank et al. (2011),
	Characteristics of algae: total lipid content = 5 wt.%; ash = 7.8 wt.%; the lower heating value (LHV) = 21.2 MJ kg^{-1}; moisture = 80%	Murphy et al. (2011), Quinn et al. (2013), Lardon et al. (2009), Alba et al. (2013),
	Electricity use: pumps = 0.35 kWh m^{-3}; CO_2 injection = 22.2 Wh kg^{-1} CO_2; CO_2 collection = 1.31 MJ kg^{-1} CO_2	Biller et al. (2011), Min et al. (2011)
	Water loss: leak = 0.2 m^3 m^{-2} per year; evaporation = 0.6 m^3 m^{-2} per year	
Algae harvesting and dewatering	Flocculation and sedimentation: increase solid content by 10 times; algae harvest efficiency is 90 wt.%; flocculant = 2.5 g/m^3 slurry	
	Centrifuge: increase solid content to 20 wt.%; algae collection efficiency is 95 wt.%; electricity use = 3.3 kWh m^{-3} slurry	
Algal biofuel production	Efficiency: homogenization = 94 wt.% algae; lipid extraction = 90 wt.% lipid; transesterification = 99 wt.% bio-oil; glycerin = 0.12 kg kg^{-1} biodiesel	Clarens et al. (2011), Frank et al. (2011), Batan et al. (2010), Vasudevan et al. (2012)
	Chemical inputs: hexane = 45.4 g kg^{-1} bio-oil; NaOH = 0.023 kg kg^{-1} bio-oil; HCl = 0.085 kg kg^{-1} biodiesel; KOH = 1.5 wt.% bio-oil; methanol = 0.117 kg kg^{-1} biodiesel	
	Energy use: homogenization = 0.246 kWh elec. kg^{-1} dry algae; lipid extraction = 0.51 kWh elec. + 6.83 MJ heat/kg bio-oil; transesterification = 0.09 kWh elec. + 1.7 MJ heat kg^{-1} diesel	

(Continued)

TABLE 4.3 (CONTINUED)
Process Design Parameters for the Integrated Waste Reuse
System in the MWTP

Unit processes	Assumptions	References
Algae residual treatment	Digestion process is used in recycling energy in residual	
Use	A passenger car with diesel engine is modeled, 28.08 mpg	GREET Model
	Sludge digestion and application	
Digestion	Digestion: CH_4 yield: 0.3 L g^{-1} TSS; CH_4 in biogas yield: 0.67 vol vol^{-1} biogas; electricity use: 0.11 kWh kg^{-1} TSS; heat use: 2.5 MJ kg^{-1} TSS	Clarens et al. (2011), GREET (2015)
	Biogas clean: electricity: 0.25 kWh m^{-3} biogas; CH_4 emissions: 3%; electricity generation: elec. conversion efficiency: 0.33; heat generation: 0.309	
	Digestate handling: CH_4 emission: 2%; N_2O emission: 0.14 kg N_2O kg^{-1} N	
Energy generation	Electricity generation: 0.331 MJ MJ^{-1} fuel; heat: 0.309 MJ MJ^{-1} fuel	GREET (2015)
Land use	Fertilizer replaced: urea: 0.25 kg N kg^{-1} N in digestate; DAP: 0.07 kg P kg^{-1} P in digestate	

When system models were set up in LCI, the diesel produced from algae was assumed to be used for plant vehicles; therefore, the diesel distribution was not accounted for. Because the allocation method of waste flows has not been widely accepted in current literature, this study still treated all waste as free or zero burden. The biodiesel and bioelectricity produced from algae and sludge can replace petroleum-based fuels, whereby their impacts are treated as credits subtracted from the total impacts. Similarly, a wastewater treatment credit was accounted for with algal biofuels, as some nutrients in centrate were removed during algae cultivation. The calculation of saving wastewater treatment was based on the assumption that the total carbon in wastewater was 0.0637 kg m^{-3} (Ecoinvent 3). Consequently, every 1 kg of carbon removed in centrate was equivalent to treating 15.7 m^3 of regular wastewater. In addition, the application of digestate to land created credits to replace the use of synthetic fertilizers, including urea and Diammonium phosphate (DAP). These credits were assigned to the digestion technology. Finally, the algae residual was sent to sludge digestion for energy conversion; as a result, all the credits for electricity generation and fertilizer replacement were assigned to digestion.

The life cycle impact assessment (LCIA) only included major environmental impact categories for analysis. They were fossil fuel use (MJ primary), GHG emissions (kg CO_2 eq.), terrestrial acidification (kg SO_2 eq.), aquatic eutrophication (kg N eq.), territorial ecotoxicity (kg triethyelene glycol soil), and carcinogens (kg C_2H_3Cl eq.).

The impact characterizations were based on Impact 2,000+ requirements. In calculating GHG emissions, the CO_2 emissions from digestion and biogas combustion were not counted, because they are biological CO_2. In contrast, CH_4 and N_2O emissions were counted.

4.5.3 LCA Results

The environmental impacts of the integrated system are shown in Figure 4.6. The fossil fuel use, GHG emissions, and acidification were all negative, indicating that the system could remove these three impacts in related ecosystems and benefit the environment. The fossil fuel use was -2.33×10^6 MJ of primary energy per day, in which the sludge digestion system contributed 95% of the total impact. This implied that the system could produce energy to replace fossil fuel–based energy and save resources. As the neutral lipid content in wastewater-based algae cultivation was not high, the biodiesel yield was not significant. In contrast, the digestion system generated much more energy as electricity than as biodiesel. Therefore, the fossil fuel saving was mainly derived from the electricity generation. The total GHG emissions were -3.49×10^4 kg CO_2 eq. per day, indicating that the system could sequester GHGs from the atmosphere. The result showed that digestion and algal biofuel production made almost the same contribution to reducing emissions by replacing petroleum-based diesel and electricity while replacing wastewater treatment of centrate. The land acidification was -879 kg SO_2 eq. per day, and the algae cultivation contributed to 83% of this acidification benefit. The acidification benefits were mainly from removal of pollutants in centrate.

The other three impacts, that is, eutrophication, territorial ecotoxicity, and carcinogens, were positive, indicating that emissions and discharges could be released to the environment from the integrated system. The aquatic eutrophication potential was extremely high, reaching 7.42×10^4 kg PO_4 eq. per day, and the

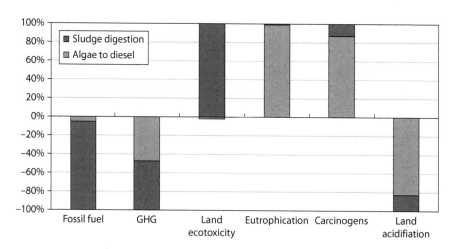

FIGURE 4.6 Environmental impacts of an integrated system that reuses centrate and sludge created in the MWTP.

algae cultivation in which centrate and supernatant were recycled to grow algae caused major eutrophication impacts. In algae cultivation PBRs, although a portion of nutrients (N and P) was absorbed, around 60 % of N and 45% of P could not be assimilated by the algae and therefore, remained in the wastewater exiting the system. This stream caused significant release of nutrients. The system also raised environmental concerns about terrestrial ecotoxicity, as the total release reached 9.3×10^8 kg TEG in soil, and the land application of digestate accounted for 99.9% of ecotoxicity. Lastly, the carcinogens released from the system were 214 kg C_2H_3Cl eq. per day, which was also high enough to raise a significant human health concern. Unlike the exotoxicity, algae cultivation contributed 87% of the total carcinogen release, which was mainly derived from using electricity for algae cultivation. The land application of digestate also raised serious concern over carcinogen release (9662 kg C_2H_3Cl eq.), but as the electricity generated in the digestion stage accounted for high credits (−9334 kg C_2H_3Cl eq.), the overall carcinogens from the digestion stage only accounted for 13% of total carcinogen release.

The study also compared the LCA of the integrated system with the single digestion + land application scenario. The results, seen in Figure 4.7, showed that the integrated system had better performance or lower impacts on fossil fuel use, GHG emissions, eutrophication, and acidification, and could reduce those impacts by 40%, 2170%, 8.5%, and 593%, respectively, compared with the single digestion system. The reduction of impacts by the integrated system was attributed mainly to the biofuel and electricity generation credits and the centrate treatment credits. However, the land ecotoxicity and carcinogens of the integrated system increased by 33% and 31% as compared with the single digestion system, because more digestate was applied to land as fertilizer and consequently increased the impacts on ecotoxicity and carcinogens.

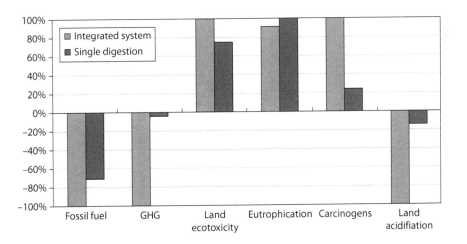

FIGURE 4.7 Comparison of the integrated waste reuse system with the conventional sludge digestion followed by land application.

4.6 CONCLUSION AND DISCUSSION

The study proposed a waste reuse system that reuses centrate and sludge created from sludge thickening to bioenergy and land nutrients. The results showed that the integrated system could bring environmental benefits as well as impacts. When compared with digestion followed by land application, the integrated system improved the environmental performance on fossil fuel use, GHG emissions, eutrophication, and acidification, but increased impacts on ecotoxicity and carcinogens. Decision makers should pay more attention to this trade-off when adding treatment facilities to MWTPs.

The study was just a primary analysis for the integrated waste management of the waste streams in an MWTP. A sensitivity analysis should be conducted to identify key factors for environmental performance where a strategy for improvement can be figured out. As the major benefits were created from the generation of bioenergy, and impacts stemmed from land application of digestate, the environmental performance can be improved by generating more bioenergy and reducing biosolids for disposal. Some feasible technologies that can be considered to further improve the integrated system are

1. Cycling CO_2 from digestion and injecting it into algae cultivation, which could improve algae yields and remove more nutrients from centrate.
2. Adding Scum-to-Biodiesel production into the integrated waste management system. More biodiesel could be produced.
3. Further treating digestate with pyrolysis or gasification or liquefaction technology, which could produce more biofuels and reduce biosolids for disposal.
4. Considering other algal biofuel production technologies, such as pyrolysis or liquefaction, to increase biofuel yields and therefore improve environmental performance.

REFERENCES

Alba, L. G., Torri, C., Fabbri, D., Kersten, S. R. A., and Brilman, D. W. F. (2013). Microalgae growth on the aqueous phase from Hydrothermal Liquefaction of the same microalgae. *Chemical Engineering Journal*. 228, 214–223.

Anderson, E., Addy, M., Ma, H., Chen, P., and Ruan, R. (2016). Economic screening of renewable energy technologies: Incineration, anaerobic digestion, and biodiesel as applied to waste water scum. *Bioresource Technology*, 222, 202–209.

Anderson, E., Addy, M., Xie, Q., Ma, H., Liu, Y., Cheng, Y., Ruan, R., et al. (2016). Glycerin esterification of scum derived free fatty acids for biodiesel production. *Bioresource Technology*, 200, 153–160. doi:10.1016/j.biortech.2015.10.018

Argonne National Laboratory (2011). Life-Cycle Analysis of Algal Biofuel with the GREET Model; ANL/ESD/11-5; Center for Transportation Research, Argonne National Laboratory: Argonne, IL; http://greet.es.anl.gov/publication-algal-lipid-fuels.

Asano, T., Burton, F. L., Leverenz, H. L., Tsuchihashi, R., and Tchobanoglous, G. (2007). *Water Reuse, Issues, Technologies, and Applications*. New York: McGraw-Hill.

Batan, L., Quinn, J., Willson, B., and Bradley, T. (2010). Net energy and greenhouse gas emission evaluation of biodiesel derived from microalgae. *Environmental Science & Technology*, 44(20), 7975–7980. doi:10.1021/es102052y.

Bi, C., Min, M., Nie, Y., Xie, Q., Lu, Q., Deng, X., and Ruan, R. (2015). Process development for scum to biodiesel conversion. *Bioresource Technology*, 185, 185–193. doi:10.1016/j. biortech.2015.01.081.

Biller, P., Riley, R., and Ross, A. (2011). Catalytic hydrothermal processing of microalgae: Decomposition and upgrading of lipids. *Bioresource Technology*, 102(7), 4841–4848. doi:10.1016/j.biortech.2010.12.113.

Buonocore, E., Mellino, S., De Angelis, G., and Liu, G. (2016). Life cycle assessment indicators of urban wastewater and sewage sludge treatment. *Ecological Indicators*. doi.org/10.1016/j.ecolind.2016.04.047, accessed 20 May.

Cao, Y. and Pawłowski, A. (2013). Life cycle assessment of two emerging sewage sludge-to-energy systems: Evaluating energy and greenhouse gas emissions implications. *Bioresource Technology*, 127, 81–91.

Chang, F., Wang, C., Wang, Q., Jia, J., and Wang, K. (2016). Pilot-scale pyrolysis experiment of municipal sludge and operational effectiveness evaluation. *Energy Sources, Part A: Recovery, Utilization, and Environmental Effects*, 38(4), 472–477.

Clarens, A. F., Nassau, H., Resurreccion, E. P., White, M. A., and Colosi, L. M. (2011). Environmental impacts of algae-derived biodiesel and bioelectricity for transportation. *Environmental Science & Technology*, 45(17), 7554–7560. doi:10.1021/es200760n.

Clarens, A. F., Resurreccion, E. P., White, M. A., and Colosi, L. M. (2010). Environmental Life Cycle Comparison of Algae to Other Bioenergy Feedstocks. *Environmental Science & Technology*, 44(5), 1813–1819. doi:10.1021/es902838n.

Davis, R. S., and Slaughter, J. B. (2006). *Biosolids Management: Options, Opportunities & Challenges*. Washington, DC: National Association of Clean Water Agencies (NACWA).

Fonts, I., Juan, A., Gea, G., Murillo, M. B., and Sanchez, J. L. (2008). Sewage Sludge Pyrolysis in Fluidized Bed, 1: Influence of Operational Conditions on the Product Distribution. *Industrial & Engineering Chemistry Research*, 47(15), 5376–5385. doi:10.1021/ie7017788.

Frank, E., Han, J., Palou-Rivera, I., Elgowainy, A., and Wang, M. (2011). *Life-Cycle Analysis of Algal Lipid Fuels with the GREET Model (United States Department of Energy, Argonne National Laboratory)*. Oakridge, TN: Argonne National Laboratory.

Gourdet, C., Girault, R., Berthault, S., Richard, M., Tosoni, J., and Pradel, M. (2017). In quest of environmental hotspots of sewage sludge treatment combining anaerobic digestion and mechanical dewatering: A life cycle assessment approach. *Journal of Cleaner Production*, 143, 1123–1136. doi:10.1016/j.jclepro.2016.12.007.

GREET (2015). *The Greenhouse Gases, Regulated Emissions, and Energy Use in Transportation Model*. Argonne, IL: Argonne National Laboratory.

Hospido, A., Moreira M., Martin, M., and Feijoo, G. (2005). Environmental evaluation of different treatment processes for sludge from urban wastewater treatments: Anaerobic digestion versus thermal processes. *The International Journal of Life Cycle Assessment*, 10(5), 336–345.

Kajitvichyanukul, P., Ananpattarachai, J., Amuda, O. S., Alade, A. O., Hung, Y., and Wang, L. K. (2008). Landfilling engineering and management. In: Wang L. K., Shammas N. K., Hung YT. (eds). *Biosolids Engineering and Management. Handbook of Environmental Engineering*, Vol. 7, Humana Press.

Lardon, L., Hélias, A., Sialve, B., Steyer, J., and Bernard, O. (2009). Life-cycle assessment of biodiesel production from microalgae. *Environmental Science & Technology*, 43(17), 6475–6481. doi:10.1021/es900705j.

McCarty, P.L., Bae, J., and Kim, J. (2011). Domestic wastewater treatment as a net energy producer can this be achieved? *Environmental Science & Technology* 45(17), 7100–7106. http://dx.doi.org/10.1021/es2014264.

Menetrez, M. Y. (2012). An overview of algae biofuel production and potential environmental impact. *Environmental Science & Technology*, 46(13), 7073–7085. doi:10.1021/es300917r.

Metcalf and Eddy, Inc. (2003). *Wastewater Engineering Treatment and Reuse*, 4th Ed. New York: Tata McGraw-Hill.

Metropolitan Council Environmental Services (2014). *Waste Water Treatment for Youngsters* [Brochure]. St Paul, MN: Metropolitan Council Environmental Services. https://metrocouncil.org/Wastewater-Water/Publications-And-Resources/MCES-INFORMATION/Educational/ES_kids_book-pdf.aspx.

Mills, N., Pearce, P., Farrow, J., Thorpe, R., and Kirkby, N. (2014). Environmental & economic life cycle assessment of current & future sewage sludge to energy technologies. *Waste Management*, 34(1), 185–195. doi:10.1016/j.wasman.2013.08.024.

Min, M., Wang, L., Li, Y., Mohr, M. J., Hu, B., and Zhou, W. G. (2011). Cultivating Chlorella sp in a pilot-scale photobioreactor using centrate wastewater for microalgae biomass production and wastewater nutrient removal. *Applied Biochemistry and Biotechnology*, 165(1), 123–137.

Mu, D., Addy, M., Anderson, E., Chen, P., and Ruan, R. (2016). A life cycle assessment and economic analysis of the Scum-to-Biodiesel technology in wastewater treatment plants. *Bioresource Technology*, 204, 89–97. doi:10.1016/j.biortech.2015.12.063.

Mu, D., Min, M., Krohn, B., Mullins, K. A., Ruan, R., and Hill, J. (2014). Life cycle environmental impacts of wastewater-based algal biofuels. *Environmental Science & Technology*, 48(19), 11696–11704. doi:10.1021/es5027689.

Murphy, F., and Allen, T. (2011). Energy-water nexus for mass cultivation of algae. *Environmental Science & Technology*, 45, 5861–5868.

National Algal Biofuels Technology Roadmap. (2010). Office of Energy Efficiency and Renewable Energy, Biomass Program, United States Department of Energy: College Park, MD.

NREL (National Renewable Energy Laboratory) (2010). Techno-economic analysis of biomass fast pyrolysis to transportation fuels; NREL/TP-6A20-46586. Washington, DC: National Renewable Energy Laboratory. www.nrel.gov/docs/fy11osti/46586.pdf.

PNNL (Pacific Northwest National Laboratory) (2016). Hydrothermal Liquefaction and Upgrading of Municipal Wastewater Treatment Plant Sludge: A Preliminary Techno-Economic Analysis. Prepared for the U.S. Department of Energy under Contract DE-AC05-76RL01830. Pacific Northwest National Laboratory Richland, Washington 99352.

Pradel, M., Aissani, L., Villot, J., Baudez, J.-C., and Laforest, V. (2016). From waste to added value product: Towards a paradigm shift in life cycle assessment applied to wastewater sludge—a review. *Journal of Cleaner Production,* 131, 60–75.

Quinn, J., Catton, K. B., Johnson, S., and Bradley, T. H. (2013). Geographical assessment of microalgae biofuels potential incorporating resource availability. *Bioenergy Research*, 6, 591–600.

Reilly, M. (2001). The Case against Land Application of Sewage Sludge Pathogens. *Canadian Journal of Infectious Diseases*, 12(4), 205–207. doi:10.1155/2001/183583

Rice, C. P., O'Keefe, P., and Kubiak, T. (2003). Sources, pathways, and effects of PCBs, dioxins, and dibenzofurans. In D. J. Hoffman, B. A. Rattner, G. A. Burton, Jr., and J. Cairns, Jr. (eds.), *Handbook of Ecotoxicology* (pp. 501–574), 2nd Ed. New York: Lewis.

Ruiz, H. A., Rodríguez-Jasso, R. M., Fernandes, B. D., Vicente, A. A., and Teixeira, J. A. (2013). Hydrothermal processing, as an alternative for upgrading agriculture residues and marine biomass according to the biorefinery concept: A review. *Renewable and Sustainable Energy Reviews*, 21, 35–51. doi:10.1016/j.rser.2012.11.069.

Singh, A. and Olsen, S. I. (2013). A critical review of biochemical conversion, sustainability and life-cycle assessment of algal biofuels. *Applied Energy*, 101, 822. doi:10.1016/j.apenergy.2011.01.052.

Smith, K. M., Fowler, G. D., Pullket, S., and Graham, N. J. D. (2009). Sewage sludge-based adsorbents: A review of their production, properties and use in water treatment applications. *Water Research*, 43(2009), 2569–2594.

Soh, L., Montazeri, M., Haznedaroglu, B. Z., Kelly, C., Peccia, J., Eckelman, M. J., and Zimmerman, J. B. (2014). Evaluating microalgal integrated biorefinery schemes: Empirical controlled growth studies and life cycle assessment. *Bioresource Technology*, 151, 19–27. doi:10.1016/j.biortech.2013.10.012.

Speight, J. G. (2008). Chapter 8 Fuels from Biomass. In: *Synthetic fuels handbook: Properties, process, and performance.* New York, NY, McGraw-Hill, 221–264.

Trinh, T. N., Jensen, P. A., Dam-Johansen, K., Knudsen, N. O., and Sorensen, H. R. (2013). Influence of the Pyrolysis Temperature on Sewage Sludge Product Distribution, Bio-Oil, and Char Properties. *Energy & Fuels*, 27(3), 1419–1427. doi:10.1021/ef301944r.

United States Environmental Protection Agency (2016). *Clean Watersheds Needs Survey 2012: Report to Congress (EPA-830-R-15005).* Washington, DC: Office of Wastewater Management Municipal Support Division, Sustainable Management Branch.

United States Environmental Protection Agency, Municipal Technology Branch (2003). *Biosolids Technology Fact Sheet: Use of Landfilling for Biosolids Management* (pp. 1–8). Washington, DC: Municipal Technology Branch.

United States Fish and Wildlife Service, New Jersey Field Office (2015). *White Paper: Recontamination of Mitigation Sites in the Meadowlands* (pp. 1–25). Galloway, NJ: U.S. Fish and Wildlife Service.

USA, City of Whitewater, WI. (2010). Wastewater treatment plan anaerobic digestion study. Strand Associates. www.whitewater-wi.gov/images/stories/public_works/wastewater/Whitewater_WWTP_ADS_Final_Report-May_2010.pdf.

Vasudevan, V., Stratton, R. W., Pearlson, M. N., Jersey, G. R., Beyene, A. G., Weissman, J. C., Rubino, M., et al (2012). Environmental performance of algal biofuel technology options. *Environ. Sci. Technol.*, 2012, 46(4), 2451–2459.

Wang, L., Min, M., Li, Y., Chen, P., Chen, Y., Liu, Y., Wang, Y., and Ruan, R. (2010). Cultivation of green algae *Chlorella sp.* in different wastewaters from municipal wastewater treatment plant. *Applied Biochemistry and Biotechnology*, 162(4), 1174–1186.

Yoshida, H., Christensen, T. H., and Scheutz, C. (2013). Life cycle assessment of sewage sludge management: A review. *Waste Management & Research*, 31(11), 2013.

Youcai, Z. (2016). Pollution control and recycling of sludge in sanitary landfill. In *Pollution Control and Resource Recovery: Sewage Sludge* (pp. 275–352). S.l.: Butterworth-Heinemann, Elsevier, Amsterdam.

5 A Life Cycle Evaluation of the Environmental Effects of a Hydrometallurgical Process for Alkaline Battery Waste Treatment
A Case Study

Angela Daniela La Rosa

CONTENTS

5.1 INTRODUCTION

Life cycle assessment (LCA) plays a key role in sustainable production and consumption policies. In 2003, the European Commission with the Communication on Integrated Product Policy (IPP) (EC 2003) described the importance of an approach to environmental sustainability based on attention to the product life cycle. This approach was subsequently confirmed in the Communication "Action Plan on Sustainable Consumption and Production and Sustainable Industrial Policy" (SCP) in 2008, which emphasizes the need for tools for the assessment and communication of the environmental performance of products, such as LCA and the different types of environmental declarations that are based on it (Environmental Product Declaration Carbon Footprint and Climate Declaration, Water Footprint, etc.).

The aim was to encourage decoupling between economic growth and the massive use of natural resources and enable the transition to a low carbon economy, as evidenced by the strategy of the European Commission "Energy 2020," which aims to increase the production of renewable energies by 20% by 2020.

This chapter deals with the description of a technological process for the treatment and recovery of secondary raw materials starting from no-lead alkaline batteries (zinc/manganese type). The primary advantage of this technological choice is environmental protection due to the reduction of pollution at the end of life of such materials. The recovery of metals from the recycling process is expected to bring benefits to the environment, although it involves greater costs compared with landfilling or incineration.

5.1.1 Background on Battery End of Life

There are different alternatives for the final disposition of batteries (Bernardes et al., 2004):

- *Landfill*: To date, most household batteries, especially primary batteries, are disposed of in MSW (municipal solid waste) and are sent to sanitary landfills.
- *Stabilization*: This process represents a pre-treatment of batteries to avoid the contact of metals with the environment in landfill. The process is not much used because of the high costs involved.
- *Incineration*: This is used when household batteries are disposed of in MSW and are sent to a municipal waste combustion facility. The incineration of batteries can cause the emission of mercury, cadmium, lead, and dioxins into the environment.
- *Recycling*: Hydrometallurgical and pyrometallurgical processes can be used to recycle metals present in the batteries. These recycling processes are currently being studied in different parts of the world.

A literature review on battery recycling (Espinosa et al., 2004) reports the following processes active nowadays:

- *Sumitomo*: A Japanese process totally based on pyrometallurgy. Its cost is very high, and it is used to recycle all types of portable batteries. It is not indicated for recycling Ni-Cd batteries.

- *Recytec*: A Swiss process that combines pyrometallurgical, hydrometal-lurgical, and physical treatments. It is used for recycling all types of por-table batteries and also fluorescent lamps and Hg-containing tubes. This process does not recycle Ni-Cd batteries. The investment for this process is smaller than that for the Sumitomo process, but the operating costs are higher (Frenay et al., 1994; Jordi, 1995; Ammann, 1995).
- *Atech*: A process based on physical treatment of scrap batteries, this has a lower cost than hydrometallurgical or pyrometallurgical processes. It is used for recycling all portable batteries.
- *Snam–Savam*: A French process for recycling Ni-Cd batteries, totally based on a pyrometallurgical method (Schweers et al., 1992).
- *Sab Nife*: A Swedish process for recycling Ni-Cd batteries, entirely based on pyrometallurgy (Landskrona et al., 1983; Anulf, 1990).
- *Inmetco*: A North American process, initially developed by the International Nickel Company (INCO) with the objective of recovering dusts from electric arc furnaces. It can also be used to recover metallic wastes from other processes, and Ni-Cd batteries can be included as one such waste (Hanewald et al., 1991, 1992).
- *Waelz*: A pyrometallurgical process to recover metals from steelmak-ing dusts. The process uses a rotary furnace and recovers metals such as Zn, Pb, and Cd from steelmaking wastes (Egocheaga-Garcia et al., 1997; Moser et al., 1992).
- *TNO*: A hydrometallurgical Dutch process for battery recycling. This process developed two recycling alternatives, one for Zn-C and alkaline household batteries and the other for Ni-Cd batteries. The alternative for household batteries was not commercially implemented (Hurd, 1993).
- *Accurec*: A German pyrometallurgical process to recycle batteries, in which Ni-Cd batteries are treated separately (ALD Vacuum Technologies AG, 2001).

The present chapter is a case study applied to a hydrometallurgy process.

Recycling through hydrometallurgy basically consists of acid or base leaching of scrap to bring the metals into solution. Once in solution, metals can be recovered by precipitation, altering the pH of the solution, adding some reaction agent, or elec-trolysis. The solution can also be separated by solvent extraction, using an organic solvent that binds to the metallic ion, separating the metal from the solution. The metal can then be recovered by electrolysis or by precipitation (Xue et al., 1992; Lyman and Palmer, 1994; Contestabile et al., 1999).

5.2   STRUCTURE OF AN LCA

According to the ISO 14140-44 standard of 2006, an LCA study shall include the following four phases:

1. Definition of the study's objectives and boundaries of the system: this is a preliminary phase, defining the objectives that the LCA has to reach.

2. Inventory analysis (or eco-inventory; in English, life cycle inventory [LCI]): at this stage, the flows of matter and energy input and output from the various phases of the life cycle are quantified.
3. Life cycle impact assessment (LCIA): the potential environmental impacts associated with certain flows in the previous inventory phase are estimated.
4. Interpretation of results: an evaluation of the output of the previous two phases is run and tested for correspondence with the study objectives defined in the first phase.

5.2.1 Software Used

The software used for the study was SimaPro 8.2, produced by the Dutch company Pré Consultant, which is among the most popular software for conducting LCA studies. Implemented for the first time in 1990, the software is used by customers in over 60 countries worldwide, especially by big industries, consulting firms, and universities, to conduct relevant assessments on the environmental performance of various products, processes, and services. It offers great flexibility by providing various modeling parameters, allows interactive analysis of results, and comes with a large database. The latest version of this professional tool, SimaPro 8.2, allows the user to collect, monitor, and interpret the environmental performance of products and services by analyzing even complex life cycles in a systematic and transparent way following the recommendations of the series standards ISO 14040.

The SimaPro software is inclusive of various databases for the inventory analysis that include thousands of processes and materials. The databases that have mainly been used are described in the following subsections.

5.2.1.1 Ecoinvent v2

The Ecoinvent database, of Swiss origin, covers nearly 4000 processes in the following industries, related mainly to activities in Switzerland and Western Europe:

- Energy
- Transport
- Building materials
- Chemical products
- Washing products
- Paper and cardboard
- Agriculture
- Treatment of pollutants

All processes are provided with extensive documentation including a description of the data (name, unit of measure, derived from the data, category and sub-category of membership) and information on uncertainty.

5.2.1.2 Industry Data v.2.0

This contains inventory data provided by industry associations. Most of this data is provided following a "cradle to grave" approach, that is, considering the product or process from source to disposal.

To assess the impacts (LCIA), the following methods are used: CML 2000 and ReCiPe endpoint (H) in the SimaPro software.

5.3 PROCESS DESCRIPTION

5.3.1 DESCRIPTION OF THE SELECTION PROCESS

Batteries and accumulators arrive at the recycling plant with different packaging, in containers of various shapes, such as big bags (Figure 5.1a), octabins (Figure 5.1b), and plastic containers 1 m³ in volume (Figure 5.1c) and equipped with a lid; in any case, the load unit is represented by a pallet of average mass $m = 500$ kg.

The material flow is characterized geometrically and chemically; with regard to the chemical characteristics, the weight fractions are estimated as alkaline batteries 80%; zinc-carbon batteries 19%; other types of batteries 1%. As regards the geometrical characterization, Table 5.1 shows the standardization used at international level.

For mechanical treatment, the loading units are moved by means of a trolley of industrial type equipped with a tilter, which allows lifting and flowing on a vibrating table; this action enables the mechanical removal of the dust accumulated during the collection and recovery of the batteries. The raw material after this treatment falls

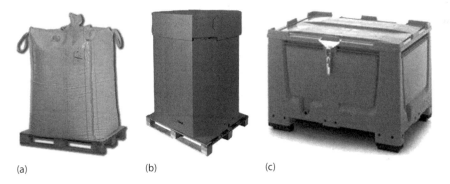

(a) (b) (c)

FIGURE 5.1 Loading unit of waste batteries: (a) big bag, (b) octabins, and (c) plastic containers (1 m³ in volume).

TABLE 5.1
Characterization of Batteries according to International Standards

Type	Length (L) (mm)	Diameter (D) (mm)	Volume (F) (mL)	Alkali (g)	Zn-C	Ni-Cd	Ni-MH	Lithium
AAAA	42.5	8.3		6.5				
AAA	44.5	10.5	0.1	11.5	9.1	12.0		
AA	49.2	13.5	0.3	23.5	19.0	29.0	26.0	24.0
A23	28.5	10.3		8.0				
C	50.0	26.2	1.1	65.0	48.0	85.0	82.0	
D	61.5	33.2	2.3	135.0	98.0	200.0	170.0	
4.5 V		17.0						

onto a conveyor belt. The production speed can be considered equal to $r = 1.0$ t/h. Downstream of the conveyor belt is placed a mechanical selection system realized by means of three sieves that intercept the batteries according to their cross-sectional diameter. The nominal power of the geometric sorting line is equal to $Pn = 11.0$ kW. Following this, chemical selection is carried out using a scanning X-ray machine to identify stacks of a certain type: first alkaline, then zinc carbon. Once selected, the batteries are cut along their longitudinal section. Batteries with the same chemical characteristics are fed into a tumbling machine together with an equal volume of water; the mechanical action and washing allow the contents, called *black paste* (BP), to be extracted from the metal. The black paste is fed into the hydrometallurgical recycling process.

5.3.2 ENERGY CONSUMPTION CALCULATION

An analysis was carried out to evaluate the energy consumption of the chemical treatment plant for alkaline batteries, assuming a steady-state operation. The total energy consumption for each batch of treated batteries was given by

$$214.845 \text{ kWh} + 69.13 \text{ kWh} + 58.56 \text{ kWh} + 97.6 \text{ kWh} = 440.14 \text{ kWh} \qquad (5.1)$$

for a total of 1584.5 MJ.

In this sum, the first term is the total consumption due to the various equipment used in the different elementary phases; the second term is the consumption of the scrubber; the third term is the consumption of the electrolytic cell; and the fourth term is the consumption of the evaporator/concentrator. The materials covered after each batch are listed below:

- Final BP, which after drying, may represent a secondary raw material to be used for the manufacture of new batteries
- High-purity zinc oxalate, which is a raw material valuable for the production of pure metallic zinc
- Manganese (IV) oxide, which is a valuable product for the manufacture of new batteries
- In summary, 66 kg of final BP is obtained after a treatment of 5 h and 53 min, with a total energy consumption of 330.15 kWh, which corresponds to 1188.5 MJ.

5.4 LCA STUDY

5.4.1 FUNCTIONAL UNIT, GOAL, AND AIM OF THE STUDY

The present LCA takes into account the process of selection of exhausted alkaline batteries and the subsequent hydrometallurgical process used to recycle the resulting mass. The approach used is "cradle to gate," as other steps of the battery life cycle (production, transport, and use) are not taken into account. The functional unit chosen for this study is one batch of BP weighing 105 kg. BP is the remaining material after the preliminary separation of iron, plastics, and paper from the collected battery wastes.

The scheme in Figure 5.2 shows the boundaries of the system examined.

The batteries are disassembled, and after the separation of iron scraps, plastic, and paper, the remaining material is BP, which makes up approximately 50% by weight of the original battery. Thereafter, the BP undergoes the recycling treatment. The metal composition of the BP mass is shown in Table 5.2.

5.4.2 INVENTORY ANALYSIS (LCI) OF THE HYDROMETALLURGICAL PROCESS

Table 5.3 shows the analysis of the flows of materials and energy related to the hydrometallurgical process. The analysis refers to the initial batch. It does not take into consideration at this stage the selection of waste battery operations or transportation of the materials.

5.4.3 ASSUMPTIONS

The sulfuric acid solution used is regenerated after each batch process. After three times, it requires an electrolytic process in the cell to be regenerated, and after 20 times, the sulfuric acid solution must be neutralized. In the reactor, 400 kg of H_2SO_4

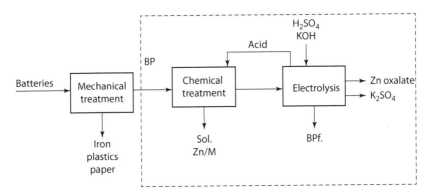

FIGURE 5.2 System boundaries consisting of the hydrometallurgical process.

TABLE 5.2
Composition of Metal BP Mixture Laboratory Data and Percentage Composition

Test Method EPA 6010C 2007	Results (mg/kg)	Percentage (%)
Manganese	405,882	53.3
Zinc	330,095	43.35
Potassium	25,140	3.3
Iron	206	0.02

TABLE 5.3

Inventory: Input and Output Flows of the Hydrometallurgical Process

Flows	Quantity	Unit	Data Source
	Input		
Batteries BP	105	kg	
H_2SO_4 (R101)	120 (sol. 30%)	kg	Sulfuric acid (100%), Ecoinvent
H_2SO_4 (R104) electrolytic treatment	438 (sol. 30%)	kg	
KOH (TK 101)	0.51 (solution	kg	Potassium hydroxide, Ecoinvent,
KOH (R104)	60 g/L, 1 M)		PM (Molecular weight) = 56 g/mol
Neutralization electrolytic treatment	1.2 (solution 60 g/L, 1 M)		
K_2CO_3 (R104) Neutralization electrolytic treatment	168	kg	Potassium carbonate, Ecoinvent
Ammonium oxalate (TK104)			
Industrial water (R101)	1100	kg	Tap water, used as substitute for
Water used to wash the zinc oxalate (R102)	300	kg	mains water, Ecoinvent, Europe, 2000
	Electricity consumption		
Electricity for the use of mixer, reactors, agitators, compressors, evaporators, pumps, filter, scrubber	440	kWh	Grid electricity, medium voltage, Italia, Ecoinvent
	Output		
Zinc oxalate	118	kg	Zinc, Ecoinvent, Europe, 1994–2003
BP final (manganese oxide and carbon black)	66	kg	Process data: About 36 kg of carbon black and 30 kg of MnO_2
K_2SO_4 solid (S102)	384	kg	
H_2O distilled	3033	kg	

in a 30% by mass solution (120 kg of sulfuric acid diluted with water to 1462 L) is charged. The following reactions take place:

$$ZnO\ (solid) + H_2SO_4\ (solution) \rightarrow ZnSO_4\ (sol) + H_2O \qquad (5.2)$$

$$Zn(OH)_2\ (solid) + H_2SO_4\ (solution) \rightarrow ZnSO_4\ (sol) + 2H_2O \qquad (5.3)$$

$$Zn\ (solid) + H_2SO_4\ (solution)\beta \rightarrow ZnSO_4\ (sol) + H_2\ (gas) \qquad (5.4)$$

$$MnO\ (solid) + H_2SO_4\ (solution) \rightarrow MnSO_4\ (sol) + H_2O \qquad (5.5)$$

$$Mn_2O_3 \ (solid) + 2H_2SO_4 \ (solution) \rightarrow 2MnSO_4 \ (sol) + 2H_2O + \tfrac{1}{2} \ O_2 \quad (5.6)$$

Regarding emissions, the chemical treatment plant has only one emission point in the air, the chimney of the scrubber, from which only hydrogen should be emitted at less than 50% of the explosive limit. Drainage of distilled water is expected as an output from the vacuum concentrator; this may be partially used within the process or discharged into the sewerage system.

5.5 LIFE CYCLE IMPACT EVALUATION

5.5.1 RESULTS AND DISCUSSION

For the evaluation of the impacts, the methods CML 2000 and ReCiPe end-point (H) present in the software SimaPro 8.2 were used. The characterization method CML 2000, developed by Leiden University of Amsterdam, is a mid-points method that focuses attention on the following categories of environmental impact: consumption of abiotic resources, greenhouse effect, depletion of the stratospheric ozone layer, human toxicity, ecotoxicity, photochemical formation of ozone, acidification, and eutrophication. The impacts of each of these categories are reported in appropriate units of measurement. For example, the greenhouse effect or global warming potential (GWP) is calculated in kilograms of CO_2.

The ReCiPe method also includes a method of assessment of the damages expressed as

- Harm to human health (unit: Disability Adjusted Life Years [DALY], which means that life years are weighted according to the disability caused by various diseases)
- Damage to ecosystems (unit: species/year, i.e., number of species in danger of extinction per year)
- Damage to resources, expressed in dollars

Figure 5.3 shows the diagram of the flows involved in the hydrometallurgical process derived from the inventory table. The impact result is indicated as a percentage of equivalent CO_2 produced and therefore, refers to the impact category GWP.

The diagram is obtained on the basis of the inventory data and refers to the recycling of 1 kg of alkaline batteries. The flows in input (indicated in red) are the constituents of BP (mainly zinc oxide, manganese oxide, and carbon black), treatment with sulfuric acid and subsequent neutralization with potash, and the power consumption. The flows in output (in green) refer to the materials recovered for possible reuse. The percentage impact of the output flows is negative, because they indicate avoided impacts.

Figure 5.4 is another useful way to present the amount of CO_2 produced or avoided. The main impacts are associated with the use of potassium carbonate and electricity consumption, while the main contribution to avoiding the impact is associated with material recovery (potassium sulfate, zinc, etc.).

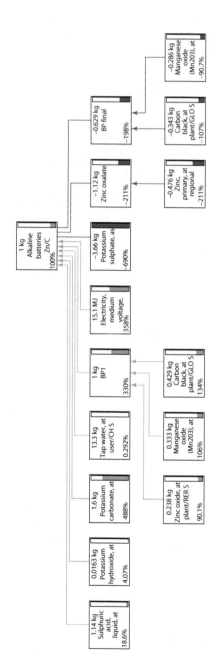

FIGURE 5.3 LCA tree diagram of the hydrometallurgical process. Percentage values refer to the contribution of each process to GWP for the treatment of 1 kg of alkaline batteries.

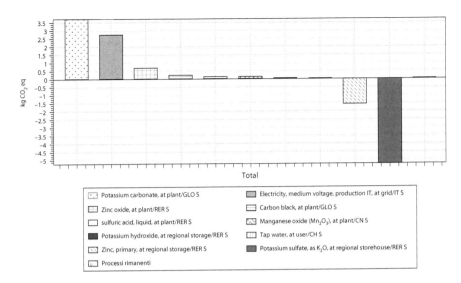

FIGURE 5.4 Equivalent CO_2 produced by each process step for the treatment of 1 kg of alkaline batteries. Impact evaluation method CML 2000.

Figure 5.5 shows the potential impacts on all other categories taken into account by the method CML 2000 (biotic and abiotic resources, greenhouse effect, ozone, human toxicity, ecotoxicity, smog, acidification, eutrophication, etc.). Negative values refer to impacts avoided thanks to the recovery of materials. There are avoided impacts in all categories.

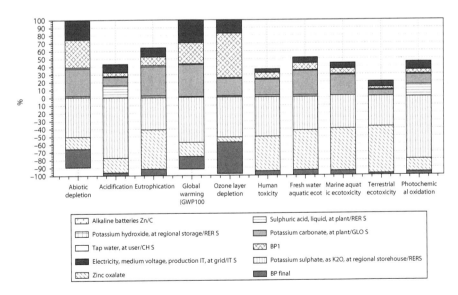

FIGURE 5.5 Potential impacts for all the categories for the treatment of 1 kg of alkaline batteries according to the CML 2000 method.

The graphs of Figures 5.6 and 5.7 were obtained using the ReCiPe endpoint method. Figure 5.6 refers to the damage caused to human health (Human Health), ecosystems (Ecosystems), and mineral resources (Resources), Figure 5.7 shows the contributions of the individual components of the process to human health.

The method takes into account different effects: greenhouse effect, thinning of the ozone layer, acidification, eutrophication, photo-chemical smog formation, and

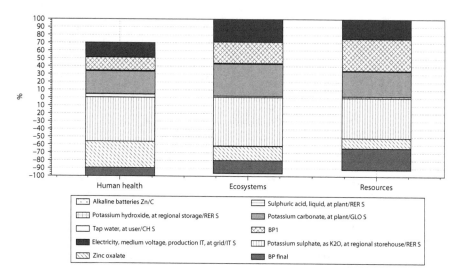

FIGURE 5.6 Percentage of damage contribution in three main categories (Human Health, Ecosystems, Resource) for the treatment of 1 kg of alkaline batteries according to the ReCiPe endpoint method.

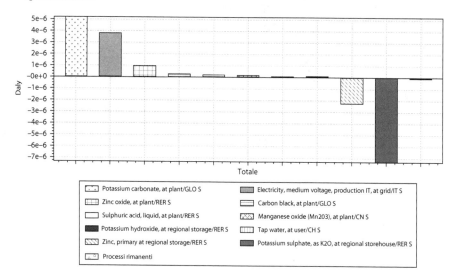

FIGURE 5.7 Damage to human health expressed as DALY for the treatment of 1 kg of alkaline batteries.

toxic substances (heavy metals, carcinogenic substances, or pesticides). The relative incidence of the various environmental effects is established through a weighting operation by which the results obtained are converted, through the use of appropriate numerical factors, to determine the levels of damage equivalent to the different environmental impacts.

The damage avoided due to the recovery of materials is remarkable. It is interesting to note that the impact caused by the electricity consumption of the hydrometallurgical process and the use of chemicals such as sulfuric acid and potassium hydroxide is largely offset by the recovery of materials (zinc, potassium, and final BP).

The diagram in Figure 5.7 shows on the ordinate, the years of life lived with disability and on the abscissa, the input and output data for the recycling of 1 kg of alkaline batteries. The highest value, which means great damage, is attributed to the use of potassium carbonate, followed by the consumption of electricity, zinc oxide, and sulfuric acid. The avoided damage (negative values) results from the recovery of metallic zinc and potassium.

Quantitative amounts of the impacts (produced and avoided) are reported in Table 5.4.

Characterization factors are calculated and expressed using reference units (e.g., kg 1,4-dichlorobenzene (1,4-DB) equivalent, kg CFC-11 (trichlorofluoromethane) equivalent, etc.).

If we take into account, for example, the impact on the category Marine aquatic ecotoxicity, the value attributed to the use of potassium carbonate (3062.47 kg 1,4-DB eq) is totally balanced by the avoided impact generated by the recovery of potassium sulfate (−4565.79 kg 1,4-DB eq). The recovery of zinc oxalate generates an avoided impact of (−6031.03 kg 1,4-DB eq).

In total, the impact reported in this category is negative (−6297.41 kg 1,4-DB eq). This means that all the generated impacts, including that caused by power consumption (786.15 kg 1,4-DB eq), are cancelled, and environmental benefits are generated. Similar trends can be seen for the other impact categories. According to a literature review, recycling scenarios demonstrate environmental benefit compared with landfilling. It is important to develop a voluntary primary battery collection and recycling program as a net positive for the environment compared with disposal to landfill with other domestic garbage (Olivetti et al., 2011).

5.6   CONCLUSIONS

The cradle-to-gate analysis carried out in this case study confirms what was already expected: the recovery of materials (potassium, zinc oxalate, zinc, and BP final) from the hydrometallurgical process and their reuse generate environmental benefits due to avoidance of the production of these compounds from raw materials. All the impacts generated to run the recycling process (especially the use of chemicals and energy consumption) are balanced by the avoided impacts due to materials recovery. The LCA method provides instruments to understand the potentialities of clean industrial technologies and processes moving toward more sustainable development.

TABLE 5.4

Quantitative Amounts of Generated and Avoided Impacts for the Hydrometallurgical Process of 1 kg of Alkaline Batteries. Method CML 2000

Impact Category	Unit	Total	Sulfuric Acid	Potassium Hydroxide	Potassium Carbonate	Tap water	BP1	Electricity	Potassium sulfate	Zinc Oxalate	BP final
Abiotic depletion	kg Sb eq	0.01	0.00	0.00	0.03	0.00	0.03	0.02	-0.04	-0.01	-0.02
Acidification	kg SO_2 eq	-0.07	0.02	0.00	0.01	0.00	0.01	0.01	-0.09	-0.02	0.00
Eutrophication	kg PO_4—eq	-0.01	0.00	0.00	0.01	0.00	0.00	0.00	-0.01	-0.01	0.00
Global warming (GWP100)	kg CO_2 eq	0.76	0.14	0.03	3.71	0.00	2.50	2.72	-5.24	-1.60	-1.50
Ozone layer depletion (ODP)	kg CFC-11 eq	0.00	0.00	0.00	0.00	0.00	0.00	0.00	0.00	0.00	0.00
Human toxicity	kg 1.4-DB eq	-8.20	0.24	0.02	2.73	0.00	1.05	0.60	-6.47	-5.67	-0.70
Fresh water aquatic ecotoxicity	kg 1.4-DB eq	-2.16	0.08	0.01	1.41	0.00	0.40	0.31	-1.86	-2.23	-0.29
Marine aquatic ecotoxicity	kg 1.4-DB eq	-6297	163.43	25.71	3062	2.08	887	786	-4565	-6031	-628
Terrestrial ecotoxicity	kg 1.4-DB eq	-0.12	0.00	0.00	0.01	0.00	0.01	0.01	-0.06	-0.09	0.00
Photochemical oxidation	kg C_2H_4 eq	0.00	0.00	0.00	0.00	0.00	0.00	0.00	0.00	0.00	0.00

REFERENCES

ALD Vacuum Technologies AG, Vacuum thermal recycling of spent batteries, ALD vacuum technologies technical information, March 2001.

Ammann P, Economic considerations of battery recycling based on Recytec process, *Journal of Power Sources* 57 (1995) 41–44.

Anulf T, SAB NIFE recycling concept for nickel–cadmium batteries—an industrialized and environmentally safe process, in: *Proceedings of the Sixth International Cadmium Conference*, Cadmium Association (UK), 1990, pp. 161–163.

Bernardes AM, Espinosa DCR, and Tenório JAS, Recycling of batteries: A review of current processes and technologies, *Journal of Power Sources* 130 (2004) 291–298.

Contestabile M, Panero S, and Scrosati B, A laboratory-scale lithium battery recycling process, *Journal of Power Sources* 83 (1999) 75–78.

Egocheaga-Garcia B, Developing the Waelz process: Some new possibilities for the preparations of the load in the Waelz process and ultradepuration of the volatile fraction obtained in this process, in: *Proceedings of the Third International Conference on the Recycling of Metals* 1997, ASM, 1997, pp. 387–402.

Espinosa DCR, Bernardes AM, and Tenório JAS, An overview on the current processes for the recycling of batteries, *Journal of Power Sources* 135 (2004) 311–319.

Frenay J, Ancia PH, and Preschia M, Minerallurgical and metallurgical processes for the recycling of used domestic batteries, in: *Proceedings of the Second International Conference on Recycling of Metals*, 1994, ASM, pp. 13–20.

Hanewald RH, Schweyer L, and Hoffman MD, High temperature recovery and reuse of specialty steel pickling materials and refractories at Inmetco, in: *Proceedings of the Electric Furnace Conference*, Toronto, Canada, November 12–15, 1991, Iron and Steel Society, Inc., 1992, pp. 141–146.

Hanewald RH, Munson WA, and Schweyer DL, Processing EAF dusts and other nickel–chromium waste materials pyrometallurgically at Inmetco, *Minerals and Metallurgical Processing*, November 1992, pp. 169–173.

Hurd DJ, Muchnik DM, and Schedler TM, Recycling of consumer dry cell batteries—pollution technology review, n.213. Notes Data Corporation, NJ, USA, 1993, pp. 210–243.

Jordi H, A financing system for battery recycling in Switzerland, *Journal of Power Sources* 57 (1995) 51–53.

Landskrona S, Melin AL, and Svensson VH, Process for the recovery of metals from the scrap from nickel–cadmium electric storage batteries, US Patent 4401463, August 30, 1983.

Lyman JW and Palmer GR, Recycling of nickel–metal hydride battery alloys, in: *Proceedings of the TMS Annual Meeting 1994*, San Francisco, California, EUA, 1994, The Minerals, Metals & Materials Society, 1994, pp. 557–573.

Moser WS, Mahier GT, Knepper RT, Kuba MR, and Pusateri FJ, Metals recycling from steelmaking and foundry wastes by horsehead resource development, in: *Proceedings of the Electric Furnace Conference 1992*, Atlanta, GA, USA, November 10–13, 1992, Iron and Steel Society, Inc., 1992, pp. 145–157.

Olivetti E, Gregory J, and Kirchain R, Life cycle impacts of alkaline batteries with a focus on end-of-life, A study conducted for the National Electrical Manufactures Association, 2011.

Schweers ME, Onuska JC, and Hanewald RK, A pyrometallurgical process for recycling cadmium containing batteries, in: *Proceedings of the HMC-South '92*, New Orleans, 1992, pp. 333–335.

Xue Z, Hua Z, Yao N, and Chen S, Separation and recovery of nickel and cadmium from spent Cd–Ni storage batteries and their process wastes, *Separation Science and Technology* 27 (2) (1992) 213–221.

6 Textile Wastewater Treatment by Advanced Oxidation Processes

A Comparative Study

*Maria Gamallo, Yolanda Moldes-Diz,
Roberto Taboada-Puig, Juan Manuel Lema,
Gumersindo Feijoo, and Maria Teresa Moreira*

CONTENTS

6.1 INTRODUCTION

6.1.1 THE ENVIRONMENTAL, HEALTH, AND ECONOMIC IMPACTS OF TEXTILE EFFLUENTS

The textile and apparel industry is a solid and relevant industry in many developed or developing countries, mainly in Asia. In this region, the production and marketing of textiles are one of the primary sources of employment for millions of people and a major contributor to export earnings and gross domestic product (Chen et al., 2017a). Nowadays, world consumption of clothing fibers has reached the level of 11 kg per capita, which represents a remarkable demand for raw materials and chemicals (Shui and Plastina, 2013).

The chemical substances used in the manufacture of textiles can be categorized into:

- Functional chemical substances: substances added to the textile fibers to provide the desired properties of the final product (colorants and crease resistant agents)
- Auxiliary chemical substances: those required in the textile manufacture process (organic solvents, surfactants, softeners, salts, acids, bases, biocides)
- Chemical substances present as impurities and not intentionally added, such as formaldehyde released from certain resins, polyaromatic hydrocarbons, metals, and so on

These chemicals can be released and result in human and environmental exposure during normal wear and tear, washing, or when disposed of as waste. Among the substances mentioned in the preceding list, colorants arise as target contaminants in textile wastewaters, because their presence is highly visible at very low concentrations (<1 mg L^{-1}), which negatively affects water transparency (Chequer et al., 2013). Decreasing light penetration into water decreases photosynthetic activity, causing oxygen deficiency and de-regulating the biological cycles of aquatic biota (Apostol et al., 2012). The term *colorant* encompasses both dyes and pigments, which can be differentiated by their solubility. While dyes are soluble compounds, mainly applied to textile materials in aqueous solution, pigments are insoluble, and they can be added in the formulation of paints, printing inks, or plastics (Christie, 2001).

Until the mid-nineteenth century, only natural dyes with poor dyeing performance were used. Therefore, as a result of the need for more resistant dyes with easier application and improved thermal behavior, synthetic dyes were developed. That is why, 50 years after the origin of the dye industry in 1856, 90% of the dyes used were manufactured (Welham, 1963). The massive consumption of dyes in the textile industry accounts for approximately 70% of the dyes used worldwide.

Colorants can be classified according to either their chemical structure or the application method. The most important reference on the classification of dyes and pigments is the *Color Index*, a report published by the Society of Dyers and Colourists in 1924 (O'Neill et al., 1999). This book series provides a complete listing of commercial dyes and known pigments, and it is periodically updated. Each colorant receives a Generic Name CI, which incorporates an application class, a hue, and a number that reflects the chronological order of dyes since they became commercially available.

6.1.2 Chemical Classification of Colorants

According to the chemical classification method, dyes are grouped depending on similar chemical structural features. The families of dyes and pigments of outstanding relevance are the azo dyes (-N=N-) followed by the carbonyl (C=O), phthalocyanine, arylcarbonium ion, sulfide, polymethine, and, finally, nitro types (Christie, 2001).

6.1.2.1 Azo Dyes and Pigments

Azo dyes and pigments are by far the most important chemical family among commercial organic dyes, and they account for about 60%–70% of the dyes used in traditional textile applications (Chequer et al., 2013). They are produced mainly in China, India, Korea, Chinese Taipei, and Argentina (OECD, 2005).

Azo dyes, as their name implies, contain the azo functional group (-N=N-), attached on either side to carbon atoms with sp^2 hybridization, and it is usually bonded to two aromatic rings. Most azo dyes of commercial importance have one single azo group and are therefore referred to as monoazo dyes, but many of them have two, three, or more groups. The versatility of the synthesis sequence of azo dyes creates an immense number of different dyes. Moreover, the synthesis process uses inexpensive and readily available organic starting materials such as aromatic amines and phenols as well as low energy.

Regarding their color properties, azo dyes can provide a broad range of tones with high color intensity. They can present satisfactory technical properties, such as fastness to light, heat, water, and other solvents, but they perform slightly worse than other chemical groups such as carbonyl dyes and phthalocyanines (Christie, 2001). However, they can be improved thanks to the mordant process, in which certain metal ions are used; nowadays, this is primarily restricted to the coupling of certain azo dyes with chromium (III) in wool.

6.1.2.2 Carbonyl Dyes and Pigments

This chemical family of colorants arises as the second in importance, and it is characterized by the presence of the carbonyl group (C=O), acting as its main chromophore. Carbonyl-type colorants are found in a far more extensive variety of structural arrangements than in the case of azo dyes and pigments. The main reason for the importance of carbonyl dyes is that they are capable of presenting absorption bands at long wavelengths with relatively short conjugated systems.

The chemistry developed in the manufacture of the carbonyl dyes is usually more elaborate than in the case of azo dyes and often involves multiple steps and the use of specific intermediates. Consequently, the number of commercial products is more limited, and these tend to be more expensive. The most relevant group of carbonyl dyes is the family of compounds known as anthraquinones (Gürses et al., 2016).

Regarding solidity properties, carbonyl dyes are often superior to their azo counterparts, which justifies their selection when a high technical behavior is required for a particular application. A particular textile application dominated by carbonyl dyes is the use of vat dyes, a group of dyes applied to cellulosic fibers such as cotton.

6.1.2.3 Phthalocyanines

Phthalocyanines are undoubtedly the most important chromophore system developed during the twentieth century. Their fortuitous discovery in 1927 from the reaction of phthalic anhydride with ammonia in a glass lined reactor enabled the isolation of an insoluble, dark blue crystalline substance, which turned out to be iron phthalocyanate (FePc; De Diesbach and Von der Weid, 1927). Although textile dyes based on phthalocyanines are of limited importance, this class of dyes provides by far the most important blue and green organic pigments and exhibits exceptional stability. The absorption band of the metal phthalocyanates depends on the nature of the central metal ion, the arrangement of the substituents on the outer rings, and the degree of condensation of the rings. The most widely investigated are the metal complexes of the first transition series: iron, cobalt, nickel, copper, and zinc (Demirbaş et al., 2017).

6.1.2.4 Polyene and Polymethine Dyes

Two families of structurally related dyes containing one or more methine groups (-CH=) belong to this group. Polyene dyes consist of a series of conjugated double bonds, usually terminating in aliphatic or alicyclic groups, which provides its color. The best-known group of polyene dyes is formed by carotenoids. Polymethine dyes are capable of providing a broad range of bright and intense colors but in general, tend to show relatively low strength properties. This feature has limited their use in fabrics, where they are mainly restricted to some dispersed dyes for polyesters, cationic dyes for acrylic fibers, and basic dyes (Langhals, 2004).

6.1.2.5 Aryl Carbonium Ion Dyes

Historically, aryl carbonium ion dyes were the first synthetic dyes developed for textile applications. In fact, the first commercial synthetic dye, Mauvein, belongs to this group (Hübner, 2006). Although their use has declined considerably, many of them still have some importance, especially for their use as basic (cationic) dyes for coloring acrylic fibers and paper. Aryl carbonium ion dyes are characterized by the central carbon atom being attached to either two or three aromatic rings. Structurally, they are closely related to polymethinic dyes, especially to cyanine-type dyes, and provide intense and vivid colors but with lower technical properties compared with the azo chemical families and phthalocyanines (Gordon and Gregory, 1987).

6.1.2.6 Sulfur Dyes

These are known to be complex mixtures of polymeric molecular species that contain a high proportion of sulfur in the form of sulfur (-S-), disulfide (-S-S-), and poly-sulfide (-Sn-) bonds, and heterocyclic rings. In general, they are highly insoluble in water and are applied to cellulosic fibers. The most outstanding advantage of sulfurized dyes is that they provide reasonable technical performance at low cost

(Nagl, 2000). However, they also present environmental issues, mainly attributed to the sulfur residues present in wastewaters (Goswami and Basak, 2000).

6.1.2.7 Nitro Dyes

Nitro dyes are a small group of dyes, with a relatively simple nitrodiphenylamine structure, used as dispersed dyes for polyester. They can provide bright shades of red, orange, and yellow, although their colors are among the weakest of commercial chromophores (Christie, 2001).

6.1.3 TEXTILE DYES

Dyes are used in the coloring of a broad set of substrates, including paper, leather, and plastics, but their most important market is in the textile sector, where they are used in an extensive variety of products. Textile fibers can be classified into three main groups: natural, semi-synthetic, and synthetic. Regarding natural fibers, wool and silk (animal origin) and cotton (cellulosic fiber) play a major role. Examples of semisynthetic fibers are viscose (regenerated cellulosic fiber) and cellulose acetate (a derivative of chemically modified cellulose). The most important synthetic fibers are polyester, polyamides (nylon), and acrylic fibers. Figure 6.1 depicts the percentage of textile use in the EU, with the outstanding positions of cotton (40%), polyester (20%), and polyamide (13%).

Depending on the type of organic polymer molecules present in textile fibers, the class of dyes to be used is different. Table 6.1 shows a summary of the diverse types of dyes used in the textile industry.

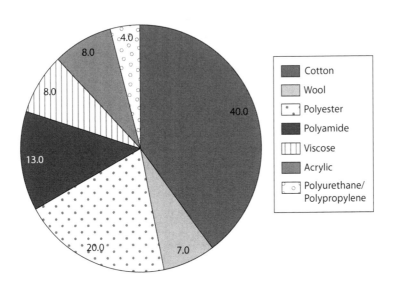

FIGURE 6.1 Percentage of textiles consumption in the EU. (From Beton et al., 2014, *Environmental Improvement Potential of Textiles (IMPRO Textiles)*, Publications Office of the European Union, Luxembourg.)

TABLE 6.1

Use of the Different Textile Fibers and Dyes

Textile fibers		Types of dyes	
Protein fibers	Wool/silk	Acid dyes	Azo type
			Carbonyl type
		Mordant dyes	Azo type
			Carbonyl type
		Premetalized dyes	Azo type
Cellulosic fibers	Cotton/viscose/flaz/	Direct dyes	Azo type
	jute/jemp/tow	Vat dyes	Carbonyl type
Synthetic fibers	Polyester	Dispersed dyes	Azo type
			Carbonyl type
	Polyamides	Acid dyes	Azo type
			Carbonyl type
		Dispersed dyes	Azo type
			Carbonyl type
	Acrylic fibers	Basic dyes	Azo type
			Aryl carbonium type
			Methine derivatives type

A potential link between exposure to sensitizing substances present in textile dyes and finishing resins and allergic skin reactions has been reported (Van der Putte et al., 2013). Approximately 10% of 2400 textile-related substances are considered to be of potential risk to human health. Under reductive conditions, azo dyes can cleave, producing aromatic amines of potential toxicity (Puntener and Page, 2004). As an example, a study identified an azo dyeing plant as one of the sources of mutagenic activity detected in the Cristais River in Brazil, a source of drinking water for 60,000 inhabitants (De Lima et al., 2007).

An EU-harmonized classification of most of the identified substances with severe health hazardous properties (potentially carcinogenic, mutagenic, or toxic) has been published. Several are restricted under the European Regulation on Registration, Evaluation and Authorization of Chemicals (REACH) or are included on the candidate list (substances of very high concern that may be found in textile articles). In this framework, azo dyes have been included in Section 43 (Azo dyes and Azocolorants) of REACH Annex XVII. Many azo dyes are suspected to cause impacts on growth, neurosensory damage, metabolic stress and death in fish, and growth and productivity in plants (Chung, 1983).

Environmental exposure to chemical substances from textile articles mainly occurs due to leaching via laundering, from release due to wear and tear, and from textile waste. The dye lost through the processes of the textile industry poses a remarkable problem for wastewater management: 200,000 tons of dye are estimated to be released into the global environment every year, and the concentration of azo dye in textile effluents can reach 500 ppm (Chequer et al., 2013). In the EU, the release of acid and direct dyes, based on the assumption of best available technologies for the textile manufacturing process, has been estimated as 2–22 tons per year from the washing of cotton and polyamide textiles (Swedish-Chemicals-Agency, 2014).

Other textile-related hazardous materials of known potential environmental risk are silver, nonylphenol, and highly fluorinated compounds. Increasing volumes of biocide-treated textile articles are marketed. For example, it has been shown that silver, triclosan, and triclocarban contained in textile articles are released from bio-cide-treated textiles by laundering. Concern is increasingly being raised about the possibility that bacteria may develop resistance to the antibacterial chemicals, which could end in resistance to antibiotics (Swedish-Chemicals-Agency, 2014).

6.2 ADVANCED OXIDATION PROCESSES (AOPS) FOR TREATMENT OF INDUSTRIAL TEXTILE EFFLUENTS

Tightening government legislation is forcing textile industries to treat their waste effluent to an increasingly high standard. In this sense, the control of water pollution in textile effluent has been investigated from different perspectives, from physicochemical to biological processes. Despite the efficiency of physicochemical alternatives, this type of method is often very costly, and the accumulation of concentrated sludge creates a disposal problem. In the search for effective removal of dyes from large volumes of effluents with low-cost systems, biological processes were addressed, but with limited efficiency.

Beyond conventional processes, advanced oxidation processes (AOPs) are considered potential alternatives, as these processes perform the oxidation of organic contaminants through reactions with hydroxyl radicals (•OH, standard redox potential $[E^0] = 2.80$ V) (Glaze et al., 1987). AOPs typically comprise two stages: (1) the formation of •OH and (2) the reaction of these oxidants with organic contaminants in water. AOPs are considered a highly competitive technology for the removal of organic pollutants that, due to their high chemical stability and/or low biodegradability, are not successfully removed by conventional techniques (Oller et al., 2011). The most common advanced oxidation processes are listed in the following subsections.

6.2.1 OZONATION

Ozone is a powerful oxidant for wastewater treatment, used as a disinfection chemical or as an oxidant to remove color and odor (2.07 V; Von Gunten, 2003). The reaction with a great number of organic compounds occurs either by direct oxidation, as the ozone molecule, or by indirect reaction, through the formation of secondary oxidants such as •OH (Baig and Liechti, 2001).

The use of ozone presents a number of drawbacks, such as its relatively low solubility and stability in water, and it requires an additional filtration step to remove biodegradable organic carbon from water. With the aim of improving the efficiency of ozonation, other AOP methods have been investigated, such as wet peroxide ozonation (O_3/H_2O_2) and catalytic ozonation. The addition of both H_2O_2 and O_3 to wastewater accelerates the decomposition of ozone and enhances the production of •OH. At pH values above 5, low concentrations of H_2O_2 will dissociate into HO_2^-, which can initiate ozone decomposition more efficiently than •OH (Staehelin and Hoigne, 1982). Catalytic ozonation can be conducted in combination with transition metals, such as divalent forms of Fe, Mn, Ni, Co, Cd, or Cu, or with metal oxides (MnO_2, TiO_2, Al_2O_3) or metals/metal oxides (Cu-Al_2O_3, Cu-TiO_2) (Kasprzyk-Hordern et al., 2003).

Operational variables such as time, pH, ozone dose, and the implementation of a biological treatment before or after ozonation have been evaluated. Ozonation under acidic conditions allows direct oxidation by molecular ozone, while at high pH, •OH production is favored, and hydroxyl oxidation starts to dominate (Hoigné and Bader, 1976). Accordingly, basic pH conditions were beneficial to achieve color removal and chemical oxygen demand (COD) elimination (Somensi et al., 2010). In another work, the ranges of ozonation time (15–60 min) and dosage (20–50 g O_3 m^{-3}) were established for significant removal of COD and nearly complete decolorization (Ciardelli and Ranieri, 2001). The ozonation of dye baths before and after a biological treatment at various time intervals was investigated by Baban et al. (2003). The authors concluded that 40 min ozonation of biologically treated wastewater yielded almost colorless effluent, whereas only 10 min of ozone oxidation after a biological treatment was sufficient to reduce the overall toxicity significantly (92%).

6.2.2   H₂O₂/UV

The H_2O_2 oxidation of textile wastewater has been demonstrated to be ineffective at both acid and alkaline values (Tünay et al., 1996). However, the use of UV irradiation photolyzes H_2O_2 to form two •OH radicals that react with organic contaminants. Its combination with other processes (H_2O_2, O_3, O_3/H_2O_2, O_3/H_2O_2/UV) was investigated for the oxidation of a synthetic textile effluent (Ledakowicz and Gonera, 1999). The authors concluded that the most suitable AOP is ozonation combined with UV radiation or the combination of AOPs (O_3/H_2O_2/UV) before biodegradation.

6.2.3   FENTON (H₂O₂/FE²⁺) AND PHOTO-FENTON PROCESSES

The Fenton process is one of the most studied AOPs due to its efficiency and short reaction time (Pignatello et al., 2006). The •OH and reactive oxidizing species are produced by the catalytic decomposition of H_2O_2 with Fe^{2+}, as described in Equations 6.1 through 6.5.

$$Fe^{2+} + H_2O_2 \xrightarrow{k=76\,Lmol^{-1}s^{-1}} Fe^{3+} + OH + OH^- \tag{6.1}$$

$$Fe^{3+} + H_2O_2 \xrightarrow{k=0.01\,Lmol^{-1}s^{-1}} Fe^{2+} + HO_2 + H^+ \tag{6.2}$$

$$RH + OH \rightarrow R + H_2O \tag{6.3}$$

$$R + Fe^{3+} \rightarrow R^+ + Fe^{2+} \tag{6.4}$$

$$R + OH \rightarrow R - OH \tag{6.5}$$

A modified configuration corresponds to the Photo-Fenton process, which can also produce •OH radicals.

$$Fe^{2+} + H_2O_2 \rightarrow Fe^{3+} + OH^- + \cdot OH \qquad (6.6)$$

$$Fe^{3+} + H_2O \xrightarrow{hv} Fe^{2+} + H^+ + \cdot OH \qquad (6.7)$$

$$OH + RH \rightarrow H_2O + R \qquad (6.8)$$

A number of advantages are associated with the Fenton for the elimination of organic pollutants: no energy necessary to activate H_2O_2, low cost and easy operation, short reaction time, and no mass transfer limitation due to its homogeneous catalytic nature. However, this process has some drawbacks that have to be considered, such as low pH requirements (2–3), consumption of ferrous salts due to ineffective regeneration, and the presence of residual iron in the effluent and chemical sludge production.

The efficiency of Fenton processes for the treatment of textile effluents is extensively reported in the literature. The combination of chemical oxidation using the Fenton reagent (H_2O_2/Fe^{2+}) achieved partial removal of color and organic content (COD) of a textile industry effluent (Papadopoulos et al., 2007). When Fenton oxidation was coupled to an aerobic sequencing batch reactor (Blanco et al., 2012) and aerobic biological treatment (Pérez et al., 2002), not only was the removal of COD and color observed but also efficient pathogen removal.

6.2.4 PHOTOCATALYSIS

Photocatalysis relies on the capacity of semiconducting materials to act as sensitizers for light-reduced redox processes due to their electronic structure. A photocatalyst is a chemical that produces electron–hole pairs by absorption of light quanta, causes chemical transformations of the reaction participants that come into contact with it, and regenerates its chemical composition after each cycle (Fox and Dulay, 1993). Photocatalysts include titanium dioxide (TiO_2), zinc oxide (ZnO), zinc sulfide (ZnS), ferric oxide (Fe_2O_3), silicon (Si), tin oxide (SnO_2), and cadmium sulfide (CdS), among others. TiO_2 has been the most widely cited photocatalyst in the literature due to its properties: considerable activity, high stability, lack of environmental impact, and low cost (Augugliaro et al., 2006).

6.2.5 SONOLYSIS

During acoustic sonication, bubbles are formed, which grow and collapse implosively, producing an unusual chemical and physical phenomenon. This collapse generates hotspots with transient temperature of about 5000 K and pressures of about 1000 atm. Under such extreme conditions, water molecules dissociate into •OH radicals and H atoms. The radical species react with other molecules to induce sonochemical degradation (Özdemir et al., 2011). The sonochemical decolorization and decomposition of azo compounds, such as C.I. Reactive Red 22 and Methyl Orange, was investigated (Okitsu et al., 2005). The results showed that the removal of both dyes was rapid (>90% in 30 min). When the scope of the dyes was broader (Methyl Orange, o-Methyl Red, and p-Methyl Red; Joseph et al., 2000), it was observed that the reaction rates for the o-Methyl Red were between 30% and 40% faster than for the other compounds. Apparently, there is a strong influence from a carboxylic group in the ortho position on the azo group.

In the pursuit of better process efficiency and color removal, it is a frequent strategy to combine different AOPs. Chung and Kim (2011) evaluated the efficiency of combined ozonation with three AOPs (O_3/H_2O_2, O_3/UV, and $O_3/H_2O_2/UV$) for the removal of dye from synthetic effluents. The conclusion of this work is that although all the processes eliminated color within a very short operational time, the $O_3/H_2O_2/UV$ processes presented the best results in terms of COD removal and increased biodegradability.

Similarly, the treatment of a textile industry effluent by different AOPs (O_3, O_3/UV, H_2O_2/UV, $O_3/H_2O_2/UV$, Fe^{2+}/H_2O_2) was investigated (Azbar et al., 2004). The results showed that AOPs provide higher color removal than conventional chemical precipitation. Among the different AOPs, the $O_3/H_2O_2/UV$ combination showed the best results (99% COD elimination and 96% color removal). Moreover, the Fenton process also showed good results (96 % COD removal and 94% color removal) but proved to be more economical. With the aim of investigating the relevance of the different AOP technologies, a search for bibliographic references published between 2012 and 2017 was conducted (Figure 6.2). Regarding the different types of dyes, azo dyes are by far the most studied, with 81% of the bibliographical references; carbonyl dyes accounted for 14%, which corresponds to the sum of two individualized searches for anthraquinone dyes and indigo dyes; and finally, phthalocyanine dyes accounted for 5%. Among the different AOPs used to remove dyes, the most studied, with 51% of the bibliographical references, is the Fenton process, followed by photocatalysis (11%) and ozonation (9%).

6.3 PROCESS INTENSIFICATION USING NANOSTRUCTURED MATERIALS FOR DYE REMOVAL

In the last two decades, the increasing interest in nanosciences and materials with enhanced functionality has culminated especially in the preparation of nanomaterials with a characteristic surface area and size, between 10 and 100 nm. Several approaches have been proposed for their use in the removal of a wide range of pollutants, comprising heavy metals, algae, organics, bacteria, viruses, nutrients, cyanide, and antibiotics (Bethi et al., 2016). In this regard, heterogeneous catalysis based on the use of nanoparticles seems to be a promising technology in the treatment of different types of industrial textile effluents using AOPs. In this section, two strategies of AOPs based on the use of nanoparticles were assessed for the removal of diverse types of dyes present in wastewater: (1) a photochemical process based on TiO_2 nanoparticles and (2) heterogeneous Fenton catalytic degradation based on magnetite nanoparticles. In addition, a review of recent studies related to the application of nanoparticles for the treatment of dye-containing effluents was conducted.

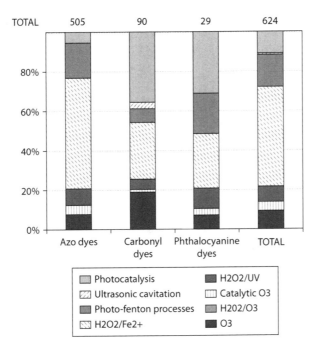

FIGURE 6.2   Scientific papers related to the application of AOPs for the removal of dyes. (From ISI Web of Knowledge; years 2012–2017.)

6.3.1   PHOTOCHEMICAL PROCESS BASED ON TiO$_2$ NANOPARTICLES

Heterogeneous photocatalysis has been reported to perform the oxidation of the target organic contaminants within a short contact time without generating any other solid wastes (Neppolian et al., 2007). Among the different types of nanomaterials, semiconductors can act as photocatalysts, sensitizing the oxidation of organic compounds (Mohapatra et al., 2014; Hoffmann et al., 1995). If a semiconductor is irradiated, photons with energy higher than the band gap energy of the semiconductor can excite electrons from the valence band into the conduction band, leaving a positively charged vacancy or hole in the valence band. The combination of both is referred to as an *electron–hole pair.* Many of the electron–hole pairs will recombine as the electron returns to its original state, emitting light or heat. However, some pairs may migrate to the catalyst surface, where they can participate in redox reactions that lead to the decomposition of compounds adsorbed at the surface (Parsons, 2004; Nguyen and Juang, 2015).

The most typical nanoparticles used in photocatalysis are metal oxide nanoparticles. Their principal advantages are the high surface area and better photolytic properties. Table 6.2 provides an overview of the recent work undertaken in the field of photocatalysis along with comprehensive information on the selected nanocatalysts and dyes.

TABLE 6.2
Application of Photochemical Advanced Oxidation Processes for Dye Decolorization

Heterogeneous Photocatalysis				
Catalysts	Dye	Concentration	Main Findings	References
ZnO	Congo Red, Methyl Orange, and Reactive Black 38	30 mg L^{-1}	Satisfactory photocatalytic properties under UV irradiation	Chen et al. (2017b)
TiO$_2$	Reactive Black 5 and Reactive Red 239	30–150 mg L^{-1}	High decolorization rates after five cycles of TiO$_2$	Saggioro et al. (2011)
Y$_2$O$_3$ composites	Xylenol Orange and Rhodamine B	20 mg L^{-1}	High potential of TiO$_2$/Y$_2$O$_3$ and Bi$_2$O$_3$/Y$_2$O$_3$ composites for dye removal	Guo et al. (2017)
V$_2$O$_5$/TiO$_2$	Toludine Blue, Safranin Orange, and Crystal Violet	20–80 μM	Superior performance of the hybrid catalyst	Rauf et al. (2007)
Au/Fe$_2$O$_3$	Disperse Blue 79	4.5 mg L^{-1}	Metallic gold particles enhance the photocatalytic activity of the iron oxide support	Wang (2007)

Heterogeneous catalysis by a wide range of nanocatalysts has been investigated for the treatment of diverse contaminants, obtaining high degradation percentages in most of the cases. In particular, TiO$_2$ and zinc oxide ZnO are preferred, because they are relatively inexpensive, non-toxic, and stable, and they present high oxidization potential (Gmurek et al., 2017; Gaya and Abdullah, 2008; Sousa-Castillo et al., 2016). In particular, TiO$_2$ has been applied in the photocatalytic treatment of polluted wastewaters (Oros-Ruiz et al., 2013; Khodja et al., 2001; Barka et al., 2013; Khataee and Kasiri, 2010).

In this section, we assess a heterogeneous photocatalytic system using TiO$_2$ nanoparticles for the treatment of matrices polluted with dyes, specifically Methyl Green, Orange II, and Reactive Blue 19, with remarkable chemical and structural differences. A number of factors that govern the kinetics of photocatalysis, such as the catalyst concentration and pH, the addition of oxidant, and the initial substrate concentration, were evaluated. The selected

TiO_2 nanoparticles were spherical with a size of 30 nm. The experimental setup for the photocatalytic decolorization consists of a custom-made photocatalytic oxidation quartz cell with a volume of 3 mL. The photodegradation activity was investigated as a function of time, at room temperature and irradiated with light at a predominant wavelength of 365 nm produced by a mercury lamp. At regular intervals, absorbance measurements in a UV-Vis spectrophotometer were carried out to monitor the color removal of Methyl Green and Orange II (5 mg L^{-1}) and Reactive Blue 19 (25 mg L^{-1}). The photocatalytic dye removal yield was determined as the rate of color disappearance, calculated using the following equation:

$$\text{Dye removal}\left(\%\right) = \frac{\left(C_0 - C_t\right)}{C_0} \times 100 \qquad (6.9)$$

where C_0 and C_t represent the dye concentration in milligrams per liter before and after reaction, respectively.

The results of the decolorization of different dyes using TiO_2 nanoparticles under UV irradiation compared with direct photolysis and dark adsorption are reported in Figure 6.3. The experimental results indicate that Methyl Green and Orange II undergo an acceptable decolorization rate, higher than 90%, whereas for Reactive Blue 19, the decolorization results after 30 min of photocatalytic reaction were lower than 10%, which could be attributed to the recalcitrant nature of the dye due to its aromatic anthraquinone structure. In a similar study, Reactive Blue 19 was found to be the most recalcitrant to decolorization by TiO_2-based photocatalysis (Lizama et al., 2002). Dye removal in parallel control experiments was negligible in the mentioned conditions.

Several authors achieved similar conclusions in the case of Methyl Green and Orange II; their experiments carried out without catalyst and with only UV irradiation did not exhibit significant degradation after 30 min of direct photolysis (May-Lozano et al., 2017; Bousnoubra et al., 2016).

In this sense, results have demonstrated the feasibility of TiO_2 as a model photocatalyst for dye decontamination in textile wastewater in the presence of UV light. These findings also suggest that various operating parameters, such as type of photocatalyst and dye and light intensity, among others, may influence the photocatalytic oxidation process. The optimization of these conditions is crucial from the perspective of efficient design and application of heterogeneous photocatalysis processes.

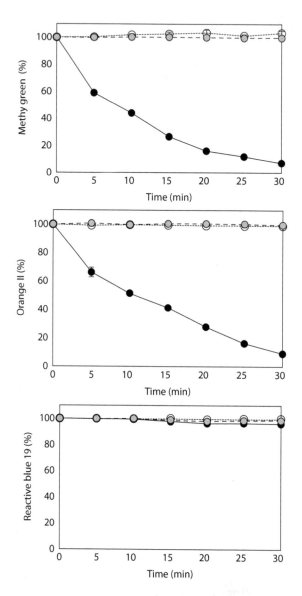

FIGURE 6.3 Decolorization percentages of Methyl Green (5 mg L^{-1}), Orange II (5 mg L^{-1}), and Reactive Blue 19 (25 mg L^{-1}) by heterogeneous photocatalysis. (●) Heterogeneous photocatalysis using TiO$_2$ (0.01 g L^{-1}). Photolysis (◉) and dark adsorption (○) were considered as controls.

6.3.2 HETEROGENEOUS FENTON CATALYTIC DEGRADATION BASED ON MAGNETITE NANOPARTICLES

The concept of the Fenton process relies on the generation of OH• from H_2O_2 with iron ions acting as a homogeneous catalyst in an acidic medium (Bautista et al., 2008; Fenton, 1894). An overview regarding the application of the heterogeneous Fenton process for the removal of dyes is provided in Table 6.3.

Beyond these processes, the use of Fe_3O_4 nanoparticles as a catalyst substituting the use of soluble Fe^{2+} may be a promising technology. The potential of these materials derives from their higher ability for degradation of recalcitrant pollutants compared with conventional iron-supported catalysts due to the presence of both Fe^{2+} and Fe^{3+} species. In addition, their magnetic properties allow their easy, fast, and inexpensive separation from the reaction medium (Munoz et al., 2015; Wang et al., 2014). Color removal of Reactive Blue 19 (25 mg L^{-1}) by the heterogeneous Fenton process was evaluated under various concentrations of Fe_3O_4 nanoparticles (100–500 mg L^{-1}), as shown in Figure 6.4.

TABLE 6.3
Application of Heterogeneous Fenton Process for Dye Decolorization

Catalysts	Dye	Concentration	References
La-Fe composite (Fe_2O_3–La_2O_3)	Methylene Blue and Rhodamine B	100 mg L^{-1}	Fida et al. (2017)
$NiFe(C_2O_4)_x$	Methyl Orange	20 mg L^{-1}	Liu et al. (2017b)
α-Fe_2O_3 composites	Methylene Blue	40 mg L^{-1}	Liu et al. (2017a)
Cu-impregnated zeolite Y	Congo Red	0.143 mM	Singh et al. (2016)

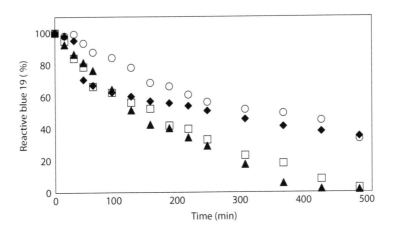

FIGURE 6.4 Effect of Fe_3O_4 nanoparticles on the decolorization yield of Reactive Blue 19. Nanoparticle concentration: 100 mg L^{-1} (O), 200 mg L^{-1} (□), 300 mg L^{-1} (▲) and 500 mg L^{-1} (◆). Initial dye concentration: 25 mg L^{-1} and H_2O_2: 100 mg L^{-1}.

Removal rates of 65%–98.3% were obtained for the range considered. The efficiency of Reactive Blue 19 removal was found to increase as the catalyst dosage was increased up to 300 mg L^{-1}. At 500 mg L^{-1}, the removal of dye decreased by 30%. Similar results were observed in previous studies, where beyond a certain dosage of Fe_3O_4 nanoparticles, the degradation efficiency dropped, which was likely caused by high nanoparticle concentrations. Peroxidase-like activity occurred on the surface of the Fe_3O_4 nanoparticles due to the presence of iron ions. An excessive dosage of nanoparticles could lead to the formation of aggregates that would reduce the available surface area and thus the density of surface-adsorbed H_2O_2 (Huang et al., 2012). The results obtained in this work point to the heterogeneous Fenton process as a promising technology for the treatment of textile wastewaters. However, further studies are needed to achieve a complete characterization and understanding of the process. As well as other variables, the evaluation of the reaction rates and the role of iron nanoparticles have to be addressed in the development of this technology.

6.3.3 NANOBIOCATALYSTS BASED ON SILICA-COATED, OLEIC ACID–COATED, AND POLYETHYLENIMINE-COATED MAGNETIC NANOPARTICLES AND THEIR USE IN DYE DECOLORIZATION

Oxidoreductases, such as laccases, have been used in oxidation processes such as dye decolorization because of their high activity, selectivity, and specificity (Wesenberg et al., 2003). Stability and recovery of the biocatalyst are the main concerns when using enzyme biocatalysts in industrial applications (Cabana et al., 2011). However, these drawbacks can be overcome through the immobilization of enzymes, which can lead to improvement of the stability, separability, and reusability of biocatalysts in continuous operations (Duran et al., 2002). Different supports providing a range of functionality, morphology, and physical properties have been studied for the immobilization of enzymes (Lloret et al., 2011; López-Gallego et al., 2005; Kunamneni et al., 2008). Recently, nanostructured materials have attracted considerable interest as supports (Zhao et al., 2011).

Magnetic nanoparticles (mNPs) are characterized by a magnetic core that can ensure their easy recovery with a simple magnet. In addition, mNPs provide a high superficial area for the binding of enzymes, low mass transfer resistance, less fouling, and a decrease in operational costs, which make them potential candidates for enzyme immobilization (Kalkan et al., 2011). The mNPs should be coated with organic molecules (surfactants, biomolecules, or polymers, among others) or inorganic layers (silica, metal oxide, etc.) to provide a proper surface and retain the stability of magnetic iron oxide nanoparticles (Wu et al., 2008).

In the present work, laccase immobilization by covalent bonding onto mNPs coated with silica, polyethylenimine (PEI), and oleic acid was developed, and the immobilized enzyme was applied for the decolorization of two different dyes: Methyl Green and Reactive Blue 19. Furthermore, a Microtox® test based on the luminescent marine bacterium *Vibrium fischeri* was performed to investigate the potential toxicity of the transformation products obtained from the enzymatic treatment.

Laccase immobilization on supports such as nanoporous silver and gold particles (Mazur et al., 2007), mesostructured foams (Rekuć et al., 2009), or epoxy-activated resins (Kunamneni et al., 2008; Lloret et al., 2012) has been successfully performed during recent years. However, few biocatalysts have been used for practical applications due to their poor robustness (DiCosimo et al., 2013). Hence, different strategies for laccase immobilization were evaluated on magnetic supports to choose a robust nano-biocatalyst for the decolorization of dyes. The covalent bonding between the support and carboxylic groups (oleic acid) did not result in satisfactory immobilization, which may be attributed to excessive linking of EDC (1-ethyl-3-(3-dimethylaminopropyl) carbodiimide hydrochloride) with the enzyme, leading to crosslinking among enzymes (due to the presence of both $-NH_2$ and $-COOH$ groups in the enzyme molecule). Activity loss due to excessive crosslinking of enzyme was also observed in other studies, with immobilization yields below 14% (Majumder et al., 2008; Kumar et al., 2014). However, other immobilization processes were successful, and the best results were obtained for silica-coated mNPs, which were selected for the decolorization of dyes.

The capacity of the immobilized enzymes to remove 20 mg L^{-1} of the triphenyl-methane dye Methyl Green and 100 mg L^{-1} of the anthraquinone dye Reactive Blue 19 was assessed in repeated batch operation for 24 h. A control with functionalized nanoparticles lacking laccase was performed. Immobilized laccase reached a decolorization percentage higher than 95% in the first cycle, which was maintained for 5 cycles, diminishing slightly to 87% after the 10th cycle (Figure 6.5).

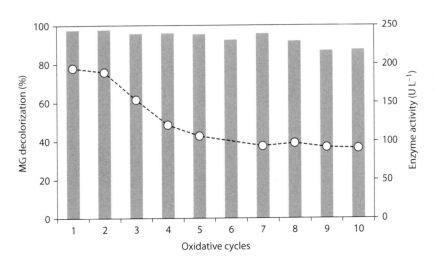

FIGURE 6.5 Methyl Green decolorization yield (%, gray bars) and residual laccase activity (O) in subsequent cycles of enzymatic treatment with laccase immobilized onto mNP.

This slight decay could be due to the reduction of the enzyme activity, since it gradually decreased during the first five cycles, and then 50% of the initial activity was maintained until the end of the operation. Kunamneni et al. (2008) studied the decolorization of Methyl Green by laccase immobilized on epoxy-activated carriers, and they found that the biocatalyst could oxidase five cycles in a fixed-bed reactor, but decolorization dropped to 20% in the sixth cycle. In the case of Reactive Blue 19 decolorization, immobilized laccase reached a decolorization percentage higher than 80% in the first cycle and was maintained for three cycles, diminishing slightly to 56% after the 10th cycle (Figure 6.6).

With the aim of evaluating the toxicity of the transformation products, a Microtox® test was carried out for controls and samples withdrawn after 24 h. The corresponding $EC_{50\%}$ (15 min) values are reported in Table 6.4.

The low $EC_{50\%}$ values for both Methyl Green and Reactive Blue 19 are related to the high toxicity of the dyes to Microtox® test bacteria, which is in agreement with results from previous reports (Casas et al., 2009; Ayed et al., 2010). After laccase treatment, toxicity decreased, which suggests that the biotransformation products are less toxic than the parent substrate. The detoxification of dyes using free

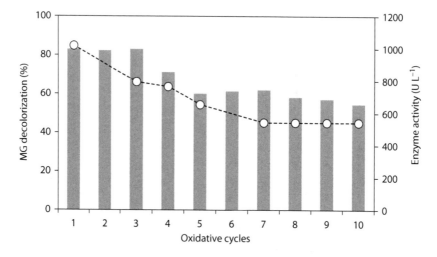

FIGURE 6.6 Reactive Blue 19 decolorization yield (%, gray bars) and residual laccase activity (○) in subsequent cycles of enzymatic treatment with laccase immobilized onto mNP.

TABLE 6.4

Microtox® Test Results of Methyl Green and Reactive Blue 19 Samples

Sample	Methyl Green $EC_{50, 15\ min}$ (%)	Reactive Blue 19 $EC_{50, 15\ min}$ (%)
Control	7	10
24-h treatment with laccase	12	28

or immobilized laccases is an important parameter to be considered, since, as it has been previously reported, transformation products from the enzymatic treatment of dyes can lead to more toxic metabolites as determined by the Microtox® assay (Champagne and Ramsay, 2010).

6.4 CONCLUSIONS

From the results obtained, the different AOPs considered here represent an attractive alternative for the treatment of contaminated ground water and industrial wastewater resources. Most of these AOPs involve the *in situ* generation of highly potent chemical oxidants such as •OH radicals, accelerating the oxidation and decolorization of dye-containing wastewaters. Furthermore, the results for stability and reusability of the used nanoparticles, used either as chemical catalysts or as support for enzyme immobilization, are promising in the development of unconventional AOPs for wastewater treatment.

ACKNOWLEDGEMENTS

This research has been supported by MINECO (CTQ2016-79461-R, program co-funded by FEDER). The authors belong to the Galician Competitive Research Group (GRC 2013-032, program co-funded by FEDER), as well as to CRETUS (AGRUP2015/02).

REFERENCES

Apostol, L.C., L. Pereira, R. Pereira, M. Gavrilescu, and M.M. Alves. 2012. Biological decolorization of xanthene dyes by anaerobic granular biomass. *Biodegradation* 23 (5): 725–737.

Augugliaro, V., M. Litter, L. Palmisano, and J. Soria. 2006. The combination of heterogeneous photocatalysis with chemical and physical operations: A tool for improving the photoprocess performance. *Journal of Photochemistry and Photobiology C: Photochemistry Reviews* 7 (4): 127–144.

Ayed, L., K. Chaieb, A. Cheref, and A. Bakhrouf. 2010. Biodegradation and decolorization of triphenylmethane dyes by *Staphylococcus epidermidis*. *Desalination* 260 (1): 137–146.

Azbar, N., T. Yonar, and K. Kestioglu. 2004. Comparison of various advanced oxidation processes and chemical treatment methods for COD and color removal from a polyester and acetate fiber dyeing effluent. *Chemosphere* 55 (1): 35–43.

Baban, A., A. Yediler, D. Lienert, N. Kemerdere, and A. Kettrup. 2003. Ozonation of high strength segregated effluents from a woollen textile dyeing and finishing plant. *Dyes and Pigments* 58 (2): 93–98.

Baig, S., and P.A. Liechti. 2001. Ozone treatment for biorefractory COD removal. *Water Science and Technology* 43 (2): 197–204.

Barka, N., I. Bakas, S. Qourzal, A. Assabbane, and Y. Ait-ichou. 2013. Degradation of phenol in water by titanium dioxide photocatalysis. *Oriental Journal of Chemistry* 29 (3): 1055–1060.

Bautista, P., A.F. Mohedano, J.A. Casas, J.A. Zazo, and J.J Rodriguez. 2008. An overview of the application of Fenton oxidation to industrial wastewaters treatment. *Journal of Chemical Technology and Biotechnology* 83 (10): 1323–1338.

Bethi, B., S.H. Sonawane, B.A. Bhanvase, and S. Gumfekar. 2016. Nanomaterials based advanced oxidation processes for waste water treatment: A review. *Chemical Engineering and Processing: Process Intensification* 109: 178–189.

Beton, A., D. Dias, L. Farrant, T. Gibon, Y. Le Guern, M. Desaxce, A. Perwueltz, et al. 2014. *Environmental Improvement Potential of Textiles (IMPRO Textiles)*. European Union, Luxembourg: Publications Office of the European Union.

Blanco, J., F. Torrades, M. De la Varga, and J. García-Montaño. 2012. Fenton and biological-Fenton coupled processes for textile wastewater treatment and reuse. *Desalination* 286: 394–399.

Bousnoubra, I., K. Djebbar, A. Abdessemed, and T. Sehili. 2016. Decolorization of methyl green and bromocresol purple in mono and binary systems by photochemical processes: Direct UV photolysis, acetone/UV and H_2O_2/UV. A comparative study. *Desalination and Water Treatment* 57 (57): 27710–27725.

Cabana, H., A. Ahamed, and R. Leduc. 2011. Conjugation of laccase from the white rot fungus *Trametes versicolor* to chitosan and its utilization for the elimination of triclosan. *Bioresource Technology* 102: 1656–1662.

Casas, N., T. Parella, T. Vicent, G. Caminal, and M. Sarrà. 2009. Metabolites from the biodegradation of triphenylmethane dyes by *Trametes versicolor* or laccase. *Chemosphere* 75 (10): 1344–1349.

Ciardelli, G. and N. Ranieri. 2001. The treatment and reuse of wastewater in the textile industry by means of ozonation and electroflocculation. *Water Research* 35 (2): 567–572.

Champagne, P.P. and J.A. Ramsay. 2010. Dye decolorization and detoxification by laccase immobilized on porous glass beads. *Bioresource Technology* 101 (7): 2230–2235.

Chen, W., C.K.M. Lau, D. Boansi, and M.H. Bilgin. 2017a. Effects of trade cost on the textile and apparel market: Evidence from Asian countries. *The Journal of the Textile Institute* 108 (6): 971–986.

Chen, X., Z. Wu, D. Liu, and Z. Gao. 2017b. Preparation of ZnO photocatalyst for the efficient and rapid photocatalytic degradation of azo dyes. *Nanoscale Research Letters* 12 (1): 143.

Chequer, F.M.D., G.A.R. de Oliveira, E.R.A. Ferraz, J.C. Cardoso, M.V.B. Zanoni, and D.P. de Oliveira. 2013. Textile dyes: Dyeing process and environmental impact. In *Eco-Friendly Textile Dyeing and Finishing*, edited by Melih Günay. Rijeka: InTech.

Christie, R. 2001. *Colour Chemistry*. Royal Society of Chemistry, London, UK.

Chung, K. 1983. The significance of azo-reduction in the mutagenesis and carcinogenesis of azo dyes. *Mutation Research/Reviews in Genetic Toxicology* 114 (3): 269–281.

Chung, J. and J. Kim. 2011. Application of advanced oxidation processes to remove refractory compounds from dye wastewater. *Desalination and Water Treatment* 25 (1–3): 233–240.

De Diesbach, H. and E. Von der Weid. 1927. Quelques sels complexes des o-dinitriles avec le cuivre et la pyridine. *Helvetica Chimica Acta* 10 (1): 886–888.

De Lima, R.O.A., A.P. Bazo, D.M.F. Salvadori, C.M. Rech, D. De Palma Oliveira, and G. De Aragão Umbuzeiro. 2007. Mutagenic and carcinogenic potential of a textile azo dye processing plant effluent that impacts a drinking water source. *Mutation Research/ Genetic Toxicology and Environmental Mutagenesis* 626 (1–2): 53–60.

Demirbaş, Ü., H.T. Akçay, A. Koca, and H. Kantekin. 2017. Synthesis, characterization and investigation of electrochemical and spectroelectrochemical properties of peripherally tetra 4-phenylthiazole-2-thiol substituted metal-free, zinc(II), copper(II) and cobalt(II) phthalocyanines. *Journal of Molecular Structure* 1141: 643–649.

DiCosimo, R., J. McAuliffe, A.J. Poulose, and G. Bohlmann. 2013. Industrial use of immobilized enzymes. *Chemical Society Reviews* 42 (15): 6437–6474.

Duran, N., M.A. Rosa, A. D'Annibale, and L. Gianfreda. 2002. Applications of laccases and tyrosinases (phenoloxidases) immobilized on different supports: A review. *Enzyme and Microbial Technology* 31: 907–931.

Fenton, H.J.H. 1894. Oxidation of tartaric acid in presence of iron. *Journal of the Chemical Society, Transactions* 65: 899–910.

Fida, H., G. Zhang, S. Guo, and A. Naeem. 2017. Heterogeneous Fenton degradation of organic dyes in batch and fixed bed using La-Fe montmorillonite as catalyst. *Journal of Colloid and Interface Science* 490: 859–868.

Fox, M.A. and M.T. Dulay. 1993. Heterogeneous photocatalysis. *Chemical Reviews* 93 (1): 341–357.

Gaya, U.I. and A.H. Abdullah. 2008. Heterogeneous photocatalytic degradation of organic contaminants over titanium dioxide: A review of fundamentals, progress and problems. *Journal of Photochemistry and Photobiology C: Photochemistry Reviews* 9 (1): 1–12.

Glaze, W.H., J. Kang, and D.H. Chapin. 1987. The chemistry of water treatment processes involving ozone, hydrogen peroxide and ultraviolet radiation. *Ozone: Science & Engineering* 9 (4): 335–352.

Gmurek, M., M. Olak-Kucharczyk, and S. Ledakowicz. 2017. Photochemical decomposition of endocrine disrupting compounds—A review. *Chemical Engineering Journal* 310: 437–456.

Gordon, P.F. and P. Gregory. 1987. *Organic Chemistry in Colour, Springer Study Edition.* Berlin/Heidelberg: Springer-Verlag.

Goswami, P. and Basak, M. 2014. Sulfur Dyes. *Kirk-Othmer Encyclopedia of Chemical Technology,* 1–31.

Guo, X., J. Liu, and G. Guo. 2017. Photocatalytic removal of dye and reaction mechanism analysis over Y_2O_3 composite nanomaterials. *MATEC Web of Conferences* 88: 02003, Chengdu, China.

Gürses, A., M. Açıkyıldız, K. Güneş, and M.S. Gürses. 2016. Dyes and pigments: Their structure and properties. In *Dyes and Pigments,* 13–29. Cham: Springer International.

Hoffmann, M.R., S.T. Martin, W. Choi, and D.W. Bahnemann. 1995. Environmental applications of semiconductor photocatalysis. *Chemical Reviews* 95 (1): 69–96.

Hoigné, J. and H. Bader. 1976. The role of hydroxyl radical reactions in ozonation processes in aqueous solutions. *Water Research* 10 (5): 377–386.

Huang, R., Z. Fang, X. Yan, and W. Cheng. 2012. Heterogeneous sono-Fenton catalytic degradation of bisphenol A by Fe_3O_4 magnetic nanoparticles under neutral condition. *Chemical Engineering Journal* 197: 242–249.

Hübner, K. 2006. 150 Jahre Mauvein. *Chemie in unserer Zeit* 40 (4): 274–275.

Joseph, J.M., H. Destaillats, H.M. Hung, and M.R. Hoffmann. 2000. The sonochemical degradation of azobenzene and related azo dyes: Rate enhancements via Fenton's reactions. *Journal of Physical Chemistry A* 104 (2): 301–307.

Kalkan, N.A., S. Aksoy, E.A. Aksoy, and N. Hasirci. 2011. Preparation of chitosan-coated magnetite nanoparticles and application for immobilization of laccase. *Journal of Applied Polymer Science* 123 (2): 707–716.

Kasprzyk-Hordern, B., M. Ziółek, and J. Nawrocki. 2003. Catalytic ozonation and methods of enhancing molecular ozone reactions in water treatment. *Applied Catalysis B: Environmental* 46 (4): 639–669.

Khataee, A.R. and M.B. Kasiri. 2010. Photocatalytic degradation of organic dyes in the presence of nanostructured titanium dioxide: Influence of the chemical structure of dyes. *Journal of Molecular Catalysis A: Chemical* 328 (1): 8–26.

Khodja, A.A., T. Sehili, J.-F. Pilichowski, and P. Boule. 2001. Photocatalytic degradation of 2-phenylphenol on TiO_2 and ZnO in aqueous suspensions. *Journal of Photochemistry and Photobiology A: Chemistry* 141 (2): 231–239.

Kumar, S., A.K. Jana, M. Maiti, and I. Dhamija. 2014. Carbodiimide-mediated immobilization of serratiopeptidase on amino-, carboxyl-functionalized magnetic nanoparticles and characterization for target delivery. *Journal of Nanoparticle Research* 16 (2): 2233.

Kunamneni, A., I. Ghazi, S. Camarero, A. Ballesteros, F.J. Plou, and M. Alcalde. 2008. Decolorization of synthetic dyes by laccase immobilized on epoxy-activated carriers. *Process Biochemistry* 43 (2): 169–178.

Langhals, H. 2004. Color chemistry. synthesis, properties and applications of organic dyes and pigments. 3rd revised edition. By Heinrich Zollinger. *Angewandte Chemie International Edition* 43 (40): 5291–5292.

Ledakowicz, S. and M. Gonera. 1999. Optimisation of oxidants dose for combined chemical and biological treatment of textile wastewater. *Water Research* 33 (11): 2511–2516.

Liu, Y., W. Jin, Y. Zhao, G. Zhang, and W. Zhang. 2017a. Enhanced catalytic degradation of methylene blue by α-Fe_2O_3/graphene oxide via heterogeneous photo-Fenton reactions. *Applied Catalysis B: Environmental* 206: 642–652.

Liu, Y., G. Zhang, S. Chong, N. Zhang, H. Chang, T. Huang, and S. Fang. 2017b. $NiFe(C_2O_4)_x$ as a heterogeneous Fenton catalyst for removal of methyl orange. *Journal of Environmental Management* 192: 150–155.

Lizama, C., J. Freer, J. Baeza, and H.D. Mansilla. 2002. Optimized photodegradation of Reactive Blue 19 on TiO_2 and ZnO suspensions. *Catalysis Today* 76: 235–246.

Lloret, L., G. Eibes, G. Feijoo, M.T. Moreira, and J.M. Lema. 2012. Continuous operation of a fluidized bed reactor for the removal of estrogens by immobilized laccase on Eupergit supports. *Journal of Biotechnology* 162 (4): 404–406.

Lloret, L., G. Eibes, G. Feijoo, M.T. Moreira, J.M. Lema, and F. Hollmann. 2011. Immobilization of laccase by encapsulation in a sol-gel matrix and its characterization and use for the removal of estrogens. *Biotechnology Progress* 27: 1570–1579.

López-Gallego, F., T. Montes, M. Fuentes, N. Alonso, V. Grazu, L. Betancor, J.M. Guisán, and R. Fernández-Lafuente. 2005. Improved stabilization of chemically aminated enzymes via multipoint covalent attachment on glyoxyl supports. *Journal of Biotechnology* 116 (1): 1–10.

Majumder, A.B., K. Mondal, T.P. Singh, and M.N. Gupta. 2008. Designing cross-linked lipase aggregates for optimum performance as biocatalysts. *Biocatalysis and Biotransformation* 26 (3): 235–242.

May-Lozano, M., V. Mendoza-Escamilla, E. Rojas-García, R. Lopez-Medina, G. Rivadeneyra-Romero, and S.A. Martinez-Delgadillo. 2017. Sonophotocatalytic degradation of Orange II dye using low cost photocatalyst. *Journal of Cleaner Production* 148: 836–844.

Mazur, M., P. Krysiński, A. Michota-Kamińska, J. Bukowska, J. Rogalski, and G.J. Blanchard. 2007. Immobilization of laccase on gold, silver and indium tin oxide by zirconium–phosphonate–carboxylate (ZPC) coordination chemistry. *Bioelectrochemistry* 71 (1): 15–22.

Mohapatra, D.P., S.K. Brar, R.D. Tyagi, P. Picard, and R.Y. Surampalli. 2014. Analysis and advanced oxidation treatment of a persistent pharmaceutical compound in wastewater and wastewater sludge—carbamazepine. *Science of the Total Environment* 470: 58–75.

Munoz, M., Z.M. de Pedro, J.A. Casas, and J.J. Rodriguez. 2015. Preparation of magnetite-based catalysts and their application in heterogeneous Fenton oxidation—A review. *Applied Catalysis B: Environmental* 176–177: 249–265.

Nagl, G. 2000. Sulfur dyes. In *Ullmann's Encyclopedia of Industrial Chemistry*. Wiley-VCH Verlag GmbH & Co. KGaA. DOI: 10.1002/14356007.a25_613.

Neppolian, B., H. Jung, and H. Choi. 2007. Photocatalytic degradation of 4-chlorophenol using TiO_2 and Pt-TiO_2 nanoparticles prepared by sol-gel method. *Journal of Advanced Oxidation Technologies* 10 (2): 369–374.

Nguyen, A.T. and R.-S. Juang. 2015. Photocatalytic degradation of p-chlorophenol by hybrid H_2O_2 and TiO_2 in aqueous suspensions under UV irradiation. *Journal of Environmental Management* 147: 271–277.

OECD. 2005. *Environmental Requirements and Market Access*. OECD Publishing, Paris, France.

Okitsu, K., K. Iwasaki, Y. Yobiko, H. Bandow, R. Nishimura, and Y. Maeda. 2005. Sonochemical degradation of azo dyes in aqueous solution: A new heterogeneous kinetics model taking into account the local concentration of OH radicals and azo dyes. *Ultrasonics Sonochemistry* 12 (4): 255–262.

Oller, I., S. Malato, and J.A. Sánchez-Pérez. 2011. Combination of advanced oxidation processes and biological treatments for wastewater decontamination—A review. *Science of the Total Environment* 409 (20): 4141–4166.

O'Neill, C., F.R. Hawkes, D.L. Hawkes, N.D. Lourenço, H.M. Pinheiro, and W. Delée. 1999. Colour in textile effluents—sources, measurement, discharge consents and simulation: A review. *Journal of Chemical Technology & Biotechnology* 74 (11): 1009–1018.

Oros-Ruiz, S., R. Zanella, and B. Prado. 2013. Photocatalytic degradation of trimethoprim by metallic nanoparticles supported on TiO_2-P25. *Journal of Hazardous Materials* 263: 28–35.

Özdemir, C., M.K. Öden, S. Şahinkaya, and D. Güçlü. 2011. The sonochemical decolorisation of textile azo dye CI Reactive Orange 127. *Coloration Technology* 127 (4): 268–273.

Papadopoulos, A.E., D. Fatta, and M. Loizidou. 2007. Development and optimization of dark Fenton oxidation for the treatment of textile wastewaters with high organic load. *Journal of Hazardous Materials* 146 (3): 558–563.

Parsons, S. 2004. *Advanced Oxidation Processes for Water and Wastewater Treatment.* IWA Publishing, London, UK.

Pérez, M., F. Torrades, X. Domènech, and J. Peral. 2002. Fenton and photo-Fenton oxidation of textile effluents. *Water Research* 36 (11): 2703–2710.

Pignatello, J.J., E. Oliveros, and A. MacKay. 2006. Advanced oxidation processes for organic contaminant destruction based on the Fenton reaction and related chemistry. *Critical Reviews in Environmental Science and Technology* 36 (1): 1–84.

Puntener, A. and C. Page. 2004. *European Ban on Certain Azo Dyes.* Edited by Quality and Environment. TFL Leather Technology Ltd, London, UK.

Rauf, M.A., S.B. Bukallah, A. Hamadi, and F. Hammadi. 2007. The effect of operational parameters on the photoinduced decoloration of dyes using a hybrid catalyst V_2O_5/TiO_2. *Chemical Engineering Journal* 129: 167–172.

Rekuć, A., J. Bryjak, K. Szymańska, and A.B. Jarzębski. 2009. Laccase immobilization on mesostructured cellular foams affords preparations with ultra high activity. *Process Biochemistry* 44 (2): 191–198.

Saggioro, E.M., A.S. Oliveira, T. Pavesi, C.G. Maia, L.F.V. Ferreira, and J.C. Moreira. 2011. Use of titanium dioxide photocatalysis on the remediation of model textile wastewaters containing azo dyes. *Molecules* 16 (12): 10370–10386.

Shui, S. and A. Plastina. 2013. *World Apparel Fiber Consumption Survey.* Washington: Food and Agriculture Organization of the United Nations and International Cotton Advisory Committee.

Singh, L., P. Rekha, and S. Chand. 2016. Cu-impregnated zeolite Y as highly active and stable heterogeneous Fenton-like catalyst for degradation of Congo red dye. *Separation and Purification Technology* 170: 321–336.

Somensi, C.A., E.L. Simionatto, S.L. Bertoli, A. Wisniewski, and C.M. Radetski. 2010. Use of ozone in a pilot-scale plant for textile wastewater pre-treatment: Physico-chemical efficiency, degradation by-products identification and environmental toxicity of treated wastewater. *Journal of Hazardous Materials* 175 (1): 235–240.

Sousa-Castillo, A., M. Comesaña-Hermo, B. Rodríguez-Gonzalez, M. Pérez-Lorenzo, Z. Wang, X-T. Kong, A.O. Govorov, and M.A. Correa-Duarte. 2016. Boosting hot electron-driven photocatalysis through anisotropic plasmonic nanoparticles with hot spots in Au-TiO_2 nanoarchitectures. *The Journal of Physical Chemistry C* 120 (21): 11690–11699.

Staehelin, J. and J. Hoigne. 1982. Decomposition of ozone in water: Rate of initiation by hydroxide ions and hydrogen peroxide. *Environmental Science & Technology* 16 (10): 676–681.

Swedish-Chemicals-Agency. 2014. *Chemicals in Textiles. Risks to Human Health and the Environment.* Report from a government assignment, Stockholm, Sweden.

Tünay, O., I. Kabdasli, G. Eremektar, and D. Orhon. 1996. Color removal from textile wastewaters. *Water Science and Technology* 34 (11): 9–16.

Van der Putte, I., S. Qi, F. Affourtit, K. De Wolf, S. Devaere, and E. Albrecht. 2013. *Study on the Link between Allergic Reactions and Chemicals in Textile Products*, PRS advies, Delft, The Netherlands.

Von Gunten, U. 2003. Ozonation of drinking water: Part I. Oxidation kinetics and product formation. *Water Research* 37 (7): 1443–1467.

Wang, C.T. 2007. Photocatalytic activity of nanoparticle gold/iron oxide aerogels for azo dye degradation. *Journal of Non-Crystalline Solids* 353 (11): 1126–1133.

Wang, W., Y. Liu, T. Li, and M. Zhou. 2014. Heterogeneous Fenton catalytic degradation of phenol based on controlled release of magnetic nanoparticles. *Chemical Engineering Journal* 242: 1–9.

Welham, R.D. 1963. The early history of the synthetic dye industry. *Journal of the Society of Dyers and Colourists* 79 (3): 98–105.

Wesenberg, D., I. Kyriakides, and S.N. Agathos. 2003. White-rot fungi and their enzymes for the treatment of industrial dye effluents. *Biotechnology Advances* 22 (1): 161–187.

Wu, W., Q. He, and C. Jiang. 2008. Magnetic iron oxide nanoparticles: Synthesis and surface functionalization strategies. *Nanoscale Research Letters* 3 (11): 397–415.

Zhao, G., J. Wang, Y. Li, X. Chen, and Y. Liu. 2011. Enzymes immobilized on superparamagnetic Fe_3O_4@clays nanocomposites: Preparation, characterization, and a new strategy for the regeneration of supports. *Journal of Physical Chemistry C* 115 (14): 6350–6359.

7 A Biological Approach for the Removal of Pharmaceutical Pollutants from Wastewater

Ponnusamy Senthil Kumar and
Anbalagan Saravanan

CONTENTS

7.1   INTRODUCTION

Pharmaceuticals are a group of compounds that have attracted increasing attention over the past decade. There are many different compound classes, which are
intended to affect specific areas of disease. Pharmaceutical items have been widely
used in many fields, for example, medicine, industry, the cultivation of domesticated
animals, aquaculture, and people's day-to day lives. They are becoming universal
in the environment because of their broad applications (Wang and Wang, 2016).
Pharmaceutical items include a wide variety of natural preparations, for example,
cleansers, salves, toothpaste, scents, sunscreens, and so forth, which are generally
used in large quantities all over the world. Pharmaceutical items are among a group
of chemicals named *contaminants of rising concern* (CRC). CRCs are not really new
toxins, as they have been available in the world for quite a while; however, their presence and importance are just now being assessed (Daughton, 2001). This extensive
variety of pharmaceuticals reaching the oceanic environment could affect exposed
organisms, more specifically in coastal and marine ecosystems (Crane et al., 2006;
Fent et al., 2006; Coetsier et al., 2009). Pharmaceutical products can be divided into
a few classes based on their different purposes. Each of these classes includes many
products. The particular classes, their related purposes, and the principal properties
of pharmaceutical products are shown in Table 7.1. A more clarified portrayal in
view of crude materials, last items, and uniqueness of plants, has been endeavoured.
The classes are based on the similarity of compound procedures and medications
and, additionally, certain classes of things. In light of the procedures required for
their manufacture, pharmaceutical enterprises can be subdivided into the following
five noteworthy subcategories:

- Fermentation plants
- Organic chemical synthesis plants
- Organic chemical fermentation/synthesis plants (generally moderate to
  large plants)
- Natural/biological product extraction plants (antibiotics/vitamins/enzymes,
  etc.)
- Drug mixing, formulation, and preparation plants (tablets, capsules, solu-
  tions, etc.)

The production of these drugs continues to increase because of the huge demand
for pharmaceutical products to prevent or cure sickness.

Regarding environmental pollution concerns, the levels of pharmaceuticals
released into the earth from wastewater treatment plants (WWTPs) may negatively affect the biological community (Niemuth and Klaper, 2015; Blair et al.,
2015). Recently, it has become clear that the elimination of certain pharmaceutical compounds during wastewater treatment processes is rather low, and as a
result, they are found in surface, ground, and drinking waters. Pharmaceutical
products can be discharged into the environment by two means: (1) direct (surface water) and (2) indirect pathways (ground water). For this reason, pharmaceuticals may be able to cause the same harmful exposure potential as persistent

TABLE 7.1
Classifications of Pharmaceutical Products

S. No	Pharmaceutical Products	Functions/Uses	Compounds and their Molecular Formulae
1	Antibiotics	Kill bacteria	Ampicillin $C_{16}H_{19}N_3O_4S$, Amoxicillin $C_{16}H_{19}N_3O_5S$, Chloramphenicol $C_{11}H_{12}Cl_2N_2O_5$, Clarithromycin $C_{38}H_{69}NO_{13}$, Sarafloxacin $C_{20}H_{17}F_2N_3O_3$, Erythromycin $C_{37}H_{67}NO_{13}$, Sulfapyridine $C_{11}H_{11}N_3O_2S$, Sulfamethoxazole $C_{10}H_{11}N_3O_3S$, Tetracycline $C_{22}H_{24}N_2O_8$
2	Anesthetics	Abdominal surgery	Lignocaine $C_{14}H_{22}N_2O$, Isoflurane $C_3H_2ClF_5O$, vasoconstrictors (primarily adrenaline)
3	Antiepileptic	Treatment of epileptic seizures	Carbamazepine $C_{15}H_{12}N_2O$, Primidone $C_{12}H_{14}N_2O_2$
4	Antineoplastics	Kill neoplastic cells	Epirubicin $C_{27}H_{29}NO_{11}$, Ifosfamide $C_7H_{15}Cl_2N_2O_2P$, Methotrexate $C_{20}H_{22}N_8O_5$
5	Antiserum	Treat envenomation	Agglutinins
6	Blood lipid regulators	Regulation of triglycerides and cholesterol in blood	Clofibrate $C_{12}H_{15}ClO_3$, Gemfibrozil $C_{15}H_{22}O_3$
7	Hormones	Influence digestion, control sexual development	Mestranol $C_{21}H_{26}O_2$, Estradiol $C_{18}H_{24}O_2$, Testosterone $C_{19}H_{28}O_2$, Androstenedione $C_{19}H_{26}O_2$
8	Paracetamol	Pain reliever and fever reducer	Acetaminophen $C_8H_9NO_2$
9	Penicillins and $C_{16}H_{18}N_2NaO_4S^+$	Kill bacteria, treat skin infections, dental infections and ear infections	6-aminopenicillanic acid with a side chain attached to the 6-amino group
10	Sulfonamides	Treat ulcerative colitis	Benzenesulfonamide $C_6H_7NO_2S$, Methanesulfonamide CH_5NO_2S
11	Enrofloxacin (veterinary medicaments)	Treat animals afflicted with certain bacterial infections	Enrofloxacin $C_{19}H_{22}FN_3O_3$
12	β-Blockers	Treat hypertension	Metoprolol $C_{15}H_{25}NO_3$, Propranolol $C_{16}H_{21}NO_2$, Nadolol $C_{17}H_{27}NO_4$, Pindolol $C_{14}H_{20}N_2O_2$, Acebutolol $C_{18}H_{29}ClN_2O_4$

pollutants, since their transformation and removal rates can be compensated by their continuous input into the environment. A few compound classes will be highlighted, either because the concentrations found in water are high, because of their (increasingly) high-volume use, or because of the persistence of these compounds. Surface spill-over and direct use in aquaculture can cause the pollution of marine environments (Huerta et al., 2013). Pharmaceutical compounds can enter surface water by means of direct release of WWTP effluents or from businesses, health centers, family units, and animal farming or aquaculture; this is initially ameliorated by dilution in surface water to trace level (nanograms to micrograms per liter; Mompelat et al., 2009; Wu et al., 2015; Roberts et al., 2016). The other potential mitigating influence on pharmaceutical products in receiving waters is land runoff in the event of biosolids spread on farmland, adsorption on suspended solids (and residue), and dissolved organic matter (DOM), which can enter groundwater by filtering or bank filtration (Osenbruck et al., 2007). Pharmaceutical compounds can be introduced into the environment through several pathways, including human sources such as deliberate disposal, the flushing of old solutions, health center waste, and WWTP effluent. The largest natural sources of pharmaceutical items are rural, since pharmaceuticals are used worldwide as supplements, not just for prophylactic and remedial purposes, but also to advance development and increase the effectiveness of nutrition. It must be noted that pharmaceuticals are processed in animals to varying degrees; in this situation, most anti-infective agents are used in the range from 30% to 90% of the permitted dosage level, and the unchanged parent compounds and their frequently active metabolites are released specifically to soil by means of the use of compost as manure, or on the other hand, by outdoor-raised animals discharging feces and urine directly onto the land. Once in the soil, the destination and impact of these substances depend on, for example, their physical-compound properties, climatic conditions, pH, and soil type, among a range of factors. The fluctuating engineered properties will impact the passage of the compound through wastewater treatment as well as its toxicity, persistence, and bioavailability in the ground.

7.2   GENERAL AND ADMINISTRATIVE NECESSITIES

As a rule, the release of toxins into water by the pharmaceutical business is controlled by natural assurance experts, such as the Environmental Protection Agency (EPA) in the United States and the European Environmental Agency (EEA) in Europe. They set release limits for pharmaceutical compounds for the manufacturing locations concerned. The U.S. EPA has issued directions for emission constraints on pharmaceutical makers, and there is a review of EPA controls for pharmaceutical producers (Watkins and Weiss, 1999).

In the EPA, there are entirely restricting points of confinement for release of pharmaceuticals into surface and ground water accessible, and the release of hazardous waste into surface water is controlled by means of the water structure mandate. To date, extraordinary efforts have been made to monitor the presence of pharmaceutical products in WWTPs. They are usually found at very low concentrations in the range of nanograms and micrograms per liter. However, because of the highly hazardous nature of pharmaceutical products, productive technologies need to be created for their removal (Yang et al., 2011; Gupta et al., 2012).

For waste administration, the main need is to counteract or reduce the entry of toxins into aquatic environments. The reuse of compounds for pharmaceutical manufacture is constrained because of the need to comply with good manufacturing practice (GMP). Figure 7.1 shows wastewater management.

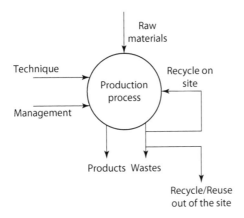

FIGURE 7.1 Wastewater management.

7.3 ENVIRONMENTAL RISKS OF PHARMACEUTICAL PRODUCTS

The environmental risk posed by pharmaceutical products in water is of extraordinary significance for the health of both people and aquatic organisms. The recognition of chemical compounds in any natural network does not necessarily imply that it is of concern or may cause harm. In any case, real concerns emerge from the location of chemicals for which there is confirmation that they may unfavorably influence oceanic life (Ebele et al., 2017). Figure 7.2 illustrates the environmental risks of pharmaceutical products.

7.3.1 BIOACCUMULATION

Pharmaceutical products are identified in the fresh water environment at moderately low concentrations. Many of them and their metabolites are naturally active and can

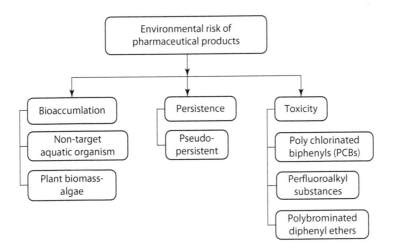

FIGURE 7.2 The environmental risks of pharmaceutical products.

affect non-target aquatic life forms. For this reason, it is important to assess their ecotoxicological effect. Currently accessible data on ecotoxins in the environment permits us to start distinguishing the dangers they may pose, albeit there has been little thought about the impacts on various species from the long-term, low-level presence of veterinary antimicrobials (Sarmah et al., 2006). Pharmaceutical products can be accumulated from soil into plants that have been treated with biosolids or compost or watered with wastewater.

7.3.1.1 Factors Determining the Bioaccumulation of Pharmaceuticals

We studied the factors determining the bioaccumulation of pharmaceuticals in freshwater mussels in a stream receiving effluent from wastewater treatment. Hence, our three targets were (1) to observe the changes in concentration of pharmaceuticals and personal care items in a stream receiving the outflow from a WWTP, (2) to compute the bioaccumulation variables (BAFs) of pharmaceuticals in mussels confined in that waterway, and (3) to determine the chemical and physical properties of pharmaceuticals that might drive the bioaccumulation.

7.3.2 PERSISTENCE

The physicochemical properties of numerous pharmaceutical products imply that many are not effortlessly removed by ordinary forms of water treatment, as is shown by their presence in drinking water. The inability to completely remove pharmaceutical products from WWTPs represents a potential hazard to aquatic life and general wellbeing. The confirmation from checking is that pharmaceutical products have also found their way into the aquatic environment and are universal. The widespread use of pharmaceutical products worldwide, combined with the increasing presentation of new pharmaceuticals to the market, is contributing significantly to the ecological presence of these chemicals and their active metabolites in the oceanic environment (Snyder, 2008; Bu et al., 2013). Also, while not all pharmaceutical products are stable, their consistent use and discharge into nature implies that many are viewed as "pseudo-persistent." Pseudo-persistent pharmaceuticals are proposed to have more prominent potential for ecological persistence than other natural contaminants such as pesticides, in light of the fact that their source is constantly being renewed even when followed by ecological procedures, for example, biodegradation, photodegradation, and particulate sorption. Thus, pharmaceuticals that might degrade would, in the long term and viably, remain as tenacious compounds in view of their constant discharge into the environment (Houtman et al., 2004).

7.3.3 TOXICITY

The real worry about the dangerous ramifications of pharmaceuticals (e.g., stable natural toxins, for example, polychlorinated biphenyls [PCBs], perfluoroalkyl substances [PFASs], and polybrominated diphenyl ethers [PBDEs]) is that they were developed specifically to increase their organic activity at low concentrations and to focus on certain metabolic, enzymatic, or cell-flagging systems. The developmental conservation of these systems in a given animal group conceivably increases the likelihood that these pharmaceuticals will be pharmacologically active in non-target living organisms. This method of activity (MoA) idea can be related to all aquatic biota that are inadvertently exposed to pharmaceuticals in their natural habitat, along these lines raising the hazard of ecotoxicological impacts.

7.4 PATHWAY FOR PHARMACEUTICAL ITEMS GOING INTO THE GROUNDWATER

In this section, the focus is on various areas where a point source has brought about the pollution: landfill, wastewater from a doctor's facility, septic systems, and wastewater effluent from WWTPs. Figure 7.3 shows in detail the pathway of pharmaceutical products entering the groundwater. Rather than point source contamination, which originates from one point, non-point source contamination is characterized by the United States Geological Survey (USGS 2015) as a diffuse release of contaminants in nature, and this could be because of the water system, precipitation, or snowmelt that conveys contaminants to the recipients.

7.4.1 LANDFILL

Around the world, the best-known path for the transfer of metropolitan strong waste is landfilling. The strong waste may discharge substances into the leachate that later enter the receiving environment and might debilitate the biological communities (Buszka et al., 2009; Eggen et al., 2010). Landfills are the final repository for various

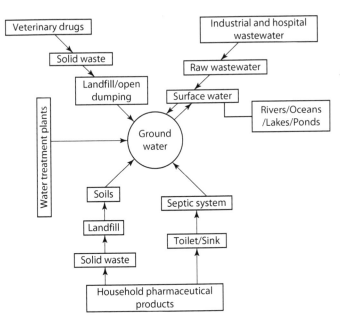

FIGURE 7.3 Pathway of pharmaceutical products entering into the groundwater.

strong and semi-strong wastes and may contain pharmaceutical items from sources such as unused medicines, soda pops, and other personal care items. Once disposed of into landfills, pharmaceutical items may be either used by microorganisms or incorporated into waste solids; however, the greater part are broken down in landfill leachate (Musson and Townsend, 2009). The anaerobic conditions in landfills and nearby receiving groundwater are likely to reduce the biodegradation of natural compounds in leachate and groundwater, resulting in the diversity and stability of pharmaceutical items in groundwater (Erses et al., 2008).

7.4.2 Septic Systems

Septic systems, or on-site wastewater treatment systems, can likewise be critical sources of pharmaceutical items entering nearby surface water and groundwater. Septic systems are a typical way of taking care of wastewater in rural areas, and the crust underneath them can stretch out tens to hundreds of meters, defiling the groundwater and the environment. They allow water used locally to be dealt with and reused to recharge nearby groundwater supplies, thus possibly allowing pharmaceuticals inadequately treated in septic frameworks to enter the underground water body.

7.4.3 Veterinary Drugs

The use of veterinary medications to prevent sickness and increase the profitability of producing domesticated animals is a developing concern and is possibly a pathway for natural pollution. For the most part, veterinary anti-microbials cannot be totally ingested or processed *in vivo*. Roughly 50%–100% of the anti-infective agents used are discharged through urine and feces, stored in waste tidal ponds, and after that discharged into the surrounding biological community, representing a potential danger to groundwater (Kim et al., 2011). The use of veterinary medication–contaminated fertilizer for preparation may likewise result in the contamination of fundamental groundwater (Hu et al., 2010).

7.4.4 Rivers and Lakes

Sewage beginning with the water treatment plants is the most significant source of pharmaceuticals in common water bodies, followed by rural release and direct release. Pharmaceuticals in the stream or lake are adsorbed onto the dirt/residue, become diluted, and experience organic or potentially photochemical changes. Categories of pharmaceuticals determined in rivers in India are listed in Table 7.2. The Cooum River, which runs through the metropolitan city of Chennai (population 8.9 million), was found to be contaminated with triclocarban (6.18 µg/L), ibuprofen (2.32 µg/L), a carboxylic acid metabolite of an antiplatelet drug (1.37 µg/L), atenolol (3.18 µg/L), and amphetamine (0.984 µg/L) (Subedi et al., 2015).

Ampicillin, ciprofloxacin, gatifloxacin, sparfloxacin, and cefuroxime were found in the Yamuna River at 13.8, 1.4, 0.48, 2.1, and 1.7 µg/L, respectively

TABLE 7.2

Groupings of Pharmaceuticals Identified in Rivers in India

S. No	River	Presence of Pharmaceutical Compounds	References
1	Cooum	Triclocarban (6.18 µg/L), Ibuprofen (2.32 µg/L), Antiplatelet carboxylic acid (1.37 µg/L), Atenolol (3.18 µg/L), Amphetamine (0.984 µg/L)	Subedi et al. (2015)
2	Yamuna	Ampicillin (13.8 µg/L), Ciprofloxacin (1.4 µg/L), Gatifloxacin (0.48 µg/L), Sparfloxacin (2.1 µg/L), Cefuroxime (1.7 µg/L)	Mutiyar and Mittal (2014)
3	Kaveri	Carbamazepine (13.0 ng/L)	Ramaswamy et al. (2011)
4	Bhavani	Triclosan (139 ng/L)	Ramaswamy et al. (2011)
5	Kaveri, Vellar, and Tamiraparani	Naproxen, Diclofenac, Ibuprofen, Ketoprofen, and acetylsalicylic acid (0.66 µg/L)	Shanmugam et al. (2013)
6	Vrishabavathi	Erythromycin, Chloramphenicol, and Trimethoprim, multidrug-resistant pathogenic microorganisms	Iyanee et al. (2013)

(Mutiyar and Mittal, 2014). A level of 13.0 ng/L of carbamazepine was found in the Kaveri River and 139 ng/L of triclosan in the Bhavani River, a branch of the Kaveri (Ramaswamy et al., 2011). Naproxen, diclofenac, ibuprofen, ketoprofen, and acetylsalicylic acid up to 0.66 µg/L were found in the Kaveri, Vellar, and Tamiraparani Rivers (Shanmugam et al., 2013). As well as erythromycin, chloramphenicol, and trimethoprim, multidrug-resistant pathogenic microorganisms were found in the Byramangala tank, supplied by the Vrishabavathi River, which contra-indicates numerous anti-infective agents recommended and consumed in India (Iyanee et al., 2013). Finally, the presence of abnormal amounts of sulfamethoxazole during rainstorms could be related to the high level of overflow from farmland in the surrounding territories.

7.4.5 Pharmaceutical Process Wastewater

Water is a basic raw material in pharmaceutical and substance-producing operations; reliable and pure water supplies are required for a range of operations including production, material preparation, and cooling. Process water quality administration is of extraordinary significance in pharmaceutical fabrication and is also a compulsory prerequisite for the cleaning of containers or therapeutic equipment in other medical services applications, including water for infusion (WFI). *Process wastewaters* is a term used to characterize wastewater in any industry that originates from the procedures taking place in the business. Process wastewaters in these terms cover any water that, at the time of assembly or preparation, interacts with the crude materials, items, intermediates, by-products, or waste items, which are used in various unit operations or procedures.

In the pharmaceutical industry, the increasing level of product development, and the related sizeable research effort will continue to create difficulties and open doors for providers of water and wastewater treatment hardware. Process water quality administration is of principal significance in pharmaceutical fabrication and is additionally an unmistakable essential for the cleaning of containers or medicinal gadgets in other healthcare applications. Extraordinary quality is the basis for meeting the exclusive requirements and controls related to items used directly for medication, such as WFI.

Moreover, at times, water may have been framed as a component of a substance response. The process wastewater created by the forms of production contains an assortment of traditional parameters (e.g., biological oxygen demand [BOD], total suspended solids [TSS], and pH) and other substance constituents. Wastewaters in the pharmaceutical industry for the most part start with the amalgamation and detailing of the medications. A large proportion of the active pharmaceutical ingredients (APIs) circulating worldwide are made by blending substances using natural, inorganic, and organic reactions. Since the reactors and separators used in a multi-item pharmaceutical industry are not planned according to the limit but are rather, regularly larger than average or used wastefully, the amount of wastewater produced is increased. There are various sub-activities happening in the pharmaceutical industry, and it is a troublesome undertaking to describe the waste from every last item.

7.4.6   Hospital Process Wastewater

Hospital effluents are likewise point sources of pharmaceutical contamination in water bodies. The concentrations of pharmaceuticals in clinic effluents are for the most part higher than in the other water treatment plant effluents.

Wastewater from hospitals around the world can generally contain traces of anything from viruses and multi-resistant bacteria to medical contrast agents and chemicals for cancer treatment. Small amounts of hormone-disrupting substances and other medicine residues are also part of the mix that passes from patients through hospital toilets and into public sewer systems. Hospital wastewater contains a complicated blend of active pharmaceutical constituents and microorganisms. Regularly, this wastewater is released to civil WWTPs with no pre-treatment. The metropolitan WWTPs are not intended to cope with a constant influx of pharmaceuticals. In addition, the hazardous wastewater may spread on occasions of flooding and consolidated sewer floods.

In hospital, the water consumed in various units, for example, inpatient wards, working rooms, research facilities, laundries, kitchens, wellbeing administrations, and administrative units, loses its physical, chemical, and organic quality and is changed into wastewater (Mahvi et al., 2009). Hospital wastewater is similar in nature to metropolitan wastewater, but the effluent wastewater from health facilities may contain unused pharmaceutical compounds, anti-toxins, disinfectants, sedatives, radioactive components, X-ray contrast agents, and other stable and hazardous compounds (Boillot, 2008). Additionally, in relation to persistent and hardly biodegradable chemicals, a few microbial pathogens resistant to anti-infective agents are present in aquatic biological systems and other common receiving bodies for contamination and can transmit serious and dangerous diseases to humans (Escher et al., 2011). The presence of chlorinated natural compounds and heavy metals, for example, Hg and Pb, has been observed in WWTP effluent from healthcare facilities (Bushra et al., 2015; Naushad et al., 2015).

Globally, there is expanding concern around the potential ecological impacts of pharmaceuticals in aquatic environments. Hospitals have been identified as a key source of pharmaceuticals that can circulate as important micro-pollutants. Painkillers, for example, diclofenac, and hormones, for instance, can have fatal effects on fish, shellfish, and green growth at extremely low levels.

7.5 TREATMENT TECHNOLOGIES

7.5.1 Water Treatment Plants

Water treatment plant outlets are the essential point sources of pharmaceutical pollution in rivers and seas. The current wastewater treatment procedures are unequipped to deal with the majority of pharmaceutical contaminants; removal efficiencies ordinarily range from 12.5% to 100% (Luo et al., 2014). Microbial metabolism, as well as deconjugation of glucuronides of the pharmaceuticals and their active metabolites, can have a negative effect on removal. Removal efficiency depends on the treatment procedure, the age of the sewage, the geology of the region, and the precipitation rate. The general pharmaceutical pollution profile is additionally reliant on the pattern of pharmaceutical production and use (Behera et al., 2011; Chen et al., 2012).

7.5.2 Treatment Methodologies For Pharmaceutical Industry Wastewater

The pharmaceutical business uses a wide collection of wastewater treatment and transfer methods. Wastewaters created by these enterprises vary in organization as well as in amount, by plant, season, and even time, contingent on the raw materials and the procedures used as a part of manufacturing different pharmaceuticals. Treatment plant area likewise acquires a variable identified with nature of accessible water. Consequently, it is exceptionally hard to determine a specific treatment framework for such an enhanced pharmaceutical industry. Numerous treatment procedure options are available to manage the wide range of wastes created by this industry; however, they are specific to the type of industry and the related waste. Figure 7.4 depicts a pharmaceutical effluent WWTP.

In any case, six general methodologies which can be used to treat pharmaceutical wastewaters:

- Recovery of individual APIs or medications that are probably going to be available in wash waters and solvents
- Physical-synthetic treatment by sedimentation or flotation
- Vigorous/anaerobic natural treatment in layer bioreactors or bioaeration
- Inactivation of dynamic substances by UV oxidation in conjunction with O_3 or H_2O_2

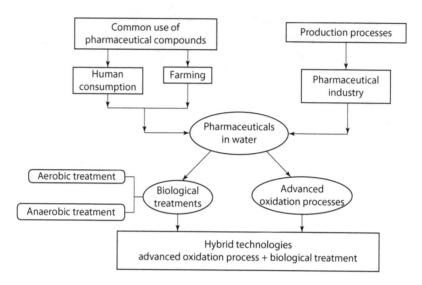

FIGURE 7.4   WWTP of pharmaceutical effluents.

- Disinfection and sterilization of resistant bioactive substances from biotechnology.
- New crossover advancements specific to the pharmaceutical business. An attempt is made here to talk about some of these issues with reference to general procedure and particular cases.

The wastewater leaving pharmaceutical units varies in content and concentration, and hence, we do not attempt a detailed treatment, since the volumes are small, and distinctive compounds are made from a similar battery of reactors and separators. Water reuse saves money through the reduction of waste transfer expenses and maintains the required water quality, balancing the operational expenses related to the preparation of waste reuse.

The pharmaceutical business requires a steady supply of excellent-quality water for generation and wastewater treatment to meet the requirements of ever-stricter administrative release limits. To meet these difficulties, organizations must question their customary intuitive and run-of-the-mill approaches and investigate new innovations and answers to stay focused.

Thus, in the present chapter, we have endeavored to

- Understand the pathway of pharmaceutical waste, beginning at the business site
- Group the specific mechanical procedures to handle waste in order
- Address the adequacy of cutting-edge methods and cross-breed innovations for the removal of 200 pharmaceuticals from the water system

Pharmaceutical wastewater streams can be difficult to treat with conventional physical/chemical and biological treatment systems. High chemical oxygen demand

(COD), variable-strength waste streams, and shock loads are just a few of the conditions that limit the effectiveness of these conventional systems. Physical/chemical systems are a common method of treating pharmaceutical wastewater; however, system results are limited due to high sludge production and relatively low efficiency of dissolved COD removal.

This study was carried out to survey and characterize physicochemical and natural oxidation of wastewater from the adjustment tank of a pharmaceutical plant. The specific aims of the review were

- To examine the impact of coagulants, for example, $FeSO_4$, $FeCl_3$, and alum, on suspended solids and COD removal efficiency
- To see the action of the oxidation dump prepare on a seat scale, to treat the pharmaceutical waste at different natural loadings
- To determine the connection between essential natural parameters including mean cell home time (MCRT), water driven maintenance time (HRT), substrate evacuation, and biomass concentration
- To determine the essential performance characteristics of natural treatment in terms of effluent quality and slime properties
- To assess the degree to which phenol is removed

To have a reasonable comprehension of the different procedures used as a part of the treatment and transfer of different sorts of waste created in the pharmaceutical business, the treatment procedures can be separated into the accompanying four classifications and subcategories.

7.5.2.1 Biological Treatment

Biological forms of treatment have been generally used as a part of pharmaceutical wastewater treatment because of their minimal effort and adequacy. They are termed either *appended development* or *suspended development*, as indicated by the living status of the microorganisms (Shi et al., 2008; Sirtori et al., 2009).

Pharmaceuticals remaining in waste streams represent an incredible risk to receiving water bodies (rivers, waterways, and seas). The release of pharmaceutical plant effluents can cause the death of fish and other marine animals. The by-product of the preparation of anti-infective agents and the remaining waste fluid is a high-quality natural wastewater that is refractory to organic forms of treatment. It is usually treated by burning in the pharmaceutical plant; however, this choice is only appropriate for very large enterprises due to the enormous investment required and the extensive running costs. Hence, an appropriate minimal-effort preparation method for treating genuine high-quality natural pharmaceutical wastewater is desirable. A biological process has better potential and requires minimal effort to treat wastewater containing pharmaceutical by-products. The release of pharmaceutical wastewater from medicine manufacturing processes stands out among the most important sources of anti-infective agents in surface water and groundwater. Ordinarily, pharmaceutical wastewater is characterized by high COD concentration, and some pharmaceutical wastewater can have COD as high as 40,000 mg/L. Although pharmaceutical wastewater may contain assorted refractory natural materials that cannot be quickly degraded, biological treatment is still a feasible decision.

The natural treatment of pharmaceutical wastewater incorporates both aerobic and anaerobic treatment systems (Raj and Anjaneyulu, 2005; Afzal et al., 2007).

7.5.2.1.1 Aerobic Treatment

Biological aerobic treatment systems are also used extensively, often with limited success due to the final clarification step. The clarifiers are susceptible to sludge bulking and variations in total dissolved solids, often associated with batch process production, which can cause destabilization of bacterial floc formation with a consequential loss of biomass in the final effluent. These systems require constant operator attention to adjust chemical dosing for the daily, even hourly, changes in influent flow. Aerobic treatment is one of the regular connected innovations. It incorporates the activated sludge (AS) process, expanded air circulation–actuated sewage handling, AS with granular enacted carbon, and layer bioreactors (Peng et al., 2004; Chen et al., 2011).

1. *Activated Sludge Process*: The activated sludge strategy is generally used as an oxygen-consuming suspended development approach. Despite the fact that the activated sludge technique is powerful for the treatment of a few types of low-quality pharmaceutical wastewater, the activated sludge process is an outstanding procedure for removing different natural contaminants and natural carbon. The activated sludge process is the best-known aerobic treatment, which has been observed to be effective for different classes of pharmaceutical wastewaters. The conventional activated sludge (CAS) treatment is an easy technique that depends basically on two parameters, the temperature and the hydraulic retention time (HRT). As well as these, the presence of natural matter, COD, BOD, pH, and the presence of non-biodegradable matter are different components that influence the effectiveness of the AS strategy.

2. *Membrane Bioreactor*: The use of membrane bioreactors (MBRs) in wastewater treatment is clearly becoming progressively more important, in light of the fact that they offer a few favorable circumstances, that is, high biodegradation efficiency, a smaller impact, and less waste generation.

An MBR was operated in parallel with the CAS procedure (air circulation tank and auxiliary settling tank). The biogenesis of the MBR was developed from inoculated sewage from the municipal WWTP (air circulation bowl) and developed over a time of around 1 month to achieve steady-state conditions. The water-driven maintenance time was set to 14 h by directing the emanating stream, and the sludge retention time (SRT) was unlimited on the grounds that no sludge was released from the reactor. In the last decade, the use of MBRs for pharmaceutical wastewater treatment has attracted much attention, as it is a financially practical option for water and wastewater treatment, particularly in view of the high SRT achieved inside smaller reactor volumes.

The MBR system helps manufacturers to meet and exceed all direct discharge regulations while simplifying the treatment process. The MBR combines membrane filtration with biological treatment. The system replaces conventional treatment and combines clarification, aeration, and filtration into a simple and cost-effective process that reduces capital and operating costs. MBR has several advantages compared with other methodologies:

- Produces reliable, high-quality effluent at all times
- Handles shock loads
- Reduces chemical costs of treatment

7.5.2.1.2 Anaerobic Treatment

Anaerobic hybrid reactors, which are a combination of suspended development and joined development systems, have recently attracted much attention. In contrast to oxygen-consuming treatment, anaerobic treatment of high-concentration natural wastewater ordinarily has the accompanying advantages: high natural stacking; low drip generation; effortlessly natural slime dewatering; fewer supplements required; without air circulation, low vitality use; can deliver biogas vitality recuperation; suitable for a more extensive range of temperature; long active anaerobic SRT. It has regularly been used for non-natural wastewater.

7.5.2.1.3 Combination Process

A biological treatment of wastewater that manages the consolidated operation of chemical and biological oxidation has been demonstrated in some studies. The impacts of chemical oxidation as a pre- or post-treatment venture in biological oxidation of wastewater have additionally been accounted for. To reduce the arrival of such pharmaceuticals into the aquatic environment or to totally remove them from wastewater, the use of propelled wastewater treatment might be required.

Be that as it may, because of the high quality, it is infeasible and uneconomical to treat this sort of wastewater using a single-stage aerobic biological treatment process. Rather, a mix of anaerobic and high-impact procedures is favored and has all the earmarks of being more successful at removing high-quality natural matter. Anaerobic and aerobic treatment strategies have both points of interest and disadvantages; the blending of the two forms together, and the specific favorable circumstances to progress this, is not receiving sufficient interest. Ordinarily, an anaerobic procedure is included to reduce high concentrations of natural matter and degrade refractory substances, followed by a vigorous treatment to oxidize the residual organic matter in the wastewater (Li et al., 2003).

In the last couple of years, the treatment of wastewaters by a modified activated sludge process, the sequencing batch reactor (SBR), has gained acknowledgment. Of interest has been the use of the system to treat wastewaters containing dangerous substances. The biomass in the reactors adapted to the wastewater and treated it effectively. The use of the SBR in the treatment of wastewaters containing toxic substances suggests that combined procedures may likewise be reasonable for the treatment of pharmaceutical wastewater, which may well contain inhibitory substances.

7.5.2.2 Advanced Oxidation Processes

Advanced oxidation processes (AOPs) can be comprehensively characterized as fluid-stage oxidation techniques in view of the intermediation of highly active species, for example, hydroxyl radicals, in the systems, resulting in the destruction of the target contaminants. The principle of AOPs is heterogeneous and homogeneous photo-catalysis by ultra-violet (UV) or sunlight-based light. They include electro-oxidation, the Fenton and photograph Fenton process, and wet air oxidation, and recent additions to this classification are ultrasound illumination and microwave treatment.

Due to the low biodegradability of numerous pharmaceuticals, the generally used treatment procedures are not sufficiently powerful for the complete of such species, and the release of treated effluents into receiving waters can cause contamination with these smaller-scale toxins. These compounds discharged into nature in this manner have ended up being sufficiently high to have a poisonous impact on natural living beings. Depending on the nature of the pharmaceutical discharge and the treatment goal (elimination or change), AOPs can be used either alone or combined with other physio-synthetic and biological processes.

7.5.2.2.1 Incineration

Burning waste to be treated with an overabundance of oxidizing air in the incinerator ignition response, so that the toxins contained in the wastewater are eliminated by high-temperature oxidative decay, is an incorporated high-temperature treatment of significant oxidation frames. Burning can significantly decrease the volume of waste water and dispose of a considerable quantity of these harmful substances with the production of heat. This strategy enables the complete oxidation of waste into safe substances with a COD removal efficiency of 99.5%. The reasonings for managing high customary substances or higher calorific estimations of misuse is thoroughly respected. Through the natural substance of the waste water is less, it might be added to the helper fuel.

7.5.2.3 Hybrid Technologies

Hybrid technologies are combinations of ordinary/progressed treatment innovations for the total destruction of pharmaceutical contaminants. The need for a combined approach emerges from the fact that none of the single treatment innovations can eliminate all compounds. There are various mixed approaches that have been used for the treatment of obstinate toxins and in addition, to reduce the cost of the treatment procedure. Basically, the approach uses regular filtration technology to remove any strong matrix, and the sludge is evacuated for burning. The resulting wastewater is then treated by the specific mix of procedures.

7.5.3 TREATMENT METHODOLOGIES FOR HOSPITAL EFFLUENTS

Hospital wastewater can represent a health hazard to people, particularly workers at WWTPs. Amid substantial rains and flooding, holding tanks in the sewer framework can flood. There is additionally a danger to marine life. Once the sewage is dealt with and discharged into the earth with its lingering content of pathogens and pharmaceuticals, the local fauna are routinely at risk.

7.5.3.1 Managing Hospital Water and Wastewater

The best way to manage hospital wastewater, as well as to satisfy local water emissions regulations, includes

- Collection of specific ingredients
- Substitutions
- Advanced treatment methods

The efficiency of a broad range of advanced technologies to treat hospital wastewater was analyzed, both in the laboratory and at pilot scale. The effectiveness of the following methods to remove persistent and toxic pollutants such as antibiotics, cytostatics, and hormones was studied.

- MBR as a pre-treatment
- Oxidation through ozonation
- Advanced oxidation process ($O^3 + H_2O_2$)
- Powdered activated carbon (PAC) and granulated activated carbon (GAC)

These treatment technologies can be used in different combinations to effectively treat hospital wastewater. After treatment, the concentration of pharmaceuticals was below the level at which organisms living in water would be adversely affected. This means that the wastewater quality was high enough to be

- Discharged directly into local water areas
- Reused as technical, cooling, or recreational water

Hospital water and wastewater systems can be enhanced by the following accompanying steps:

- Integrating water quality measures into the distribution network design to ensure consistently high water quality
- Integrating optimal wastewater treatment into hospital construction plans
- Getting water authority permission to discharge treated wastewater into the local water area
- Planning and identifying possibilities for reusing treated wastewater

7.6 FUTURE RESEARCH NEEDS

The pharmaceutical waste stream is of a varied nature, and hence, the wastewater must be treated for environmentally friendly disposal. Reduction of the waste stream at the source, alongside reuse of the water or recovery of some part of the waste, is among the most attractive choices.

On-site reduction can be accomplished by profoundly understanding the procedure, examination of all the stages, and recognizing the contaminations to be discharged. With this profound understanding, an innovative recovery method such as films can be connected at the source of toxin entry, the recovered material can be used, and the concentrated waste can be treated using other treatment advances for safe transfer. Along these lines, noteworthy regard can be given to the waste made and along these lines making the system useful.

There are two possible reasons for the low occurrence of pharmaceutical products in the effluent of WWTPs. One is the absence of microbes equipped for metabolizing them or their low metabolic activity. Another is the low bioavailability of pharmaceutical products when their concentration has diminished to a certain level. Thus, more reviews are needed to determine the correct reasons and to provide potential techniques to manage the removal of low levels of pharmaceutical products from wastewater.

7.7   CONCLUSION

To ensure the quality of surface water, there is a great necessity to enhance the waste-water treatment rate to reduce immediate release. At times, pharmaceutical products can be identified in the source water of drinking water treatment plants or even in tap water. It is urgent to secure the water sources from pollution by pharmaceutical by-products.

The removal of pharmaceutical products during a biological treatment system can be affected by many factors, such as the composition of the pharmaceutical products, the pre-treatment system, the microbial community existing in the activated sludge system, and the environmental conditions as well as operational conditions. Thus, the biological treatment system, as the most widely used treatment process, should be improved to meet the demand for dealing with wastewater containing pharmaceutical products. To increase the ability of WWTPs to treat wastewater containing pharmaceutical products, more attention should be paid to the following points.

- The impact of the composition of pharmaceutical products on the removal efficiency of pharmaceutical products ought to be considered in light of the fact that these products have distinctive physiochemical properties and biodegradability. In light of the arrangement of pharmaceutical products, a specific process ought to be used. For example, a long pressure-driven maintenance time can be used for wastewater containing high levels of biodegradable pharmaceutical products.
- It is important to better understand the degradation pathway, which would be helpful in the enhancement of the biodegradation of pharmaceutical products during wastewater treatment. Besides, despite the centralization of pharmaceutical items, the coordinated effort among biomass and pharmaceutical items should investigate the location of the key segment that controls the biodegradation of pharmaceutical items.

- The biodegradation of pharmaceutical products is a minimal-effort and powerful technique for removing the pharmaceutical products. In the activated sludge system, biosorption and biodegradation at the same time contributed to the removal of pharmaceutical products. Regardless, the duties of biosorption and biodegradation in the departure of pharmaceutical items changed.
- Biodegradation can totally remove pharmaceutical products from soil, yet the pharmaceutical products evacuated by biosorption remain in the treated sewage, and can enter nature with the reuse of treated sewage, causing possible pollution. With the specific end goal of reducing the likelihood of possible pollution, more reviews are needed to clear up the particular contributions of biosorption and biodegradation to the removal of pharmaceutical products.

REFERENCES

Afzal, M., Iqbal, S., Rauf, S. et al. 2007. Characteristics of phenol biodegradation in saline solutions by monocultures of *Pseudomonas aeruginosa* and *Pseudomonas pseudomallei*. *Journal of Hazardous Materials* 149: 60–66.

Behera, S., Kim, H., Oh, J. et al. 2011. Occurrence and removal of antibiotics, hormones and several other pharmaceuticals in wastewater treatment plants of the largest industrial city of Korea. *Science of the Total Environment* 409: 4351–4360.

Blair, B., Nikolaus, A., Hedman, C. et al. 2015. Evaluating the degradation, sorption, and negative mass balances of pharmaceuticals and personal care products during wastewater treatment. *Chemosphere* 134: 395–401.

Boillot, C. 2008. Daily physicochemical, microbiological and ecotoxicological fluctuation of a hospital effluent according to technical and care activities. *Science of the Total Environment* 403: 113–129.

Bu, Q., Wang, B., Huang, S. et al. 2013. Pharmaceuticals and personal care products in the aquatic environment in China: A review. *Journal of Hazardous Materials* 262: 189–211.

Bushra, R., Naushad, M., Adnan, R., Alothman, Z. A., Rafatullah, M., 2015. Polyaniline supported nanocomposite cation exchanger: Synthesis, characterization and applications for the efficient removal of Pb2+ ion from aqueous medium. *Journal of Industrial and Engineering Chemistry* 21: 1112–1118.

Buszka, P., Yeskis, D., Kolpin, D. W. et al. 2009: Waste-indicator and pharmaceutical compounds in landfill-leachate affected ground water near Elkhart, Indiana, 2000–2002. *Bulletin of Environmental Contamination and Toxicology* 82: 653–659.

Chen, H., Li, X., and Zhu, S. 2012. Occurrence and distribution of selected pharmaceuticals and personal care products in aquatic environments: A comparative study of regions in China with different urbanization levels. *Environmental Science and Pollution Research* 19: 2381–2389.

Chen, Z., Wang, H., Ren, N. et al. 2011. Simultaneous removal and evaluation of organic substrates and NH 3 -N by a novel combined process in treating chemical synthesis-based pharmaceutical wastewater. *Journal of Hazardous Materials* 197: 49–59.

Coetsier, C. M., Spinelli, S., Lin, L. et al. 2009. Discharge of pharmaceutical products (PPs) through a conventional biological sewage treatment plant: MECs vs PECs? *Environmental International* 35: 787–792.

Crane, M., Watts, C., and Boucard, T. 2006. Chronic aquatic environmental risks from exposure to human pharmaceuticals. *Science of the Total Environment* 367: 23–41.

Daughton, C. G. 2001. Pharmaceuticals and personal care products in the environment: Overarching issues and overview. *ACS Symposium Series* 791: 2–38.

Ebele, A. J., Abdallah, M. A-E., and S. Harrad. 2017. Pharmaceuticals and personal care products (PPCPs) in the freshwater aquatic environment. *Emerging Contaminants* 4: 1–16.

Eggen, T., Moeder, M., and Arukwe, A. 2010. Municipal landfill leachates: A significant source for new and emerging pollutants. *Science of the Total Environment* 408: 5147–5157.

EPA, United States Environmental Protection Agency, Polluted Runoff: Nonpoint Source Pollution. https://www.epa.gov/nps

Erses, A. S., Onay, T. T., and Yenigun, O. 2008. Comparison of aerobic and anaerobic degradation of municipal solid waste in bioreactor landfills. *Bioresource Technology* 99: 5418–5426.

Escher, B. I., Baumgartner, R., Koller, M. et al. 2011. Environmental toxicity and risk assessment of pharmaceuticals from hospital wastewater. *Water Research* 45: 75–92.

Fent, K., Weston, A. A., and Caminada, D. 2006. Ecotoxicology of human pharmaceuticals. *Aquatic Toxicology* 76: 122–159.

Gupta, V. K., Ali, I., Saleh, T. A. et al. 2012. Chemical treatment technologies for wastewater recycling—an overview. *RSC Advances*. 2: 6380–6388.

Houtman, C. J., Oostveen, A. M. V., Brouwer, A. et al. 2004. Identification of estrogenic compounds in fish bile using bioassay-directed fractionation. *Environmental Science and Technology* 38: 6415–6423.

Hu, X., Zhou, Q., and Luo, Y. 2010. Occurrence and source analysis of typical veterinary antibiotics in manure, soil, vegetables and groundwater from organic vegetable bases, northern China. *Environmental Pollution* 158: 2992–2998.

Huerta, B., Rodriguez-Mozaz, S., and Barcelo, D. 2013. Analysis removal, effects and risk of pharmaceuticals in the water Cycle. *Comprehensive Analytical Chemistry* 62: 169–193.

Iyanee, F. S., Simamura, K., Prabhasankar, V. P. et al. 2013. Occurrence of antibiotics in river water: A case study of Vrishabhavathi River near Bangalore, India. *33rd International Symposium on Halogenated Persistent Organic Pollutants*, DIOXIN 2013, August 25–30, 2013, Daegu, Korea.

Kim, K. R., Owens, G., Kwon, S. T. et al. 2011. Occurrence and environmental fate of veterinary antibiotics in the terrestrial environment. *Water Air Soil Pollution* 214: 163–174.

Li, Y. M., Gu, G. W., Zhao, I. et al. 2003. Treatment of coke-plant wastewater by biofilm systems for removal of organic compounds and nitrogen. *Chemosphere* 52: 997–1005.

Luo, Y., Guo, W., Ngo, H. H. et al. 2014. A review on the occurrence of micro pollutants in the aquatic environment and their fate and removal during waste water treatment. *Science of the Total Environment* 473–474: 619–641.

Mahvi, A., Rajabizadeh, A., Yousefi, N. et al. 2009. Survey of wastewater treatment condition and effluent quality of Kerman Province hospitals. *World Applied Sciences Journal* 7(12): 1521–1525.

Mompelat, S., Bot, B. L., and Thomas, O. 2009. Occurrence and fate of pharmaceutical products and by-products, from resource to drinking water. *Environmental International* 35: 803–814.

Musson. S. E. and Townsend, T. G. 2009. Pharmaceutical compound content of municipal solid waste. *Journal of Hazardous Materials* 162: 730–735.

Mutiyar, P. and Mittal, A. 2014. Occurrences and fate of selected human antibiotics in influents and effluents of sewage treatment plant and effluent-receiving river Yamuna in Delhi (India). *Environmental Monitoring and Assessment* 186: 541–557.

Naushad, M., AL Othman, Z. A., Awual, M. R., Alam, M. M., Eldesoky, G. E., 2015. Adsorption kinetics, isotherms, and thermodynamic studies for the adsorption of Pb^{2+} and Hg^{2+} metal ions from aqueous medium using Ti(IV) iodovanadate cation exchanger. *Ionics (Kiel)* 21: 2237–2245.

Niemuth, N. J. and Klaper, R. D. 2015. Emerging wastewater contaminant metformin causes intersex and reduced fecundity in fish. *Chemosphere* 135: 38–45.

Osenbruck, K., Glaser, H. R., Knoller, K. et al. 2007. Sources and transport of selected organic micropollutants in urban groundwater underlying the city of Halle (Saale), Germany. *Water Research* 41: 3259–3270.

Peng, Y. Z., Li, Y. Z., Peng, C. Y. et al. 2004. Nitrogen removal from pharmaceutical manufacturing wastewater with high concentration of ammonia and free ammonia via partial nitrification and denitrification. *Water Science and Technology* 50: 31–36.

Raj, D. S. S. and Anjaneyulu, Y. 2005. Evaluation of biokinetic parameters for pharmaceutical wastewaters using aerobic oxidation integrated with chemical treatment. *Process Biochemistry* 40: 165–175.

Ramaswamy, B., Shanmugam, G., Velu, G. et al. 2011. GC–MS analysis and ecotoxicological risk assessment of triclosan, carbamazepine, and parabens in Indian rivers. *Journal of Hazardous Materials* 186: 1586–1593.

Roberts, J., Kumar, A., Du, J. et al. 2016. Pharmaceuticals and personal care products (PPCPs) in Australia's largest inland sewage treatment plant, and its contribution to a major Australian river during high and low flow. *Science of the Total Environment* 541: 1625–1637.

Sarmah, A. K., Meyer, M. T., and Boxall, A. B. A. 2006. A global perspective on the use, sales, exposure pathways, occurrence, fate and effects of veterinary antibiotics (VAs) in the environment. *Chemosphere* 65: 725–759.

Shanmugam, G., Sampath, S., Selvaraj, K. et al. 2013. Nonsteroidal anti-inflammatory drugs in Indian rivers. *Environmental Science and Pollution Research* 21: 921–931.

Shi, H., Zheng, H. J., Zhao, D. H. et al. 2008. Treatment of nisin production wastewater by biochemical method. *Chinese Journal of Environment Engineering* 2(10): 1369–1372.

Sirtori, C., Zapata, A., Oller, I. et al. 2009. Decontamination industrial pharmaceutical wastewater by combining solar photo-Fenton and biological treatment. *Water Research* 43(3): 661–668.

Subedi, B., Balakrishna, K., Sinha, R. et al. 2015. Mass loading and removal of pharmaceuticals and personal care products, including psychoactive and illicit drugs and artificial sweeteners, in five sewage treatment plants in India. *Journal of Environmental Chemical Engineering* 3: 2882–2891.

Synder, S. A. 2008. Occurrence, treatment, and toxicological relevance of EDCs and pharmaceuticals in water. *Ozone Science and Engineering* 30: 65–69.

Wang, J. and Wang, S. 2016. Removal of pharmaceuticals and personal care products (PPCPs) from wastewater: A review. *Journal of Environmental Management* 182: 620–640.

Watkins, D. and Weiss, K. 1999. New air and wastewater compliance challenges for the pharmaceutical industry. *Pharmaceutical Engineering* 19: 6.

Wu, M., Xiang, J., Que, C. et al. 2015. Occurrence and fate of psychiatric pharmaceuticals in the urban water system of Shanghai, China. *Chemosphere* 138: 486–493.

Yang, S. F., Lin, C. F., Lin, A. C. et al. 2011. Sorption and biodegradation of sulfonamide antibiotics by activated sludge: Experimental assessment using batch data obtained under aerobic conditions. *Water Research* 45: 3389–3397.

8 Fungal Treatment of Pharmaceuticals in Effluents
Current State, Perspectives, Limitations, and Opportunities

Arash Jahandideh, Sara Mardani, Rachel McDaniel, Bruce Bleakley, and Gary Anderson

CONTENTS

8.1 INTRODUCTION: BACKGROUND AND DRIVING FORCES

The presence of pharmacologically active organic micropollutants (PhACs), including pharmaceuticals and personal care products (PPCPs) as well as endocrine-disrupting chemicals (EDCs), is an emerging issue and an increasing concern both in the effluent of wastewater treatment plants (WWTPs) and in natural water courses (Lienert et al., 2007; Marco-Urrea et al., 2009; Jones et al., 2005). These active substances are often highly persistent compounds and come from various sources, including human and veterinary medicine (Richardson et al., 2005), chemicals used as fragrances in perfumes (Simonich et al., 2002), industrial wastewater, and other household products (Kosjek et al., 2007). In fact, most of these compounds are designed to affect physiological functions in humans and animals; therefore, the permanent presence of these pharmaceuticals, even at low concentration, is of great importance (Rodarte-Morales et al., 2012). Increased aquatic toxicity (Santos et al., 2010) and endocrine disruption (Kim et al., 2007) have been previously reported due to the presence of pharmaceuticals in water resources and in sewage-impacted water bodies.

The severity of the potential risk from PhACs to the ecosystems, the biotic community, and human health is dependent on several factors, including the concentrations, adsorption efficiencies, time of exposure, bioaccumulation, and nature of the PhACs and the vulnerability of the contaminated ecosystem (Rodarte-Morales et al., 2012; Daughton and Ternes, 1999; Kümmerer, 2008). The environmental impact of PhACs is not restricted to the aquatic ecosystem. PhACs are also a great potential risk to soil matrices. The soil can become contaminated by diffusion of the PhACs through the soil from landfill leachates or by contaminated sewage sludge biosolids (a byproduct generated during wastewater treatment, which is used as a soil improver) (Rodríguez-Rodríguez et al., 2011; Drillia et al., 2005). On the one hand, PhACs are resistant products by nature, which are not easily degraded or removed at WWTPs (Carballa et al., 2004), and on the other hand, the current WWTPs are not designed to remove these components, and in most cases, they fail to remove PhACs effectively. The extensive contamination potential of PhACs, along with their contribution to severe ecotoxicity and human health problems, necessitates the development of cost-effective and efficient methods for the elimination of PhACs from wastewater or contaminated soil (Prieto et al., 2011; Asgher et al., 2008; Yang et al., 2013).

Conventional WWTPs are based on activated sludge technologies and are not designed for or capable of the removal of PhACs, which are present at low concentrations (nanograms per liter to milligrams per liter). Consequently, PhACs often pass through, untouched or partially transformed, and reach environmental compartments (air, water, and soil) (Verlicchi et al., 2012b; Cruz-Morató et al., 2013). Currently, three potential treatment technologies have been suggested for the efficient degradation and removal of PhACs in wastewater streams: conventional treatments (physical–chemical methods: photodegradation [Liu et al., 2009], activated sludge [Nakada et al., 2006], nitrification–denitrification [Suarez et al., 2010]); advanced treatments (UV, ozone, hydrodynamic cavitation [Zupanc et al., 2013], anammox reactors [Tang et al., 2011], activated carbon [Snyder et al., 2007], photo-Fenton systems [Shemer et al., 2006], and membrane bioreactors [MBRs] [Clara et al., 2005]); and bioremediation methods based on the white-rot fungi and their oxidative enzymes (Rodarte-Morales et al., 2012). The complete removal of some recalcitrant PhACs, including fragrances (i.e., galaxolide and tonalide), tranquilizers (i.e., diazepam), carbamazepine, and clofibric acid has been reported based on conventional treatments (Doll and Frimmel, 2004). However, the overall degradation of anti-epileptics (i.e., carbamazepine) and the main anti-inflammatory compounds has been reported to range from 10% to 70% in the literature (Rodarte-Morales et al., 2012; Carballa et al., 2004; Ikehata et al., 2006). The other drawback of conventional methods is the formation of undesirable and often toxic transformation products (Cruz-Morató et al., 2013). On the other hand, higher or complete removal (degradation) of anti-inflammatory compounds is reported for the advanced technologies; that is, up to 100% removal of naproxen (NPX) by ozone treatment and 80%–100% for ibuprofen (IBP) by the photo-Fenton system (Rodarte-Morales et al., 2012). These systems often fail at effectively removing other PhACs (partial removal of 50%–70%) (Gagnon et al., 2008; Mendez-Arriaga et al., 2010). In addition, the implementation of ozonation treatment, for example, is still far from being economical, which makes this method unfeasible (Ternes et al., 2003). MBR and activated carbon–based methods are also expensive methods, which are effective in the removal of some PhACs but fail to remove others (Snyder et al., 2007).

The other emerging alternative treatment of pharmaceutical compounds in WWTP effluent is a biological method based on the employment of bioremediation techniques. Bioremediation is an attractive alternative technique, which employs the metabolic potential of biological agents such as fungi to remove contaminants from soil or water (Keharia and Madamwar, 2003). Recently, the use of white-rot fungi (WRF) for the effective removal of pharmaceuticals from wastewater has attracted attention. WRF are capable of degrading lignin, dyes, polycyclic aromatic hydrocarbons (PAHs), and several PhACs with degradation of up to 100% employing their nonspecific enzymatic system, including extracellular lignin-modifying enzymes and intracellular enzymes (Rodarte-Morales et al., 2012; Wesenberg et al., 2003; Field et al., 1992; Asgher et al., 2008; Prieto et al., 2011).

8.2 PHARMACEUTICS AND PHARMACEUTICAL COMPONENTS IN EFFLUENTS

8.2.1 POTENTIAL SOURCES OF PHARMACEUTICAL POLLUTANTS IN THE ENVIRONMENT

Persistent organic pollutants (POPs) are toxic substances that can be released into the environment by the application of agrochemicals in agricultural areas, through agroindustry applications, or by the application of biosolids for soil improvement. Agrochemicals are chemicals used in agriculture, such as pesticides or fertilizers. Huge agro-industrial processes, including oil extraction procedures, bleaching (i.e., of cotton for the pulp and paper industries), and distilleries, produce several billion liters annually of colored and toxic effluents, often rich in persistent compounds (i.e., phenol compounds, chlorinated lignin, and dyes), which are a potential risk to the environment (Adhoum and Monser, 2004; Pokhrel and Viraraghavan, 2004). Another source of persistent pollutants in agricultural practice is the employment of sewage sludge from WWTPs, which is used as a soil improver and known as *biosolids*. Currently, no legislation exists regarding the limits of pharmaceuticals and PhACs in biosolids. They may contain high concentrations of PhACs, which contaminate the soil and ground water (Rodríguez-Rodríguez et al., 2011; Tilman et al., 2002; Kinney et al., 2006). The resultant POPs, which are persistent in nature, can remain in the soil, even long after their use, and can enter the human food cycle directly or by percolation to the ground-water table (Gavrilescu, 2005). In addition, large amounts of pharmaceuticals annually are used in livestock farms and fisheries as well as shrimp hatcheries (Uddin and Kader, 2006). The release of veterinary pharmaceuticals, especially antibiotics, which are used to prevent disease in animals, treat infections, and promote animal growth, is another main contributor of pharmaceuticals to the environment (Sim et al., 2011; Kim et al., 2013).

After pharmaceutical intake, the active compounds can be excreted by animals into water bodies as the parent compound, conjugates, or metabolites, which have become a potential and emerging environmental issue (Marco-Urrea et al., 2010c; Langford and Thomas, 2009). It is claimed that up to 90% of an administered dose of antibiotics is excreted through urine and feces (Drillia, Stamatelatou, and Lyberatos, 2005). It is believed that the PhACs are spread into the environment mostly via human or animal consumption and subsequent excretion in the feces and urine, and by direct disposal of expired or unused pharmaceuticals by patients and customers (Halling-Sørensen et al., 1998; Cruz-Morató et al., 2013). Mass flow analysis of PhACs has shown that the concentrations of these compounds in the environment are in the range of nanograms to micrograms per liter; therefore, acute toxic effects are unlikely (Ikehata et al., 2006). However, although little is known about the chronic environmental toxic effect of PhACs, the potential risk of chronic effects cannot be neglected (Rodarte-Morales et al., 2011; Crane et al., 2006; Boxall et al., 2003).

Hospital effluent can be considered as one of the main contributors to the presence of PhACs in the influent of WWTPs. The concentration of PhACs in hospital effluent is considerably higher (up to milligrams per liter) compared with their concentrations in the WWTP influents (Verlicchi et al., 2012a). Currently, hospital effluents are not treated separately before being discharged into public sewer networks to be treated along with urban wastewater; therefore, some researchers believe that hospital wastewater is the main contributor to PhAC concentrations in WWTP influent (Verlicchi et al., 2012b; Langford and Thomas, 2009). Nevertheless, some other authors do not concur; they believe that the amount of PhACs contributed by hospital effluent is negligible compared with the large inflow of PhACs introduced by municipal wastewater (Le Corre et al., 2012). Compared with the effluent from hospitals, effluent from pharmaceutical and industrial manufacturers would probably have fewer PhAC compounds, but in more significant concentrations, which could potentially overload the aquatic environment. Therefore, it is likely that hospital effluent and pharmaceutical manufacturers' effluent contribute to some extent to the PhAC load in WWTP influent (Langford and Thomas, 2009). Pharmaceutical production facilities are another source of PhACs in the environment. Despite the strict environmental standards and a different set of regulations for the manufacturers' effluents, Larsson et al. reported concentrations above 1 µg L^{-1} for 21 pharmaceutical compounds out of 59 initially screened for in the wastewater effluent from pharmaceutical manufacturers. There was a heavy antibiotic load, in which the reported concentrations greatly exceeded the standards (i.e., Ciprofloxacin 28,000–31,000 µg L^{-1}, Enrofloxacin 780–900 µg L^{-1}, Norfloxacin 390–420 µg L^{-1}, Lomefloxacin 150–300 µg L^{-1}, Enoxacin 150–300 µg L^{-1}, and Ofloxacin 150–160 µg L^{-1} in the effluent) (Larsson et al., 2007). The presence of a potential risk of high concentrations of PhACs in manufacturing effluent has also been reported by other authors (Fick et al., 2009; Sim et al., 2011; Phillips et al., 2010).

8.2.2 Types of Pharmaceuticals That Can Be Degraded by Fungi

Currently, more than 4000 PhACs are produced in large amounts annually. However, the contribution of some of them to WWTP influent is not significant due to their relatively small consumption rates, ease of removal, or efficient degradation. On the other hand, in some cases, the successful employment of fungi has not been reported. In this section, the main pharmaceuticals present in wastewaters that can be degraded or eliminated by fungal species are assessed. The pharmaceuticals were grouped based on their therapeutic function into the following classes: anti-inflammatory drugs, analgesic drugs, psychotropic drugs, lipid regulators, antibiotics, β-blockers, estrogens, and iodinated contrast media.

Nonsteroidal *anti-inflammatory* compounds are a group of nonprescription pharmaceuticals with a large annual consumption. Unchanged anti-inflammatory compounds or their transformed derivatives are almost ubiquitous in all WWTP influents at concentrations of up to micrograms per liter. Many fungal species have been investigated for the degradation of anti-inflammatory compounds, and the results have shown the promising capability of fungi to degrade these compounds or their derivatives. To date, the successful use of fungi for removal of the following anti-inflammatory drugs has been reported: IBP (Rodarte-Morales et al., 2011), Diclofenac (Rodarte-Morales et al., 2011), NPX (Rodarte-Morales et al., 2011), Fenoprofen (Tran et al., 2010), Indomethacin (Tran et al., 2010), Ketoprofen (Marco-Urrea et al., 2010b), Mefenamic (Hata et al., 2010), Caffeine (Cruz-Morató et al., 2014), and Propyphenazone (Tran et al., 2010).

Psychiatric drugs are frequently prescribed psychoactive compounds with the capability to affect the brain and nervous system. Psychiatric drugs are used for the treatment of mental disorders, epilepsy, trigeminal neuralgia, and other psychiatric diseases (Rose, 2001). Often, these recalcitrant compounds are present in nanogram per liter concentrations in the influent of WWTPs and are not removed or degraded by currently employed wastewater treatment techniques (Miao et al., 2005). To date, the complete or partial degradation of these persistent compounds employing different fungi has been reported. Fungal species have been used to assess the degradation of different psychotropic drugs, including Carbamazepine, Diazepam, Fluoxetine, and Citalopram.

Antibiotics are a class of antimicrobial drugs that are widely used for the treatment or prevention of bacterial infections. Antibiotics are extensively used for the treatment of human and livestock wastes as well as for agricultural activities (Gelband et al., 2015). The occurrence of these compounds in aquatic cultures is a potential risk to the environment, as they can alter the ecosystem by promoting antibiotic-resistant bacterial species. Low concentrations of antibiotics have been reported in the effluent of WWTPs in

Canada, Europe, and the United States (Batt et al., 2007). To date, the successful use of different fungal species has been reported for the efficient removal of low concentrations of different antibiotics, including Cinoxacin (Cruz-Morató et al., 2012), Ciprofloxacin (Parshikov et al., 2001b), Enrofloxacin (Martens et al., 1996), Erythromycin (Accinelli et al., 2010), Norfloxacin (Parshikov et al., 2001b), Flumequine (Čvančarová et al., 2013), Sarafloxacin (Parshikov et al., 2001a), Sulfamethazine (García-Galán et al., 2011), Sulfathiazole (Rodríguez-Rodríguez et al., 2012b), Sulfapyridine (Rodríguez-Rodríguez et al., 2012b), and Sulfamethoxazole (Rodarte-Morales et al., 2011).

Lipid regulators are a class of chemicals that are prescribed to control elevated levels of different forms of lipids in patients with hyperlipidemia. Lipid regulators include statins, which lower cholesterol, and fibrates which regulate fatty acids and triglycerides (Pahan, 2006). The removal efficiency of these compounds and their main derivative, clofibric acid, is known to be ~28% in conventional WWTPs (Marco-Urrea et al., 2010c). A large amount of lipid regulators is consumed annually, and the failure of current WWTPs to efficiently remove them makes lipid regulators a potential hazard to the environment. Several fungal species have been tested for the removal of this class of pharmaceuticals. Successful removal (near and up to 100%) has been reported for clofibric acid (Marco-Urrea et al., 2010c), and partial removal of 70% was reported for Gemfibrozil (Tran et al., 2010).

β-blockers or β-adrenergic blocking agents are a group of pharmaceuticals used for the treatment of cardiac arrhythmias and high blood pressure (hypertension). They act by blocking the effects of the hormone epinephrine (adrenaline). Conventional WWTPs are not capable of removing these compounds, and consequently, β-blockers are found ubiquitously in the effluent of WWTPs with concentrations in the range of nanograms to milligrams per liter. Several fungal species have demonstrated the capability to eliminate β-blocker compounds, including atenolol and propranolol (Marco-Urrea et al., 2010c).

Estrogens are a class of pharmaceuticals that influence the endocrine hormonal system and are prescribed in the treatment of certain hormone-sensitive cancers, such as prostate cancer and breast cancer (Ternes et al., 1999). Conventional WWTPs are not capable of efficiently eliminating these compounds; therefore, many steroids are eventually discharged into the environment via WWTP effluents or as accumulated compounds in sewage sludge, which is used for land improvement (Koh et al., 2008). The data regarding the degradation of these compounds by fungi is scarce in the literature. However, the successful removal of 17β-estradiol (Auriol et al., 2008) and 17α-ethinylestradiol compounds (Blánquez and Guieysse, 2008) due to degradation by several fungal species has been reported.

8.3   USE OF FUNGI FOR TREATMENT OF PHARMACEUTICALS IN THE EFFLUENT

8.3.1   Introduction to Fungi

The eukaryotic and largely aerobic nature of fungi, combined (except for yeasts) with their filamentous growth habit, distinguishes them in many ways as different from bacteria (Glazer and Nikaido, 1995; Morton, 2005; Harms et al., 2011). Whereas bacteria typically break down the organic compounds needed for carbon and/or energy sources using enzymes with relatively high substrate specificity, fungi are often more flexible in their ability to act on pollutants. The typically smaller size of the bacterial cells also makes them less sensitive to mechanical disturbance and shear stress than fungal hyphae (Harms et al., 2011).

Most of the fungi that have been documented to either metabolize or bind pharmaceutical pollutants are from two fungal phyla, Ascomycota (ascomycetes) and Basidiomycota (basidiomycetes) (Harms et al., 2011). This includes a few single-celled yeast genera in the Ascomycota, but mostly involves filamentous fungi forming hyphae. The ability of cell walls or other components of some fungi to biosorb pollutants such as heavy metals (Bishnoi, 2005; Harms et al., 2011; Siddiquee et al., 2015) or dyes (Kabbout and Taha, 2014; Yagub et al., 2014; Rybczyńska-Tkaczyk and Korniłłowicz-Kowalska, 2016; Lu et al., 2017) makes them useful in biosorption remediation of some pharmaceutical wastes. Mycoremediation of many pharmaceutical wastes depends on fungal enzymes. The WRF in the Basidiomycota have received perhaps the most attention due to the non-specificity and efficacy of their extracellular enzymes to break down a variety of organic compounds (Harms et al., 2011; Kües, 2015; Tortella et al., 2015).

8.3.2 FUNGAL ENZYMES

Some fungi produce a variety of enzymes that can chemically alter and sometimes even mineralize a variety of naturally occurring recalcitrant molecules, such as lignin, and many synthetic xenobiotics. Fungal groups that have evolved to interact with lignocellulose, soil humus, and/or the defense mechanisms of green plants seem to be most equipped with these enzymes (Harms et al., 2011; Kües, 2015). Many of them produce a variety of extracellular oxidoreductases that are relatively nonspecific regarding which organic substrates they can bind and act on. WRF have received much attention (Tišma et al., 2010; Marco-Urrea and Reddy, 2012; Asif et al., 2017) and are especially well equipped, producing a variety of extracellular enzymes, including several lignin-modifying class II heme peroxidases (lignin peroxidase [LiP], manganese peroxidase [MnP], and versatile peroxidase [VP]), as well as dye-decolorizing peroxidases, all being active at acidic pH. Peroxidases require hydrogen peroxide (H_2O_2) for enzymatic activity (Harms et al., 2011; Kües, 2015).

The lignin peroxidases, ranging in molecular mass from 37 to 50 kDa (Hirai et al., 2005; Asgher et al., 2007, 2008, 2006), are secreted during the secondary metabolism of WRF, and are capable of mineralizing a variety of recalcitrant substances, including aromatic compounds and lignin (Shrivastava et al., 2005). The pH and temperature activity profiles of LiPs depend on the enzyme's source; however, optimum activities have been reported between pH 2 and 5 and between 35 and 55°C (Yang, 2004; Asgher et al., 2007). Veratryl alcohol (VA) and 2-chloro-1,4-dimethoxybenzene (fungal secondary metabolites) stimulate the LiP-catalyzed oxidation of substrates, acting as redox mediators. The LiP oxidation rate is dependent on the molar ratio of H_2O_2 to the pollutant. While low concentrations of H_2O_2 activate the *Phanerochaete chrysosporium* LiP (Pc-LiP), higher concentrations rapidly inhibit enzyme activity by enzyme deactivation. LiPs can oxidize a variety of aromatic compounds, including some pharmaceuticals (Harms et al., 2011; Falade et al., 2016).

MnPs are extracellular glycoproteins with a molecular weight range of 32–62.5 kDa and optimum activity from pH 4 to 7 and from 40 to 60°C (Ürek and Pazarlioğlu, 2004; Baborová et al., 2006). Fungi secrete multiple isoforms of MnPs in carbon- and nitrogen-limited media supplemented with Mn^{2+}, which plays the role of mediator for MnP and VA (Hakala et al., 2005; Cheng et al., 2007). Immobilization with sodium alginate, gelatin, or chitosan (as carriers) and glutaraldehyde crosslinking agent can increase the stability of the enzyme (Cheng et al., 2007). The activity of MnP increases in the presence of co-oxidants, such as glutathione, unsaturated fatty acids, and Tween 80, and is inhibited by NaN_3, ascorbic acid, β-mercaptoethanol, and dithreitol (Hofrichter, 2002; Ürek and Pazarlioğlu, 2005). The MnPs generate Mn(III) ions, which can act to oxidize and degrade a variety of complex substrates, including lignin, and specifically oxidize phenolic structures and aromatic amines (Harms et al., 2011; Falade et al., 2016). Fungal MnPs have been used to break down a variety of common pharmaceutical wastes (Shin et al., 2007; Wen et al., 2010; Golan-Rozen et al., 2011; Rodarte-Morales et al., 2011).

VP enzymes are a class of hemoprotein peroxidases with broad substrate specificity, capable of direct oxidization of Mn^{2+}, methoxybenzenes, phenols, and aromatic compounds. VPs oxidize both phenolic and nonphenolic lignin model dimers, as well as the other substrates, in the absence of manganese (Ruiz-Duenas et al., 2007), due to having multiple active sites and different oxidative abilities with or without manganese (Knop et al., 2016). These features make them interesting for biotechnological applications and the bioremediation of recalcitrant pollutants (Tsukihara et al., 2006; Wong, 2009). They have the traits of both LiP (oxidation of non-phenolic aromatics) and MnP (oxidation of phenolics) (Falade et al., 2016; Eibes et al., 2011). Some studies have shown their use in treating pharmaceutical waste (Taboada-Puig et al., 2011; Eibes et al., 2011; Salame et al., 2012).

Dye-decolorizing peroxidases (DyPs) produced by some basidiomycetes have little sequence similarity to the other peroxidases described in this section. They can oxidize a variety of dyes, including xenobiotic anthraquinone derivatives that other peroxidases rarely oxidize (Hofrichter et al., 2010; Harms et al., 2011; Salvachúa et al., 2013).

Whereas the enzymes described above are all peroxidases requiring H_2O_2, laccases are N-glycosylated extracellular copper-containing oxidases, with a molecular weight range of 58 to 90 kDa (Wells et al., 2006; Murugesan et al., 2006; Mishra and Bisaria, 2006; Zouari-Mechichi et al., 2006; Quaratino et al., 2007), that require O_2 gas but not H_2O_2 for activity. They are produced by a variety of bacteria, green plants, and fungi (Upadhyay et al., 2016; Yang et al., 2017) with widespread occurrence in ascomycetes and basidiomycetes (Harms et al., 2011). Fungal laccases' pH and temperature activity profiles are dependent on the WRF species; however, optimum activities have been reported between pH 2 and 10 and between 40 and 65°C for laccases from different sources (Lu et al., 2017; Ullrich et al., 2005; Zouari-Mechichi et al., 2006; D'Souza et al., 2006; Quaratino et al., 2007; Murugesan et al., 2006). LacI and LacII are two laccase isozymes that have been identified in different fungal species, including *Physisporinus rivulosus*, *Trametes trogii, Cerrena unicolor* 137, and *Panustigrinus.* The isozymes secreted from different species have specific isoelectric points (pI), molecular weights (MW), and pH and temperature optima (Cadimaliev et al., 2005; Lorenzo et al., 2006; Mäkelä et al., 2006; Michniewicz et al., 2006; Zouari-Mechichi et al., 2006). Laccases are able to oxidize phenolic compounds, but in the presence of certain mediator compounds, they can also oxidize non-phenolic aromatic compounds (Upadhyay et al., 2016; Harms et al., 2011). Studies have indicated the potential of fungal laccases to decolorize and detoxify both dye-containing effluents and EDCs (Harms et al., 2011). Studies with the laccases of *Trametes versicolor* and other fungi have shown promise in the bioremediation of several pharmaceutical pollutants (Junghanns et al., 2005; Lloret et al., 2010, 2012; Tran et al., 2010; Strong and Claus, 2011; Margot et al., 2013 Rodríguez-Rodríguez et al., 2012b; Ba et al., 2014a,b; Macellaro et al., 2014; Rodríguez-Delgado et al., 2016; Tahmasbi et al., 2016; Becker et al., 2017).

Other extracellular fungal enzymes also exist and may be important in the metabolism of some pharmaceutical pollutants (Harms et al., 2011; Kües, 2015). A living fungus will use its extracellular enzymes in coordination with its intracellular enzymes to metabolize many organic substrates. For that reason, there may be advantages to using live fungi instead of isolated enzymes from the fungi to remediate pharmaceutical wastes. Using fungi for bioremediation is still an infant field of endeavor, and many things remain to be learned. There may be situations in which immobilized fungal enzymes are more useful for remediating organic wastes than are living fungi (Kües, 2015).

Fungi, including WRF, need intracellular enzymes to function at least as well as their extracellular enzymes. The extracellular and intracellular enzymes coordinate with one another to accomplish different tasks. Among the intracellular enzymes, the O_2-requiring P450 monooxygenases (also called mixed function cytochrome P450 oxidases) are especially interesting. Found in ascomycetes, basidiomycetes, and many other fungal groups, they carry out both biosynthetic and catabolic tasks. They have low substrate specificity and catalyze epoxidations, hydroxylations, and other modifications of many types of organic compounds. They may be at least as important as extracellular enzymes in some bioremediation situations. Cytochrome P450 oxidases have been implicated in enabling fungi to metabolize a variety of pharmaceuticals, including anti-inflammatory drugs, anti-epileptics, anti-analgesics, and lipid regulators (Kües, 2015; Harms et al., 2011). Several studies have established the involvement of cytochrome P450 oxidases in the metabolism of various pollutants, including pharmaceutical waste (Marco-Urrea et al., 2009; Subramanian and Yadav, 2009; Črešnar and Petrič, 2011; Golan-Rozen et al., 2011; Wang et al., 2013; Vasiliadou et al., 2016; Durairaj et al., 2016).

8.4 MODES OF ACTION—TECHNIQUES EMPLOYED TO DATE

The PhACs in the influent of WWTPs present a broad variety of compounds with different concentrations ranging from several milligrams per liter to ultra-trace concentrations of nanograms per liter, as discussed in the previous section. The method employed for analyzing the compounds depends on several factors, including the type and nature of the analyte, the concentration, the type of the employed organism, the source of water, and metabolites, as well as the probable interactions between different interfering compounds in the medium (Comerton et al., 2009). Although the standard methodologies are well established with standard protocols for analyzing the "classic" contaminations, nevertheless, the need for standardized analytical methods exists for trace concentrations of less characterized emerging pharmaceuticals, especially for compounds with greater polarities, which may have toxicological relevance (Snyder et al., 2003). Different direct and indirect analytical methods have been applied for quantification of PhACs and their degradation products in water or polluted soil. Indirect analytical methods involve an extraction procedure (i.e., solid-phase extraction) followed by instrumental analyses.

Different methods have been employed for the extraction of the analyte, including solid-phase extraction and solvent extraction methods, which have been described elsewhere (Rodríguez-Rodríguez et al., 2011, 2012; Marco-Urrea et al., 2009). Several instrumental analyses employed for the determination and quantification of PhAC concentration in the effluent after fungal treatment include gas chromatography coupled with mass spectrometric detection, gas chromatography combustion isotope ratio mass spectrometry (GC-CIRMS), high performance liquid chromatography (HPLC) coupled with UV detector or mass spectrometric detection, ultra-performance liquid chromatography (UPLC), HPLC with diode-array detection (HPLC-DAD), liquid chromatography/electrospray ionization tandem mass spectrometry (LC-ESI-QqQ–MS/MS), nuclear magnetic resonance (NMR) analysis, and a combination of techniques (Marco-Urrea et al., 2009; Snyder et al., 2003; Cruz-Morató et al., 2013; Prieto et al., 2011; Marco-Urrea et al., 2010a). Enzyme activity represents the quantity of the available active enzyme and is measured in enzyme units (U). One unit of enzyme is defined as the amount of enzyme with enzymatic activity capable of catalyzing the conversion of 1 μmol of the substrate per min at a specific temperature (Cajthaml et al., 2009). Enzymatic activities of different enzymes, including MnP, LiP, and laccase, are often measured based on established protocols (Camarero et al., 1999; Rodarte-Morales et al., 2012; Yang et al., 2013; Tien and Kirk, 1988; Zhang and Geißen, 2012).

The toxicity of the medium is often measured by the *Photobacterium phosphoreum* luminescence reduction test (Microtox test). In this method, the bioluminescent bacterium *Vibrio fischeri*, which liberates energy in the form of visible light, is employed for the determination of toxicity. Toxic compounds disrupt the respiratory process of the bacteria, leading to a reduction in the light output. The change in luminescence is proportional to the percentage inhibition of *Vibrio fischeri* and can be directly correlated to toxicity. Microtox tests can be employed for measuring the toxicity of both solid-phase and aqueous substances for chronic or acute toxicity testing (Marco-Urrea et al., 2009; ISO, 2007;Cruz-Morató et al., 2013). The toxicity of sewage sludge or a biosolid can also be measured employing the seed germination toxicity test. In this test, the phytotoxicity of the solid sample is evaluated by seed germination, in which the germination percentage and root length of different incubated seeds are measured and compared with the germination percentage and root length of the same seeds incubated with distilled water for an identical period of time at the same temperature (Rodríguez-Rodríguez et al., 2011).

8.4.1 REACTOR DESIGN FOR PHACS REMOVAL FROM WASTEWATER VIA FUNGAL TREATMENT

Various nutrient media and reactor configurations have been explored for using WRF as a reliable, efficient, and rapid way of removing PhACs from wastewater effluent. The choice of the reactor configuration and design parameters depend on several factors affecting the performance and the activity of the employed organism. Detailed study of the process is required to determine growth parameters of the fungi, such as the optimal temperature, operational pH, dissolved oxygen concentration of the effluent, fixed nitrogen concentration, reaction kinetics, reaction rate and yield, and construction materials. The control over the process and the process economic factors are also substantial factors that are important for scaling-up purposes.

To date, most of the practices on removal of PhACs via fungal treatment have been carried out in *in vitro* sterilized conditions, employing synthetic liquid media and controlled conditions of pH and temperature. Often, spiked concentrations of the compounds (milligrams per liter) have been used, which might not reflect the conditions in real WWTP influent. A few studies have been performed under non-sterile conditions where many contaminants and microorganisms are also present, and consequently, the removal efficiency of fungi may be hindered. Zhang and Geiben (2012) have reported 80% removal of carbamazepine (spiked at 5 mg L^{-1}) in non-sterile conditions, employing *P. chrysosporium* immobilized in polyether foam . Cruz-Morató et al. (2013) reported successful removal of 7 out of 10 PhACs initially screened in non-sterile conditions. However, the authors reported comparably lower biomass concentrations in non-sterile conditions and concluded that the practical use of fungi for real wastewater treatment practices needs a supply of nutrients (glucose and nitrogen). Control over pH to ensure the optimal biological activities and enzymatic production of the fungi is also needed.

In addition, efficient dye removal from textile wastewaters via fungal treatment has been reported in non-sterile conditions by different authors (Hai et al., 2009; Anastasi et al., 2010). For fungal treatment of wastewater, three types of reactors can be identified with regard to the mode of reactant contact: batch reactors, semi-batch reactors, and flow reactors. Descriptions of their characteristics and operation are outside the scope of this chapter but may be found in many texts on chemical reaction engineering (Levenspiel and Levenspiel, 1972). The use of different reactor configurations, including continuous and batch stirred reactors (Rodarte-Morales et al., 2012), fluidized bed, packed bed, and perfusion basket reactors for the removal of different pharmaceuticals from wastewater via fungal treatment has proved the possibility of scaling up the process (Cruz-Morató et al., 2012). In fluidized bed modes, often the fungal biomass is fluidized by the aid of air or oxygen pulses generated by an electro-valve (Cruz-Morató et al., 2013).

It is believed that the aeration regime also plays a role in the performance of fungal reactors. Moreira et al. (1996) demonstrated that the pulsation of oxygen controls the shape and the growth of the fungal pellets in the fluidized bed reactor. Rodarte-Morales et al. (2012) have studied the effect of two aeration regimes, air and oxygen pulses, on the performance and morphology of *P. chrysosporium* pellets in a fed-batch reactor. The authors concluded that oxygen pulsation results in high elimination of pharmaceuticals (up to 99%), controls the growth of the pellets, and facilitates the operation of the reactor. This aeration regime resulted in the satisfactory elimination of different PhAC compounds. However, excessive growth of the pellets has also been reported. The result shows the possibility of employing less expensive aeration regimes in the development of viable reactors for fungal treatment (Rodarte-Morales et al., 2012). The successful employment of an aeration regime has also been reported for other industrial wastewaters, that is, dye removal, via fungal treatment (Cruz-Morató et al., 2014).

Batch reactor configuration has been employed extensively for PhAC removal from wastewater via fungal treatment for decades. Rodarte-Morales et al. (2011) reported complete removal of diclofenac, NPX, and IBP (spike concentration of 1 mg L^{-1}) by *P. chrysosporium* in a fed-batch stirred reactor with an air and oxygen supply by 24 h. The successful removal of triclosan (5 mg L^{-1}) in batch bed-packed reactors has also been reported for immobilized laccase from *Coriolopsis polyzona* on a diatomaceous earth support (Celite® R-633—contact time of 200 min—five cycles [Cabana et al., 2009]), and *T. versicolor* (conjugation of laccase with the biopolymer chitosan, 100% removal after 6 h [Cabana et al., 2011]). Immobilized laccase from *Myceliophthora thermophila* on a sol-gel matrix has been tested for removal of estrogens in both batch stirred tank reactors (BSTRs) (operating in cycles) and a continuous photobioreactor (PBR). The removal of estrogen was reported as >85% in the BSTR and in the range of 55–75% in the continuous PBR (Lloret et al., 2011; Taboada-Puig et al., 2011). However, a short removal duration of 1 h has also been reported by Auriol et al. for the complete removal of estrogens in municipal wastewater employing immobilized laccase with activities of 20,000 U L^{-1} in batch reactors (Auriol et al., 2008). The batch reactor configuration has also been employed for the fungal treatment with near to complete removal efficiencies of different wastewater effluents, including domestic sewage (Thanh and Simard, 1973), starch processing effluent (Coulibaly et al., 2003), metal-contaminated effluent (Kapoor and Viraraghavan, 1998), distillery wastewaters (Kumar et al., 1998), and pulp and paper processing wastewater (Sumathi and Phatak, 1999). Continuous or flow reactor configuration is less explored compared with the batch configuration, but examples of employing this configuration for the decolorization of domestic sewage (Miyata et al., 2000), distillery wastewater (Miyata et al., 2000), and wood processing wastewater (Manzanares et al., 1995) have been reported.

8.5   ADVANCED TECHNIQUES FOR TREATMENT OF PHARMACEUTICALS IN EFFLUENTS

Sewage sludge is a byproduct of wastewater treatment, which results from the biological treatment of municipal sewage and is often used for cement production, composting, or land improvement (Rodríguez-Rodríguez et al., 2012a). The last option, called *biosolids*, is often considered as the most preferable choice, as it contributes to the recycling of nutrients as well as increasing the fertility of the land. Raw sewage sludge is treated to meet specific regulations for microbial pathogens and other classic pollutants before being applied as a land improver (Henry and Cole, 1997). However, the current regulations mostly focus on pathogens and metal reduction, and only a limited number of organopollutants are considered (i.e., halogenated compounds and dioxins), as shown in the EU's Working Document on Sludge (Rodríguez-Rodríguez et al., 2012a; Fytili and Zabaniotou, 2008). On the one hand, it is well known that pharmaceuticals can accumulate in sewage sludge at high concentrations, and legislation for most groups of emerging pollutants (i.e., pharmaceuticals and polybrominated diphenyl ethers [PBDEs]) does not exist (Rodríguez-Rodríguez et al., 2012a). Therefore, municipal biosolids may contain large amounts of PhACs, which can be a potential risk to the environment, counteracting the expected benefits of the biosolids applied to the land (Kinney et al., 2006; Rodríguez-Rodríguez et al., 2011).

Rodríguez-Rodríguez et al. (2011, 2012a) analyzed the pharmaceuticals in raw sewage sludge. The authors reported the presence of 40 out of 43 screened pharmaceuticals in the sludge. Some laboratory-scale studies have shown the possibility of employing naturally degrading microorganisms, specifically WRF, for the removal of spiked pharmaceuticals and organic pollutants from solid lignocellulosic wastes and biosolids (García-Galán et al., 2011; Rodríguez-Rodríguez et al., 2012a). For example, Rodríguez-Rodríguez et al. employed *T. versicolor* for the treatment of raw sewage sludge and reported the complete removal of phenazone, bezafibrate, fenofibrate, cimetidine, clarithromycin, sulfamethazine, and atenolol and partial removal of other pharmaceuticals (between 42% and 80%). In addition, the toxicity of the sludge was substantially reduced after the treatment (Rodríguez-Rodríguez et al., 2011).

Advanced oxidation processes (AOPs), or aqueous phase oxidation methods, are emerging technologies that work because of the intermediacy of highly reactive radicals, such as $\cdot OH$. By nature, radicals are compounds with unpaired electrons that oxidize other substances to acquire electrons. Often, these radicals are produced via chemical, photochemical, photocatalytical, or electrochemical methods (Marco-Urrea et al., 2010c). Recently, the ability of several WRFs, including *T. versicolor* and *Pleurotus eringyi*, to produce these radicals through a quinone redox cycling mechanism, which is called *bio-oxidation* (Rodríguez-Rodríguez et al., 2012a), has been shown. In this novel method, the fungi are incubated with a lignin-derived quinone and chelated ferric ion. Subsequently, the fungi catalyze the conversion of the quinone into hydroquinone, and oxidize it to produce semiquinone radicals by their lignin-modifying enzymes. By autoxidation, which has been catalyzed by Fe^{3+}, Fenton's reagent is produced, which results in Fe^{2+} and O_2^- radicals. O_2^- then reacts, and $\cdot OH$ radicals and/or other oxidizing species (ferryl ion, under certain conditions of pH and concentration of organic and inorganic ligands [Hug and Leupin, 2003]) are released (Marco-Urrea et al., 2010c; Gómez-Toribio et al., 2009a,b). Marco-Urrea et al. have reported the successful use of this method for more than 80% removal of a spiked concentration of 10 mg L^{-1} of clofibric acid, carbamazepine, atenolol, and propranolol after 6 h of incubation (Marco-Urrea et al., 2010c).

A combination of fungal treatment and filtration yields higher PhAC removal and is feasible by employing fungal treatment with MBRs. These novel MBRs have a granular activated carbon-packed anaerobic zone beneath their aerobic zone, which contains the membrane module and the microorganisms (Hai et al., 2011).

This single-step technique results in a compact process with high flexibility, which can be fine-tuned for different biological applications. Recently, the removal of different trace organic pollutants employing a fungal MBR inoculated with a pure culture of *T. versicolor* has been tested, and the complete removal of bisphenol A and diclofenac in sterile batch tests was reported (Yang et al., 2013).

8.6   PERSPECTIVES, DRAWBACKS, AND LIMITATIONS

The study of the fate of PhACs in the environment is still a fairly new topic. PhACs are known as recalcitrant compounds, and the current WWTPs fail to remove them effectively (Carballa et al., 2004). The concentration of these compounds in marine ecosystems is increasing around the globe. The term *biopharmaceuticals* refers to drugs produced by means other than direct extraction from a natural source. Very few studies have been performed on these compounds, as the product is often considered to be degraded quickly. Thus, the environmental relevance of these compounds is not yet clear. The contamination potential of PhACs, along with their contribution to severe ecotoxicity and human health problems, necessitates the development of cost-effective and efficient methods for the elimination of PhACs from the environment (Prieto et al., 2011; Asgher et al., 2008; Yang et al., 2013). The severity of the potential risks of the PhACs to the ecosystem is dependent on the vulnerability of the contaminated ecosystem (Rodarte-Morales et al., 2012; Daughton and Ternes, 1999; Kümmerer, 2008). Therefore, high-priority PhACs and vulnerable ecosystems must be identified (Hester and Harrison, 2015).

It is believed that antibiotic resistance genes develop when bacteria are exposed to sublethal doses of antibiotics. The presence of antibiotics in wastewater discharges and the marine ecosystem is believed to be associated with the development of these antibacterial-resistant pathogen bacteria. Bioremediation methods based on the WRF and their oxidative enzymes are a green alternative to conventional sewage sludge wastewater treatment technologies, and it appears that they may achieve higher overall degradation yields of antibiotics than the conventional technologies. Fungal treatment has been shown to be very promising in terms of the removal of antibiotics in the influent. In addition, the successful use of WRF has been reported for treating contaminated solid substrates. Treated sewage sludge that is applied for land improvement must meet specific regulations for microbial pathogens and other classic pollutants before being used in land improvement applications (Henry and Cole, 1997). However, the current regulations are mostly focused on the reduction of pathogens and metals (Rodríguez-Rodríguez et al., 2012a; Fytili and Zabaniotou, 2008). It is well known that pharmaceuticals can be accumulated in sewage sludge in high concentrations. Legislation related to pharmaceuticals does not exist (Rodríguez-Rodríguez et al., 2012a). Similar legislation must be developed to address the safe content of pharmaceuticals in biosolids. WRF are promising candidates for the removal of pharmaceuticals from sewage sludge, especially from industrial or hospital WWTPs.

It is well known that molecules resulting from the parent compounds due to structural change may have environmental relevance. The presence of structural transformations (due to partial degradation) or metabolites has been reported after wastewater treatment practices. Little is known about the occurrence, fate, or activity of transformed compounds and metabolites in the environment. However, these transformed compounds, which are generally considered to be less toxic, might be even more toxic (in the case of pro-drugs or transformations after oxidation processes) than the parent compound. However, in some cases, structural transformations or metabolites are not even analyzed or reported (Kümmerer, 2009) as the transformed compounds. The new molecule might not be recognized by the employed analytical method and simply reported as removed. Therefore, special care must be taken when these compounds are present. Coupling toxicity tests with the analytical test is therefore necessary when analyzing the effectiveness of treatment practices, including fungal treatment.

To date, most of the practices for the removal of PhACs via fungal treatment have been carried out in *in vitro* sterilized conditions employing synthetic liquid media and controlled conditions of pH and temperature. The results of these studies might be irrelevant for real wastewater practices, in which a community of microorganisms is growing and possibly interfering with the fungal treatment. In addition, the pH and temperature that are optimal for the fungi may change the structure of the parent compound and make these applications impractical. Spiked concentrations of the compounds (milligrams per liter) have been used, which might not reflect the conditions in real WWTP influent. Therefore, the degradation yields that are reported in those studies might not be achieved for real WWTP effluents, in which the concentration of the substance is in the range of nanograms per liter.

The other concern regarding the successful implementation of fungal treatment relates to scale-up considerations and reactor design factors. Various reactor configurations have been suggested for the use of WRF for reliable, efficient, and rapid removal of PhACs from the wastewater effluent. The choice of the reactor configuration and design parameters depends on several factors affecting the performance and activity of the employed organism. A detailed study of the process is required for determination of parameters such as the optimal temperature, operational pH, dissolved oxygen concentration of the effluent and fixed nitrogen concentration, reaction kinetics, effect of physical phenomena, reaction rate and yield, and materials of construction. The control over the process and the process economic factors are also substantial factors, which are important for the purpose of scaling up. Combinations and interactions of these factors make the scale-up study more complex. In addition, on the one hand, the flow of the municipal WWTPs' influent is huge, and on the other hand, the fungal treatment of the effluents would be a slow process, requiring a high hydraulic retention time in the fungal bioreactor. In addition, the oxygen supply or aeration that is required in the reactors imposes extra costs on the process. Therefore, the implementation of fungal treatment for the removal of pharmaceuticals from municipal wastewater at the current WWTPs might not be practical. However, for the effluent of pharmaceutical manufacturers or hospitals, or industrial sewage, the use of fungal treatment could be practical, as the concentration of the PhACs in the influent is high, and the volume is comparably lower than the influent of the WWTPs.

8.7 SUSTAINABILITY, ENVIRONMENTAL CONCERNS, RISK ASSESSMENTS, AND RISK MANAGEMENT

Currently, fungal wastewater treatment practices are being studied for further reduction in the pathogenic effects as well as the eco-toxicity of the effluents. However, from the sustainability point of view, if one of the main functions of WWTPs is to minimize the impact on the environment, other factors should be considered and studied as well as the PhAC removal efficiency of the applied method (Pasqualino et al., 2009). It is believed that further improvements in the quality of WWTPs' effluents impose an inevitable environmental load on the process, and eventually, at some point, there will be a "break-even" between the environmental benefits to be gained by the removal of PhACs and the impacts induced by this treatment. This environmental break-even point has been reported for some studied treatment scenarios, in which the imposed environmental impact of the treatment exceeded its benefits (Wenzel et al., 2008). Different wastewater treatment systems have different performance characteristics and consequently, impose certain loads on the environment. On the other hand, increasingly sophisticated improvements in treatment methods result in more environmental load (Wang et al., 2012), which means that the environmental profiles are case specific. In addition to seeking further improvements in water quality, we should look for strategies to reduce the environmental impacts. Therefore, the environmental performance of the wastewater practice should be carefully assessed before proposing its implementation in the WWTP to reduce the resource use, energy consumption, and environmental emissions attributed to the process. Often, the environmental loads are neglected due to the regulatory push for acquiring higher water quality. To establish an ecologically sustainable WWTP, a comprehensive assessment of available options is necessary to meet different standards from a life cycle perspective (Wang et al., 2012), and the environmental performance of the employed wastewater treatment method can be assessed by life cycle assessment (LCA) methodology (Coats et al., 2011).

8.7.1 LCA Factors

The economic, environmental, and social dimensions of sustainability need to be addressed in assessing the sustainability of a project or product. A scientifically based sustainability analysis also involves value judgments, assumptions, scenarios, and uncertainties. LCA is a method that is applied to compare, contrast, and evaluate environmental impacts of processes by identifying energy and resource use and various emissions resulting from a particular life cycle (Coats et al., 2011). Generally, LCA consists of four steps: "goal and scope," "life cycle inventory (LCI)," "life cycle impact assessment," and "interpretation". LCA also provides valuable information to identify the "priority" areas for improvements, where actions would have the greatest effect on reducing the corresponding environmental burdens (Hossain et al., 2008; Rebitzer et al., 2004). LCA is typically restricted to environmental aspects, while sustainable assessment (SA) is a broader concept and covers more dimensions or aspects than LCA. Therefore, to "broaden" the scope of LCA, the social and economic dimensions are added to the environmental aspects. The life cycle sustainability assessment (LCSA) covers these aspects and includes LCA, life cycle costing (LCC), and social life cycle analysis (SLCA) (Hossainet al., 2008; Rebitzer et al., 2004). Typical environmental effects that can be assessed using LCA include global warming, the formation of photo-chemical oxidants, acidification, eutrophication, and fossil fuel use. The photochemical oxidant formation impact category reflects the effects of photo-oxidant emissions derived during the process. Acidification reflects the emission of acidifying substances, and eutrophication is defined as the ecosystem response to the addition of artificial or natural substances, such as nitrates and phosphates, through fertilizers or sewage, to an aquatic system; that is, ground-water reservoirs or streams (Coats et al., 2011; Mu et al., 2010; Spatari et al., 2010).

The environmental impacts of the conventional wastewater treatment methods with different configurations have been investigated thoroughly using the LCA technique (Foley et al., 2010; Buonocore et al., 2016; Corominas et al., 2013; Yıldırım and Topkaya, 2012). Assessing different technology configurations, several authors reported that the energy consumption has the main influence on the overall environmental profile (Foley et al., 2010), followed by the impact of water discharge to rivers and sludge application to land (Pasqualino et al., 2009). To the best of the authors' knowledge, there is no LCA study existing so far on the fungal treatment of the wastewater. In addition, the LCA studies on similar biologic treatments are very scarce and limited in scope in terms of the defined boundaries, excluding solids handling, and the process configurations considered (Foley et al., 2010; Zang et al., 2015). The environmental performance of wastewater sludge treatment alternatives, considering the sludge handling, disposal processes, and potential benefits resulting from the replacement of synthetic fertilizers by biosolids, has also been investigated (Yıldırım and Topkaya, 2012; Foley et al., 2010; Yoshida et al., 2013; Concepción et al., 2012). However, other sludge management techniques have also been suggested for sewage sludge, including incineration with the final destruction of organic substances and deactivation of the sludge (Wenzel et al., 2008).

8.7.2   SUGGESTIONS FOR IMPLEMENTATIONS OF SUCCESSFUL LCA ON FUNGAL TREATMENT OF PHARMACEUTICALS

Significant uncertainties exist in LCA studies of emerging technologies, which are very likely to increase the complexity of decision-making processes. In LCA, serious difficulties can arise during the evaluation phase; that is, when the effect scores of different impact categories are weighed against each other. In addition, LCA is a highly data-intensive method, and its success depends on the availability of precise data, which is still an issue in state-of-the-art processes such as fungal treatment of wastewater. LCI data are still not available for many PhAC chemicals, particularly for transformed chemicals. To conduct a successful LCA on the fungal treatment of pharmaceuticals, and due to the process complexity, several simplifications and assumptions are necessary. The first concern would be the LCI step. LCI, which is typically the most time-consuming phase of an LCA study, imposes the greatest uncertainty.

There are different strategies to simplify the inventory analysis, depending on the goal and scope of the study, the required level of detail, the acceptable level of uncertainty, and the available resources (time, human resources, know-how, and budget). Another simplification strategy is to reduce the effort for the LCI phase by applying different cut-offs (i.e., deliberately excluding the effect of transformation chemicals from the inventory analysis). Eventually, the simplified LCA should still give the same insights for a given study as a detailed LCA, although at lower resolution. To conduct a successful LCA, it is necessary to acquire the following information in the following areas: process information, materials information, equipment information, and water-sewage management. Process information includes the data from process modeling, flow through the fungal bioreactor, process parameters such as hydraulic retention time, kinetic data and removal efficiencies, correlation between the parameters (assumptions), and the energy use in different sectors. Material information presents the data regarding the physical and chemical properties of the employed materials, toxicity of materials and regulatory limits, life cycle toxicity, and eventually, energy use and the waste management techniques employed. Equipment information can be obtained from dynamic simulation, current techno-economic and design reports of similar technologies, and energy balances.

To sum up, the implementation of fungal treatment in WWTPs is interesting from different sustainability aspects. Generally, employing biological treatments (for phosphorus removal) results in a considerably lower amount of sludge to handle. In addition, the biological sludge is considered to be less toxic compared with the chemical sludge, with lower environmental impact (Coats et al., 2011). However, fungal treatment is not necessarily environmentally benign. It is dependent on so many factors, most of which are unknown. On the other hand, in areas of high population with water scarcity, the importance of supplying water with high quality (for drinking) might supersede the environmental relevance impacts. As a part of water management strategies, the environmental criteria must be considered to ensure that the water is used rationally (either reclaimed water or fresh water) and is returning to the environment in an acceptable condition (Amores et al., 2013; Meneses et al., 2010).

8.9   CONCLUSION

Pharmaceuticals and pharmaceutical compounds in WWTP effluent come from different sources, including agroindustrial processes, pharmaceutical manufacturing, livestock farms, fisheries, shrimp hatcheries, and hospitals. Acute toxic effects of these compounds are unlikely, as the concentrations of these compounds in the environment are in the range of nanograms to several milligrams per liter. However, the potential risk of chronic effects cannot be neglected. The extensive contamination potential of PhACs necessitates the development of cost-effective and efficient methods for the elimination of PhACs from effluent. Conventional WWTPs fail to efficiently remove PhACs. Fungal biological treatment is an emerging alternative treatment for these compounds in WWTP effluent. WRF have tremendous potential for the removal of a broad spectrum of pharmaceutical and industrial pollutants. Their degradation capability, which varies between different WRF species, is due to the extracellular and non-specific nature of the enzyme system of WRF: LiP,

MnP, VP, and laccase along with other accessory enzymes. The PhACs in the influent of WWTPs present a broad variety of compounds with different concentrations. Different groups of pharmaceuticals, including anti-inflammatory drugs, analgesic drugs, psychotropic drugs, lipid regulators, antibiotics, β-blockers, estrogens, and iodinated contrast media can be degraded effectively by WRF (see Table 8.1) with various reactor configurations.

TABLE 8.1
Structure of Assessed Pharmaceuticals

Drug Family	Drug	Molar Mass (g mol^{-1})	Solubility in Water (mg L^{-1})	Reference
Anti-inflammatory	Ibuprofen $C_{13}H_{18}O_2$	206.29	21 (at 25°C)	Yalkowsky and Dannenfelser (1992)
	Diclofenac $C_{14}H_{11}Cl_2NO_2$	296.15	2.37 (at 25°C)	Fini et al. (1986)
	Naproxen $C_{14}H_{14}O_3$	230.26	15.9 (at 25°C)	Yalkowsky and Dannenfelser (1992)
	Fenoprofen $C_{15}H_{14}O_3$	242.27	<0.005 (at 25°C)	Yalkowsky and Dannenfelser (1992)
	Indomethacin $C_{19}H_{16}ClNO_4$	357.79	0.937 (at 25°C)	Yalkowsky and Dannenfelser (1992)
	Ketoprofen $C_{16}H_{14}O_3$	254.28	51 (at 22°C)	Yalkowsky and Dannenfelser (1992)
	Mefenamic $C_{15}H_{15}NO_2$	241.29	20 (at 30°C)	Yalkowsky and Dannenfelser (1992)
	Caffeine $C_8H_{10}N_4O_2$	194.19	21,600 (at 25°C)	Yalkowsky and Dannenfelser (1992)
	Propyphenazone $C_{14}H_{18}N_2O$	230.31	51,900 (at 25°C)	Yalkowsky and Dannenfelser (1992)
Psychiatric drugs	Carbamazepine $C_{15}H_{12}N_2O$	236.27	152 (at 25°C)	Yalkowsky and Dannenfelser (1992)
	Diazepam $C_{16}H_{13}ClN_2O$	284.71	50 (at 25°C)	Yalkowsky and Dannenfelser (1992)
	Fluoxetine $C_{17}H_{18}F_3NO$	309.33	50,000 at 25°C	Yalkowsky and Dannenfelser (1992)
	Citalopram $C_{20}H_{21}FN_2O$	324.39	4,000 (at 25°C)	Yalkowsky and Dannenfelser (1992)
Antibiotics	Cinoxacin $C_{12}H_{10}N_2O_5$	262.22	961 (at 25°C)	Yalkowsky and Dannenfelser (1992)
	Ciprofloxacin $C_{17}H_{18}FN_3O_3$	331.35	1,350 (at 25°C)	Yalkowsky and Dannenfelser (1992)
	Enrofloxacin $C_{19}H_{22}FN_3O_3$	359.41	146 (at 25°C)	Yalkowsky and Dannenfelser (1992)
	Erythromycin $C_{37}H_{67}NO_{13}$	733.94	2,000 (at 28°C)	Pharmacopoeia (2002)

(Continued)

TABLE 8.1 (CONTINUED)
Structure of Assessed Pharmaceuticals

Drug Family	Drug	Molar Mass (g mol^{-1})	Solubility in Water (mg L^{-1})	Reference
	Norfloxacin $C_{16}H_{18}FN_3O_3$	319.33	1.78×10^5 (at 25°C)	Yalkowsky and Dannenfelser (1992)
	Flumequine $C_{14}H_{12}FNO_3$	261.25	1,240 (at 25°C)	Yalkowsky and Dannenfelser (1992)
	Sarafloxacin $C_{20}H_{17}F_2N_3O_3$	385.36	105 (at 25°C)	Yalkowsky and Dannenfelser (1992)
	Sulfamethazine $C_{12}H_{14}N_4O_2S$	278.33	1,500 (at 29°C)	Herzfeldt and Kümmel (1983)
	Sulfathiazole $C_9H_9N_3O_2S_2$	255.32	373 (at 25°C)	Yalkowsky and Dannenfelser (1992)
	Sulfapyridine $C_{11}H_{11}N_3O_2S$	249.29	268 (at 25°C)	Yalkowsky and Dannenfelser (1992)
	Sulfamethoxazole $C_{10}H_{11}N_3O_3S$	253.28	610 (at 37°C)	Yalkowsky and Dannenfelser (1992)
Lipid regulators	Clofibric acid $C_{10}H_{11}ClO_3$	214.65	29 (at 25°C)	Pharmacopoeia (2002)
	Octyl salicylate $C_{15}H_{22}O_3$	250.33	10,000 (at 25°C)	Pharmacopoeia (2002)
β-blockers	Atenolol $C_{14}H_{22}N_2O_3$	266.34	13,300 (at 25°C)	McFarland et al. (2001)
	Propranolol $C_{16}H_{21}NO_2$	259.34	61.7 (at 25°C)	McFarland et al. (2001)
Estrogens	17β-estradiol $C_{18}H_{24}O_2$	277.41	3.6 (at 27°C)	Yalkowsky and Dannenfelser (1992)
	17α-ethinylestradiol $C_{20}H_{24}O_2$	296.40	11.3 (at 27°C)	Yalkowsky and Dannenfelser (1992)

The choice of the reactor configuration and design parameters depends on several factors affecting the performance and the activity of the employed organism. Developments in the emerging fungal treatment of pharmaceuticals in WWTP effluent are spectacular from a technological point of view. However, these biological methods still suffer from different limitations and technological gaps and have a long way to go. Little is known about the fate of PhACs and biopharmaceuticals in the environment. In addition, the occurrence, fate, or activity of transformed compounds and metabolites (which might be more toxic) in the environment is also not well known. The other concern is that the results of fungal treatment practices have often been obtained from *in vitro* sterilized conditions, employing synthetic liquid media and controlled pH and temperature; these results might be irrelevant for real wastewater treatment practices, in which a community of microorganisms is growing and possibly interfering with the fungal treatment, and may not be practical for scaling up systems.

REFERENCES

Accinelli, C., M. L. Saccà, I. Batisson, J. Fick, M. Mencarelli, and R. Grabic. 2010. Removal of oseltamivir (Tamiflu) and other selected pharmaceuticals from wastewater using a granular bioplastic formulation entrapping propagules of Phanerochaete chrysosporium. *Chemosphere* 81 (3): 436–443.

Adhoum, N. and L. Monser. 2004. Decolourization and removal of phenolic compounds from olive mill wastewater by electrocoagulation. *Chemical Engineering and Processing: Process Intensification* 43 (10): 1281–1287.

Amores, M. J., M. Meneses, J. Pasqualino, A. Antón, and F. Castells. 2013. Environmental assessment of urban water cycle on Mediterranean conditions by LCA approach. *Journal of Cleaner Production* 43: 84–92.

Anastasi, A., F. Spina, V. Prigione, V. Tigini, P. Giansanti, and G. C. Varese. 2010. Scale-up of a bioprocess for textile wastewater treatment using Bjerkandera adusta. *Bioresource Technology* 101 (9): 3067–3075.

Asgher, M., S. A. Shah, M. Ali, and R. L. Legge. 2006. Decolorization of some reactive textile dyes by white rot fungi isolated in Pakistan. *World Journal of Microbiology and Biotechnology* 22 (1): 89–93.

Asgher, M., M. J. Asad, H. N. Bhatti, and R. L Legge. 2007. Hyperactivation and thermostabilization of Phanerochaete chrysosporium lignin peroxidase by immobilization in xerogels. *World Journal of Microbiology and Biotechnology* 23 (4): 525–531.

Asgher, M., H. N. Bhatti, M. Ashraf, and R. L. Legge. 2008. Recent developments in biodegradation of industrial pollutants by white rot fungi and their enzyme system. *Biodegradation* 19 (6): 771.

Asif, M. B., F. I. Hai, L. Singh, W. E. Price, and L. D. Nghiem. 2017. Degradation of pharmaceuticals and personal care products by white-rot fungi—A critical review. *Current Pollution Reports* 3(2): 1–16.

Auriol, M., Y. Filali-Meknassi, C. D. Adams, R. D. Tyagi, T.-N. Noguerol, and B. Pina. 2008. Removal of estrogenic activity of natural and synthetic hormones from a municipal wastewater: Efficiency of horseradish peroxidase and laccase from Trametes versicolor. *Chemosphere* 70 (3): 445–452.

Ba, S., L. Haroune, C. Cruz-Morató, C. Jacquet, I. E. Touahar, J.-P. Bellenger, C. Y. Legault, J. P. Jones, and H. Cabana. 2014a. Synthesis and characterization of combined cross-linked laccase and tyrosinase aggregates transforming acetaminophen as a model phenolic compound in wastewaters. *Science of the Total Environment* 487: 748–755.

Ba, S., J. P. Jones, and H. Cabana. 2014b. Hybrid bioreactor (HBR) of hollow fiber microfilter membrane and cross-linked laccase aggregates eliminate aromatic pharmaceuticals in wastewaters. *Journal of Hazardous Materials* 280: 662–670.

Baborová, P., M. Möder, P. Baldrian, K. Cajthamlová, and T. Cajthaml. 2006. Purification of a new manganese peroxidase of the white-rot fungus Irpex lacteus, and degradation of polycyclic aromatic hydrocarbons by the enzyme. *Research in Microbiology* 157 (3): 248–253.

Batt, A. L., S. Kim, and D. S. Aga. 2007. Comparison of the occurrence of antibiotics in four full-scale wastewater treatment plants with varying designs and operations. *Chemosphere* 68 (3): 428–435.

Becker, D., S. Rodriguez-Mozaz, S. Insa, R. Schoevaart, D. Barceló, M. De Cazes, M.-P. Belleville, J. Sanchez-Marcano, A. Misovic, and J. Oehlmann. 2017. Removal of endocrine disrupting chemicals in wastewater by enzymatic treatment with fungal laccases. *Organic Process Research & Development* 21 (4): 480–491.

Bishnoi, N. R. 2005. Fungus—an alternative for bioremediation of heavy metal containing wastewater: A review. *Journal of Scientific & Industrial Research* (64): 93–100.

Blánquez, P. and B. Guieysse. 2008. Continuous biodegradation of 17β-estradiol and 17α-ethynylestradiol by Trametes versicolor. *Journal of Hazardous Materials* 150 (2): 459–462.

Boxall, A., D. Kolpin, B. Holling-Sorensen, and J. Tolls. 2003. Are veterinary medicines causing environmental risks? *Environmental Science & Technology* 37(15): 286A–294A.

Buonocore, E., S. Mellino, G. De Angelis, G. Liu, and S. Ulgiati. 2016. Life cycle assessment indicators of urban wastewater and sewage sludge treatment. *Ecological Indicators*, http://dx.doi.org/10.1016/j.ecolind.2016.04.047.

Cabana, H., J. Peter Jones, and S. N. Agathos. 2009. Utilization of cross-linked laccase aggregates in a perfusion basket reactor for the continuous elimination of endocrine-disrupting chemicals. *Biotechnology and Bioengineering* 102 (6): 1582–1592.

Cabana, H., A. Ahamed, and R. Leduc. 2011. Conjugation of laccase from the white rot fungus Trametes versicolor to chitosan and its utilization for the elimination of triclosan. *Bioresource Technology* 102 (2): 1656–1662.

Cadimaliev, D. A., V. V. Revin, N. A. Atykyan, and V. D. Samuilov. 2005. Extracellular oxidases of the lignin-degrading fungus Panus tigrinus. *Biochemistry (Moscow)* 70 (6): 703–707.

Cajthaml, T., Z. Křesinová, K. Svobodová, and M. Möder. 2009. Biodegradation of endocrine-disrupting compounds and suppression of estrogenic activity by ligninolytic fungi. *Chemosphere* 75 (6): 745–750.

Camarero, S., S. Sarkar, F. J. Ruiz-Dueñas, M. J. Martínez, and Á. T. Martínez. 1999. Description of a versatile peroxidase involved in the natural degradation of lignin that has both manganese peroxidase and lignin peroxidase substrate interaction sites. *Journal of Biological Chemistry* 274 (15): 10324–10330.

Carballa, M., F. Omil, J. M. Lema, M. Llompart, C. García-Jares, I. Rodríguez, M. Gomez, and T. Ternes. 2004. Behavior of pharmaceuticals, cosmetics and hormones in a sewage treatment plant. *Water Research* 38 (12): 2918–2926.

Cheng, X-B., R. Jia, P-S. Li, Q. Zhu, S-Q. Tu, and W-Z. Tang. 2007. Studies on the properties and co-immobilization of manganese peroxidase. *Chinese Journal of Biotechnology* 23 (1): 90–96.

Clara, M., B. Strenn, O. Gans, E. Martinez, N. Kreuzinger, and H. Kroiss. 2005. Removal of selected pharmaceuticals, fragrances and endocrine disrupting compounds in a membrane bioreactor and conventional wastewater treatment plants. *Water Research* 39 (19): 4797–4807.

Coats, E. R., D. L. Watkins, and D. Kranenburg. 2011. A comparative environmental life-cycle analysis for removing phosphorus from wastewater: Biological versus physical/chemical processes. *Water Environment Research* 83 (8): 750–760.

Comerton, A. M., R. C. Andrews, and D. M. Bagley. 2009. Practical overview of analytical methods for endocrine-disrupting compounds, pharmaceuticals and personal care products in water and wastewater. *Philosophical Transactions of the Royal Society of London A: Mathematical, Physical and Engineering Sciences* 367 (1904): 3923–3939.

Concepción, H., M. Meneses, and R. Villanova. 2012. Environmental analysis of wastewater treatment plants control strategies. Paper read at Control & Automation (MED), 2012 20th Mediterranean Conference on Control & Automation (MED 2012) 2012 Jul 3 (pp. 752–757). IEEE. Barcelona, Spain.

Corominas, L., J. Foley, J. S. Guest, A. Hospido, H. F. Larsen, S. Morera, and A. Shaw. 2013. Life cycle assessment applied to wastewater treatment: State of the art. *Water Research* 47 (15): 5480–5492.

Coulibaly, L., G. Gourene, and N. Spiros Agathos. 2003. Utilization of fungi for biotreatment of raw wastewaters. *African Journal of Biotechnology* 2 (12): 620–630.

Crane, M., C. Watts, and T. Boucard. 2006. Chronic aquatic environmental risks from exposure to human pharmaceuticals. *Science of the Total Environment* 367 (1): 23–41.

Črešnar, B., and Š. Petrič. 2011. Cytochrome P450 enzymes in the fungal kingdom. *Biochimica et Biophysica Acta (BBA)-Proteins and Proteomics* 1814 (1): 29–35.

Cruz-Morató, C., C. E. Rodríguez-Rodríguez, E. Marco-Urrea, M. Sarrà, G. Caminal, T. Vicent, A. Jelić, M. J. García-Galán, S. Pérez, and M. S. Díaz-Cruz. 2012. Biodegradation of pharmaceuticals by fungi and metabolites identification. In *Emerging Organic Contaminants in Sludges*, Vincent, T, Caminal, G, Eljarrat, E, and Barceló, D (ed). 165–213. Springer, Berlin, Germany.

Cruz-Morató, C., L. Ferrando-Climent, S. Rodriguez-Mozas, D. Barceló, E. Marco-Urrea, T. Vicent, and M. Sarrà. 2013. Degradation of pharmaceuticals in non-sterile urban wastewater by Trametes versicolor in a fluidized bed bioreactor. *Water Research* 47 (14): 5200–5210.

Cruz-Morató, C., D. Lucas, M. Llorca, S. Rodriguez-Mozas, M. Gorga, M. Petrovic, D. Barceló, T. Vicent, M. Sarrà, and E. Marco-Urrea. 2014. Hospital wastewater treatment by fungal bioreactor: Removal efficiency for pharmaceuticals and endocrine disruptor compounds. *Science of the Total Environment* 493: 365–376.

Cvančarová, M., M. Moeder, A. Filipová, T. Reemtsma, and T. Cajthaml. 2013. Biotransformation of the antibiotic agent flumequine by ligninolytic fungi and residual antibacterial activity of the transformation mixtures. *Environmental Science & Technology* 47 (24): 14128–14136.

Daughton, C. G., and T. A. Ternes. 1999. Pharmaceuticals and personal care products in the environment: Agents of subtle change? *Environmental Health Perspectives* 107 (Suppl 6): 907.

Doll, T. E., and F. H. Frimmel. 2004. Kinetic study of photocatalytic degradation of carbamazepine, clofibric acid, iomeprol and iopromide assisted by different TiO 2 materials—determination of intermediates and reaction pathways. *Water Research* 38 (4): 955–964.

Drillia, P., K. Stamatelatou, and G. Lyberatos. 2005. Fate and mobility of pharmaceuticals in solid matrices. *Chemosphere* 60 (8): 1034–1044.

D'Souza, D. T., R. Tiwari, A. K. Sah, and C. Raghukumar. 2006. Enhanced production of laccase by a marine fungus during treatment of colored effluents and synthetic dyes. *Enzyme and Microbial Technology* 38 (3): 504–511.

Durairaj, P., J.-S. Hur, and H. Yun. 2016. Versatile biocatalysis of fungal cytochrome P450 monooxygenases. *Microbial Cell Factories* 15 (1): 125.

Eibes, G., G. Debernardi, G. Feijoo, M. T. Moreira, and J. M. Lema. 2011. Oxidation of pharmaceutically active compounds by a ligninolytic fungal peroxidase. *Biodegradation* 22 (3): 539–550.

Falade, A. O., U. U. Nwodo, B. C. Iweriebor, E. Green, L. V. Mabinya, and A. I. Okoh. 2016. Lignin peroxidase functionalities and prospective applications. *Microbiology Open* 6 (1): 1–14.

Fick, J., H. Söderström, R. H. Lindberg, C. Phan, M. Tysklind, and D. G. Larsson. 2009. Contamination of surface, ground, and drinking water from pharmaceutical production. *Environmental Toxicology and Chemistry* 28 (12): 2522–2527.

Field, J. A., E. De Jong, G. Feijoo Costa, and J. A. De Bont. 1992. Biodegradation of polycyclic aromatic hydrocarbons by new isolates of white rot fungi. *Applied and Environmental Microbiology* 58 (7): 2219–2226.

Fini, A., M. Laus, I. Orienti, and V. Zecchi. 1986. Dissolution and partition thermodynamic functions of some nonsteroidal anti-inflammatory drugs. *Journal of Pharmaceutical Sciences* 75 (1): 23–25.

Foley, J., D. De Haas, K. Hartley, and P. Lant. 2010. Comprehensive life cycle inventories of alternative wastewater treatment systems. *Water Research* 44 (5): 1654–1666.

Fytili, D. and A. Zabaniotou. 2008. Utilization of sewage sludge in EU application of old and new methods—a review. *Renewable and Sustainable Energy Reviews* 12 (1): 116–140.

Gagnon, C., A. Lajeunesse, P. Cejka, F. Gagne, and R. Hausler. 2008. Degradation of selected acidic and neutral pharmaceutical products in a primary-treated wastewater by disinfection processes. *Ozone: Science and Engineering* 30 (5): 387–392.

García-Galán, M. J., C. E. Rodríguez-Rodríguez, T. Vicent, G. Caminal, M. S. Díaz-Cruz, and D. Barceló. 2011. Biodegradation of sulfamethazine by Trametes versicolor: Removal from sewage sludge and identification of intermediate products by UPLC–QqTOF-MS. *Science of the Total Environment* 409 (24): 5505–5512.

Gavrilescu, M. 2005. Fate of pesticides in the environment and its bioremediation. *Engineering in Life Sciences* 5 (6): 497.

Gelband, H., P. M. Miller, S. Pant, S. Gandra, J. Levinson, D. Barter, A. White, and R. Laxminarayan. 2015. The state of the world's antibiotics 2015. *Wound Healing Southern Africa* 8 (2): 30–34.

Glazer, A. N. and Nikaido, H. 1995. *Microbial Biotechnology; Chapter 2: Microbial Diversity.* W.H. Freeman and Company, New York, NY: pp. 49–87.

Golan-Rozen, N., B. Chefetz, J. Ben-Ari, J. Geva, and Y. Hadar. 2011. Transformation of the recalcitrant pharmaceutical compound carbamazepine by Pleurotus ostreatus: Role of cytochrome P450 monooxygenase and manganese peroxidase. *Environmental Science & Technology* 45 (16): 6800–6805.

Gómez-Toribio, V., A. B. García-Martín, M. J. Martínez, Á. T. Martínez, and F. Guillén. 2009a. Enhancing the production of hydroxyl radicals by Pleurotus eryngii via quinone redox cycling for pollutant removal. *Applied and Environmental Microbiology* 75 (12): 3954–3962.

Gómez-Toribio, V., A. B. García-Martín, M. J. Martínez, Á. T. Martínez, and F. Guillén. 2009b. Induction of extracellular hydroxyl radical production by white-rot fungi through quinone redox cycling. *Applied and Environmental Microbiology* 75 (12): 3944–3953.

Hai, F. I., K. Yamamoto, F. Nakajima, and K. Fukushi. 2009. Factors governing performance of continuous fungal reactor during non-sterile operation—the case of a membrane bioreactor treating textile wastewater. *Chemosphere* 74 (6): 810–817.

Hai, F. I., K. Yamamoto, F. Nakajima, and K. Fukushi. 2011. Bioaugmented membrane bioreactor (MBR) with a GAC-packed zone for high rate textile wastewater treatment. *Water Research* 45 (6): 2199–2206.

Hakala, T. K., T. Lundell, S. Galkin, P. Maijala, N. Kalkkinen, and A. Hatakka. 2005. Manganese peroxidases, laccases and oxalic acid from the selective white-rot fungus Physisporinus rivulosus grown on spruce wood chips. *Enzyme and Microbial Technology* 36 (4): 461–468.

Halling-Sørensen, B., S. N. Nielsen, P. F. Lanzky, F. Ingerslev, H. C. H. Lützhøft, and S. E. Jørgensen. 1998. Occurrence, fate and effects of pharmaceutical substances in the environment—A review. *Chemosphere* 36 (2): 357–393.

Harms, H., D. Schlosser, and L. Y. Wick. 2011. Untapped potential: Exploiting fungi in bioremediation of hazardous chemicals. *Nature Reviews Microbiology* 9 (3): 177–192.

Hata, T., S. Kawai, H. Okamura, and T. Nishida. 2010. Removal of diclofenac and mefenamic acid by the white rot fungus Phanerochaete sordida YK-624 and identification of their metabolites after fungal transformation. *Biodegradation* 21 (5): 681–689.

Henry, C. L., and D. W. Cole. 1997. Use of biosolids in the forest: Technology, economics and regulations. *Biomass and Bioenergy* 13 (4–5): 269–277.

Herzfeldt, C. D., and R. Kümmel. 1983. Dissociation constants, solubilities and dissolution rates of some selected nonsteroidal antiinflammatories. *Drug Development and Industrial Pharmacy* 9 (5): 767–793.

Hester, R. E. and R. M. Harrison. 2015. *Pharmaceuticals in the Environment.* Vol. 41: Royal Society of Chemistry, Cambridge, UK.

Hirai, H., M. Sugiura, S. Kawai, and T. Nishida. 2005. Characteristics of novel lignin peroxidases produced by white-rot fungus Phanerochaete sordida YK-624. *FEMS Microbiology Letters* 246 (1): 19–24.

Hofrichter, M. 2002. Review: Lignin conversion by manganese peroxidase (MnP). *Enzyme and Microbial Technology* 30 (4): 454–466.

Hofrichter, M., R. Ullrich, M. J. Pecyna, C. Liers, and T. Lundell. 2010. New and classic families of secreted fungal heme peroxidases. *Applied Microbiology and Biotechnology* 87 (3): 871–897.

Hossain, K. A., F. I. Khan, and K. Hawboldt. 2008. Sustainable development of process facilities: State-of-the-art review of pollution prevention frameworks. *Journal of Hazardous Materials* 150 (1): 4–20.

Hug, S. J. and O. Leupin. 2003. Iron-catalyzed oxidation of arsenic (III) by oxygen and by hydrogen peroxide: pH-dependent formation of oxidants in the Fenton reaction. *Environmental Science & Technology* 37 (12): 2734–2742.

Ikehata, K., N. Jodeiri Naghashkar, and M. G. El-Din. 2006. Degradation of aqueous pharmaceuticals by ozonation and advanced oxidation processes: A review. *Ozone: Science and Engineering* 28 (6): 353–414.

ISO, EN. 2007. Water Quality—Determination of the inhibitory effect of water samples on the light emission of Vibrio fischeri—Part-3: Method using freeze-dried bacteria. 2 (11348-3, 2007): 21. https://www.iso.org/standard/40518.html

Jones, O. A., J. N. Lester, and N. Voulvoulis. 2005. Pharmaceuticals: A threat to drinking water? *TRENDS in Biotechnology* 23 (4): 163–167.

Junghanns, C., M. Moeder, G. Krauss, C. Martin, and D. Schlosser. 2005. Degradation of the xenoestrogen nonylphenol by aquatic fungi and their laccases. *Microbiology* 151 (1): 45–57.

Kabbout, R., and S. Taha. 2014. Biodecolorization of textile dye effluent by biosorption on fungal biomass materials. *Physics Procedia* 55: 437–444.

Kapoor, A., and T. Viraraghavan. 1998. Removal of heavy metals from aqueous solutions using immobilized fungal biomass in continuous mode. *Water Research* 32 (6): 1968–1977.

Keharia, H., and D. Madamwar. 2003. Bioremediation concepts for treatment of dye containing wastewater: A review. *Indian Journal of Experimental Biology* 41: 1068–1075.

Kim, S. D., J. Cho, I. S. Kim, B. J. Vanderford, and S. A. Snyder. 2007. Occurrence and removal of pharmaceuticals and endocrine disruptors in South Korean surface, drinking, and waste waters. *Water Research* 41 (5): 1013–1021.

Kim, H., Y. Hong, J.-E. Park, V. K. Sharma, and S.-I. Cho. 2013. Sulfonamides and tetracyclines in livestock wastewater. *Chemosphere* 91 (7): 888–894.

Kinney, C. A., E. T. Furlong, S. D. Zaugg, M. R. Burkhardt, S. L. Werner, J. D. Cahill, and G. R. Jorgensen. 2006. Survey of organic wastewater contaminants in biosolids destined for land application. *Environmental Science & Technology* 40 (23): 7207–7215.

Knop, D., D. Levinson, A. Makovitzki, A. Agami, E. Lerer, A. Mimran, O. Yarden, and Y. Hadar. 2016. Limits of versatility of versatile peroxidase. *Applied and Environmental Microbiology* 82 (14): 4070–4080.

Koh, Y. K. K., T. Y. Chiu, A. Boobis, E. Cartmell, M. D. Scrimshaw, and J. N. Lester. 2008. Treatment and removal strategies for estrogens from wastewater. *Environmental Technology* 29 (3): 245–267.

Kosjek, T., E. Heath, and B. Kompare. 2007. Removal of pharmaceutical residues in a pilot wastewater treatment plant. *Analytical and Bioanalytical Chemistry* 387 (4): 1379–1387.

Kües, U. 2015. Fungal enzymes for environmental management. *Current Opinion in Biotechnology* 33: 268–278.

Kumar, V., L. Wati, P. Nigam, I. M. Banat, B. S. Yadav, D. Singh, and R. Marchant. 1998. Decolorization and biodegradation of anaerobically digested sugarcane molasses spent wash effluent from biomethanation plants by white-rot fungi. *Process Biochemistry* 33 (1): 83–88.

Kümmerer, K. 2008. *Pharmaceuticals in the Environment: Sources, Fate, Effects and Risks.* Springer Science & Business Media, Springer: Berlin, Germany.

Kümmerer, K. 2009. The presence of pharmaceuticals in the environment due to human use—present knowledge and future challenges. *Journal of Environmental Management* 90 (8): 2354–2366.

Langford, K. H., and K. V. Thomas. 2009. Determination of pharmaceutical compounds in hospital effluents and their contribution to wastewater treatment works. *Environment International* 35 (5): 766–770.

Larsson, D. G. J., C. de Pedro, and N. Paxeus. 2007. Effluent from drug manufactures contains extremely high levels of pharmaceuticals. *Journal of Hazardous Materials* 148 (3): 751–755.

Le Corre, K. S., C. Ort, D. Kateley, B. Allen, B. I. Escher, and J. Keller. 2012. Consumption-based approach for assessing the contribution of hospitals towards the load of pharmaceutical residues in municipal wastewater. *Environment International* 45: 99–111.

Levenspiel, O., and C. Levenspiel. 1972. *Chemical Reaction Engineering.* Vol. 2: Wiley, New York, NY.

Lienert, J., T. Bürki, and B. I. Escher. 2007. Reducing micropollutants with source control: Substance flow analysis of 212 pharmaceuticals in faeces and urine. *Water Science and Technology* 56 (5): 87–96.

Liu, Z.-H., Y. Kanjo, and S. Mizutani. 2009. Removal mechanisms for endocrine disrupting compounds (EDCs) in wastewater treatment—physical means, biodegradation, and chemical advanced oxidation: A review. *Science of the Total Environment* 407 (2): 731–748.

Lloret, L., G. Eibes, T. A. Lú-Chau, M. T. Moreira, G. Feijoo, and J. M. Lema. 2010. Laccase-catalyzed degradation of anti-inflammatories and estrogens. *Biochemical Engineering Journal* 51 (3): 124–131.

Lloret, L., G. Eibes, G. Feijoo, M. T. Moreira, J. M. Lema, and F. Hollmann. 2011. Immobilization of laccase by encapsulation in a sol–gel matrix and its characterization and use for the removal of estrogens. *Biotechnology Progress* 27 (6): 1570–1579.

Lloret, L., G. Eibes, G. Feijoo, M. T. Moreira, and J. M. Lema. 2012. Degradation of estrogens by laccase from Myceliophthora thermophila in fed-batch and enzymatic membrane reactors. *Journal of Hazardous Materials* 213: 175–183.

Lorenzo, M., D. Moldes, and M. Á. Sanromán. 2006. Effect of heavy metals on the production of several laccase isoenzymes by Trametes versicolor and on their ability to decolourise dyes. *Chemosphere* 63 (6): 912–917.

Lu, T., Q. Zhang, and S. Yao. 2017. Removal of dyes from wastewater by growing fungal pellets in a semi-continuous mode. *Frontiers of Chemical Science and Engineering*: 11 (3): 338–345.

Macellaro, G., C. Pezzella, P. Cicatiello, G. Sannia, and A. Piscitelli. 2014. Fungal laccases degradation of endocrine disrupting compounds. *BioMed Research International* 2014, 614038: 1–8.

Mäkelä, M. R., K. S. Hildén, T. K. Hakala, A. Hatakka, and T. K. Lundell. 2006. Expression and molecular properties of a new laccase of the white rot fungus Phlebia radiata grown on wood. *Current Genetics* 50 (5): 323–333.

Manzanares, P., S. Fajardo, and C. Martin. 1995. Production of ligninolytic activities when treating paper pulp effluents by Trametes versicolor. *Journal of Biotechnology* 43 (2): 125–132.

Marco-Urrea, E., and C. A. Reddy. 2012. Degradation of chloro-organic pollutants by white rot fungi. In *Microbial Degradation of Xenobiotics*, Singh, S. H. (ed.) 31–66. Springer, Berlin, Germany..

Marco-Urrea, E., M. Pérez-Trujillo, T. Vicent, and G. Caminal. 2009. Ability of white-rot fungi to remove selected pharmaceuticals and identification of degradation products of ibuprofen by Trametes versicolor. *Chemosphere* 74 (6): 765–772.

Marco-Urrea, E., M. Pérez-Trujillo, P. Blánquez, T. Vicent, and G. Caminal. 2010a. Biodegradation of the analgesic naproxen by Trametes versicolor and identification of intermediates using HPLC-DAD-MS and NMR. *Bioresource Technology* 101 (7): 2159–2166.

Marco-Urrea, E., M. Pérez-Trujillo, C. Cruz-Morató, G. Caminal, and T. Vicent. 2010b. White-rot fungus-mediated degradation of the analgesic ketoprofen and identification of intermediates by HPLC–DAD–MS and NMR. *Chemosphere* 78 (4): 474–481.

Marco-Urrea, E., J. Radjenović, G. Caminal, M. Petrović, T. Vicent, and D. Barceló. 2010c. Oxidation of atenolol, propranolol, carbamazepine and clofibric acid by a biological Fenton-like system mediated by the white-rot fungus Trametes versicolor. *Water Research* 44 (2): 521–532.

Margot, J., C. Bennati-Granier, J. Maillard, P. Blánquez, D. A. Barry, and C. Holliger. 2013. Bacterial versus fungal laccase: Potential for micropollutant degradation. *AMB Express* 3 (1): 63.

Martens, R., H.-G. Wetzstein, F. Zadrazil, M. Capelari, P. Hoffmann, and N. Schmeer. 1996. Degradation of the fluoroquinolone enrofloxacin by wood-rotting fungi. *Applied and Environmental Microbiology* 62 (11): 4206–4209.

McFarland, J. W., A. Avdeef, C. M. Berger, and O. A Raevsky. 2001. Estimating the water solubilities of crystalline compounds from their chemical structures alone. *Journal of Chemical Information and Computer Sciences* 41 (5): 1355–1359.

Mendez-Arriaga, F., S. Esplugas, and J. Gimenez. 2010. Degradation of the emerging contaminant ibuprofen in water by photo-Fenton. *Water Research* 44 (2): 589–595.

Meneses, M., J. C. Pasqualino, and F. Castells. 2010. Environmental assessment of urban wastewater reuse: Treatment alternatives and applications. *Chemosphere* 81 (2): 266–272.

Miao, X.-S., J.-J. Yang, and C. D. Metcalfe. 2005. Carbamazepine and its metabolites in wastewater and in biosolids in a municipal wastewater treatment plant. *Environmental Science & Technology* 39 (19): 7469–7475.

Michniewicz, A., R. Ullrich, S. Ledakowicz, and M. Hofrichter. 2006. The white-rot fungus Cerrena unicolor strain 137 produces two laccase isoforms with different physico-chemical and catalytic properties. *Applied Microbiology and Biotechnology* 69 (6): 682–688.

Mishra, S., and V. S. Bisaria. 2006. Production and characterization of laccase from Cyathus bulleri and its use in decolourization of recalcitrant textile dyes. *Applied Microbiology and Biotechnology* 71 (5): 646–653.

Miyata, N., T. Mori, K. Iwahori, and M. Fujita. 2000. Microbial decolorization of melanoidin-containing wastewaters: Combined use of activated sludge and the fungus Coriolus hirsutus. *Journal of Bioscience and Bioengineering* 89 (2): 145–150.

Moreira, M. T., A. Sanroman, G. Feijoo, and J. M. Lema. 1996. Control of pellet morphology of filamentous fungi in fluidized bed bioreactors by means of a pulsing flow. Application to Aspergillus niger and Phanerochaete chrysosporium. *Enzyme and Microbial Technology* 19 (4): 261–266.

Morton, J. B. 2005. Chapter 6: Protozoa and Nematodes. In D. M. Sylvia, J. J. Fuhrmann, P. G. Hartel, and D. A. Zuberer (ed.), *Fungi, Principles and Applications of Soil Microbiology*. Upper Saddle River, NJ: Pearson Education, Inc., pp. 141–161.

Mu, D., T. Seager, P. Suresh Rao, and F. Zhao. 2010. Comparative life cycle assessment of lignocellulosic ethanol production: Biochemical versus thermochemical conversion. *Environmental Management* 46 (4): 565–578.

Murugesan, K., M. Arulmani, I.-H. Nam, Y.-M. Kim, Y.-S. Chang, and P. Thangavelu Kalaichelvan. 2006. Purification and characterization of laccase produced by a white rot fungus Pleurotus sajor-caju under submerged culture condition and its potential in decolorization of azo dyes. *Applied Microbiology and Biotechnology* 72 (5): 939–946.

Nakada, N., T. Tanishima, H. Shinohara, K. Kiri, and H. Takada. 2006. Pharmaceutical chemicals and endocrine disrupters in municipal wastewater in Tokyo and their removal during activated sludge treatment. *Water Research* 40 (17): 3297–3303.

Pahan, K. 2006. Lipid-lowering drugs. *Cellular and Molecular Life Sciences CMLS* 63 (10): 1165–1178.

Parshikov, I. A., J. P. Freeman, J. O. Lay Jr, J. D. Moody, A. J. Williams, R. D. Beger, and J. B. Sutherland. 2001a. Metabolism of the veterinary fluoroquinolone sarafloxacin by the fungus Mucor ramannianus. *Journal of Industrial Microbiology and Biotechnology* 26 (3): 140–144.

Parshikov, I. A., T. M. Heinze, J. D. Moody, J. P. Freeman, A. J. Williams, and J. B. Sutherland. 2001b. The fungus Pestalotiopsis guepini as a model for biotransformation of ciprofloxacin and norfloxacin. *Applied Microbiology and Biotechnology* 56 (3): 474–477.

Pasqualino, J. C., M. Meneses, M. Abella, and F. Castells. 2009. LCA as a decision support tool for the environmental improvement of the operation of a municipal wastewater treatment plant. *Environmental Science & Technology* 43 (9): 3300–3307.

Pharmacopoeia, European. 2002. *European Pharmacopoeia*. Strasbourg: Council of Europe.

Phillips, P. J., S. G. Smith, D. W. Kolpin, S. D. Zaugg, H. T. Buxton, E. T. Furlong, K. Esposito, and B. Stinson. 2010. Pharmaceutical formulation facilities as sources of opioids and other pharmaceuticals to wastewater treatment plant effluents. *Environmental Science & Technology* 44 (13): 4910–4916.

Pokhrel, D and T. Viraraghavan. 2004. Treatment of pulp and paper mill wastewater—a review. *Science of the Total Environment* 333 (1): 37–58.

Prieto, A., M. Möder, R. Rodil, L. Adrian, and E. Marco-Urrea. 2011. Degradation of the antibiotics norfloxacin and ciprofloxacin by a white-rot fungus and identification of degradation products. *Bioresource Technology* 102 (23): 10987–10995.

Quaratino, D., F. Federici, M. Petruccioli, M. Fenice, and A. D'Annibale. 2007. Production, purification and partial characterisation of a novel laccase from the white-rot fungus Panus tigrinus CBS 577.79. *Antonie van Leeuwenhoek* 91 (1): 57–69.

Rebitzer, G., T. Ekvall, R. Frischknecht, D. Hunkeler, G. Norris, T. Rydberg, W-P. Schmidt, S. Suh, B. Pennington Weidema, and D. W. Pennington. 2004. Life cycle assessment: Part 1: Framework, goal and scope definition, inventory analysis, and applications. *Environment International* 30 (5): 701–720.

Richardson, B. J., P. K. S. Lam, and M. Martin. 2005. Emerging chemicals of concern: Pharmaceuticals and personal care products (PPCPs) in Asia, with particular reference to Southern China. *Marine Pollution Bulletin* 50 (9): 913–920.

Rodarte-Morales, A. I., G. Feijoo, M. T. Moreira, and J. M. Lema. 2011. Degradation of selected pharmaceutical and personal care products (PPCPs) by white-rot fungi. *World Journal of Microbiology and Biotechnology* 27 (8): 1839–1846.

Rodarte-Morales, A. I., G. Feijoo, M. T. Moreira, and J. M. Lema. 2012. Biotransformation of three pharmaceutical active compounds by the fungus Phanerochaete chrysosporium in a fed batch stirred reactor under air and oxygen supply. *Biodegradation* 23 (1): 145–156.

Rodríguez-Rodríguez, C. E, A. Jelić, M. Llorca, M. Farré, G. Caminal, M. Petrović, D. Barceló, and T. Vicent. 2011. Solid-phase treatment with the fungus Trametes versicolor substantially reduces pharmaceutical concentrations and toxicity from sewage sludge. *Bioresource Technology* 102 (10): 5602–5608.

Rodríguez-Rodríguez, C. E., E. Barón, P. Gago-Ferrero, A. Jelić, M. Llorca, M. Farré, M. S. Díaz-Cruz, E. Eljarrat, M. Petrović, and G. Caminal. 2012a. Removal of pharmaceuticals, polybrominated flame retardants and UV-filters from sludge by the fungus Trametes versicolor in bioslurry reactor. *Journal of Hazardous Materials* no. 233: 235–243.

Rodríguez-Rodríguez, C. E, M. J. García-Galán, P. Blánquez, M. S. Díaz-Cruz, D. Barceló, G. Caminal, and T. Vicent. 2012b. Continuous degradation of a mixture of sulfonamides by Trametes versicolor and identification of metabolites from sulfapyridine and sulfathiazole. *Journal of Hazardous Materials* 213: 347–354.

Rodríguez-Delgado, M., C. Orona-Navar, R. García-Morales, C. Hernandez-Luna, R. Parra, J. Mahlknecht, and N. Ornelas-Soto. 2016. Biotransformation kinetics of pharmaceutical and industrial micropollutants in groundwaters by a laccase cocktail from Pycnoporus sanguineus CS43 fungi. *International Biodeterioration & Biodegradation* 108: 34–41.

Rose, N. 2001. Historical changes in mental health practice. *Textbook of Community Psychiatry*, Thornicroft, G. (ed.). Oxford University Press, New York, NY. pp. 13–27.

Ruiz-Duenas, F. J., M. Morales, M. Pérez-Boada, T. Choinowski, M. J. Martínez, K. Piontek, and Á. T. Martínez. 2007. Manganese oxidation site in pleurotus eryngii versatile peroxidase: A site-directed mutagenesis, kinetic, and crystallographic study†. *Biochemistry* 46 (1): 66–77.

Rybczyńska-Tkaczyk, K. and T. Korniłłowicz-Kowalska. 2016. Biosorption optimization and equilibrium isotherm of industrial dye compounds in novel strains of microscopic fungi. *International Journal of Environmental Science and Technology* 13 (12): 2837–2846.

Salame, T. M, D. Knop, D. Levinson, S. J. Mabjeesh, O. Yarden, and Y. Hadar. 2012. Release of Pleurotus ostreatus versatile-peroxidase from Mn 2+ repression enhances anthropogenic and natural substrate degradation. *PloS One* 7 (12): e52446.

Salvachúa, D., A. Prieto, Á. T. Martínez, and M. J. Martínez. 2013. Characterization of a novel dye-decolorizing peroxidase (DyP)-type enzyme from Irpex lacteus and its application in enzymatic hydrolysis of wheat straw. *Applied and Environmental Microbiology* 79 (14): 4316–4324.

Santos, L. H. M. L. M., A. N. Araújo, A. Fachini, A. Pena, C. Delerue-Matos, and M. C. B. S. M. Montenegro. 2010. Ecotoxicological aspects related to the presence of pharmaceuticals in the aquatic environment. *Journal of Hazardous Materials* 175 (1): 45–95.

Shemer, H., Y. K. Kunukcu, and K. G. Linden. 2006. Degradation of the pharmaceutical metronidazole via UV, Fenton and photo-Fenton processes. *Chemosphere* 63 (2): 269–276.

Shin, E., H. T. Choi, and H. Song. 2007. Biodegradation of endocrine-disrupting bisphenol A by white rot fungus Irpex lacteus. *Journal of Microbiology and Biotechnology* 17 (7): 1147.

Shrivastava, R., V. Christian, and B. R. M. Vyas. 2005. Enzymatic decolorization of sulfonphthalein dyes. *Enzyme and Microbial Technology* 36 (2): 333–337.

Siddiquee, S., K. Rovina, S. Al Azad, L. Naher, S. Suryani, and P. Chaikaew. 2015. Heavy metal contaminants removal from wastewater using the potential filamentous fungi biomass: A review. *Journal of Microbial and Biochemical Technology* no. 7: 384–393.

Sim, W-J., J-W. Lee, E.-S. Lee, S-K. Shin, S-R. Hwang, and J-E. Oh. 2011. Occurrence and distribution of pharmaceuticals in wastewater from households, livestock farms, hospitals and pharmaceutical manufactures. *Chemosphere* 82 (2): 179–186.

Simonich, S. L., T. W. Federle, W. S. Eckhoff, A. Rottiers, S. Webb, D. Sabaliunas, and W. De Wolf. 2002. Removal of fragrance materials during US and European wastewater treatment. *Environmental Science & Technology* 36 (13): 2839–2847.

Snyder, S. A., P. Westerhoff, Y. Yoon, and D. L. Sedlak. 2003. Pharmaceuticals, personal care products, and endocrine disruptors in water: Implications for the water industry. *Environmental Engineering Science* 20 (5): 449–469.

Snyder, S. A., S. Adham, A. M. Redding, F. S. Cannon, J. DeCarolis, J. Oppenheimer, E. C. Wert, and Y. Yoon. 2007. Role of membranes and activated carbon in the removal of endocrine disruptors and pharmaceuticals. *Desalination* 202 (1–3): 156–181.

Spatari, S., D. M. Bagley, and H. L. MacLean. 2010. Life cycle evaluation of emerging ligno-cellulosic ethanol conversion technologies. *Bioresource Technology* 101 (2): 654–667.

Strong, P. J. and H. Claus. 2011. Laccase: A review of its past and its future in bioremediation. *Critical Reviews in Environmental Science and Technology* 41 (4): 373–434.

Suarez, S., J. M. Lema, and F. Omil. 2010. Removal of pharmaceutical and personal care products (PPCPs) under nitrifying and denitrifying conditions. *Water Research* 44 (10): 3214–3224.

Subramanian, V., and J. S. Yadav. 2009. Role of P450 monooxygenases in the degradation of the endocrine-disrupting chemical nonylphenol by the white rot fungus Phanerochaete chrysosporium. *Applied and Environmental Microbiology* 75 (17): 5570–5580.

Sumathi, S. and V. Phatak. 1999. Fungal treatment of bagasse based pulp and paper mill wastes. *Environmental Technology* 20 (1): 93–98.

Taboada-Puig, R., C. Junghanns, P. Demarche, M. T. Moreira, G. Feijoo, J. M. Lema, and S. N. Agathos. 2011. Combined cross-linked enzyme aggregates from versatile peroxidase and glucose oxidase: Production, partial characterization and application for the elimination of endocrine disruptors. *Bioresource Technology* 102 (11): 6593–6599.

Tahmasbi, H., M. R. Khoshayand, M. Bozorgi-Koushalshahi, M. Heidary, M. Ghazi-Khansari, and M. A. Faramarzi. 2016. Biocatalytic conversion and detoxification of imipramine by the laccase-mediated system. *International Biodeterioration & Biodegradation* no. 108: 1–8.

Tang, C-J., P. Zheng, T-T. Chen, J-Q. Zhang, Q. Mahmood, S. Ding, X-G. Chen, J-W. Chen, and D-T. Wu. 2011. Enhanced nitrogen removal from pharmaceutical wastewater using SBA-ANAMMOX process. *Water Research* 45 (1): 201–210.

Ternes, T. A., J. Stüber, N. Herrmann, D. McDowell, A. Ried, M. Kampmann, and B. Teiser. 2003. Ozonation: A tool for removal of pharmaceuticals, contrast media and musk fragrances from wastewater? *Water Research* 37 (8): 1976–1982.

Ternes, T. A., M. Stumpf, J. Mueller, K. Haberer, R-D. Wilken, and M. Servos. 1999. Behavior and occurrence of estrogens in municipal sewage treatment plants—I. Investigations in Germany, Canada and Brazil. *Science of the Total Environment* 225 (1): 81–90.

Thanh, N. C. and R. E. Simard. 1973. Biological treatment of domestic sewage by fungi. *Mycopathologia et Mycologia Applicata* 51 (2–3): 223–232.

Tien, M. and T. K. Kirk. 1988. Lignin peroxidase of Phanerochaete chrysosporium. *Methods in Enzymology* 161: 238–249.

Tilman, D., K. G. Cassman, P. A. Matson, R. Naylor, and S. Polasky. 2002. Agricultural sustainability and intensive production practices. *Nature* 418 (6898): 671–677.

Tišma, M., B. Zelić, and Đ. Vasić-Rački. 2010. White-rot fungi in phenols, dyes and other xenobiotics treatment—a brief review. *Croatian Journal of Food Science and Technology* 2 (2): 34–47.

Tortella, G., N. Durán, O. Rubilar, M. Parada, and M. C. Diez. 2015. Are white-rot fungi a real biotechnological option for the improvement of environmental health? *Critical Reviews in Biotechnology* 35 (2): 165–172.

Tran, N. H., T. Urase, and O. Kusakabe. 2010. Biodegradation characteristics of pharmaceutical substances by whole fungal culture Trametes versicolor and its laccase. *Journal of Water and Environment Technology* 8 (2): 125–140.

Tsukihara, T., Y. Honda, R. Sakai, T. Watanabe, and T. Watanabe. 2006. Exclusive overproduction of recombinant versatile peroxidase MnP2 by genetically modified white rot fungus, Pleurotus ostreatus. *Journal of Biotechnology* 126 (4): 431–439.

Uddin, S. A., and Md A. Kader. 2006. The use of antibiotics in shrimp hatcheries in Bangladesh. *Journal of Fisheries and Aquatic Science* 1 (1): 64–67.

Ullrich, R., N. L. Dung, and M. Hofrichter. 2005. Laccase from the medicinal mushroom Agaricus blazei: Production, purification and characterization. *Applied Microbiology and Biotechnology* 67 (3): 357–363.

Upadhyay, P., R. Shrivastava, and P. K. Agrawal. 2016. Bioprospecting and biotechnological applications of fungal laccase. *3 Biotech* 6 (1): 15.

Ürek, R. Ö., and N. K. Pazarlioğlu. 2004. Purification and partial characterization of manganese peroxidase from immobilized Phanerochaete chrysosporium. *Process Biochemistry* 39 (12): 2061–2068.

Ürek, R. Ö., and N. K. Pazarlioğlu. 2005. Production and stimulation of manganese peroxidase by immobilized Phanerochaete chrysosporium. *Process Biochemistry* 40 (1): 83–87.

Vasiliadou, I. A., R. Sánchez-Vázquez, R. Molina, F. Martínez, J. A. Melero, L. F. Bautista, J. Iglesias, and G. Morales. 2016. Biological removal of pharmaceutical compounds using white-rot fungi with concomitant FAME production of the residual biomass. *Journal of Environmental Management* no. 180: 228–237.

Verlicchi, P., M. Al Aukidy, A. Galletti, M. Petrovic, and D. Barceló. 2012a. Hospital effluent: Investigation of the concentrations and distribution of pharmaceuticals and environmental risk assessment. *Science of the Total Environment* no. 430: 109–118.

Verlicchi, P., M. Al Aukidy, and E. Zambello. 2012b. Occurrence of pharmaceutical compounds in urban wastewater: Removal, mass load and environmental risk after a secondary treatment—a review. *Science of the Total Environment* no. 429: 123–155.

Wang, J., R. Yamamoto, Y. Yamamoto, T. Tokumoto, J. Dong, P. Thomas, H. Hirai, and H. Kawagishi. 2013. Hydroxylation of bisphenol A by hyper lignin-degrading fungus Phanerochaete sordida YK–624 under non-ligninolytic condition. *Chemosphere* 93 (7): 1419–1423.

Wang, X., J. Liu, N-Q. Ren, and Z. Duan. 2012. Environmental profile of typical anaerobic/anoxic/oxic wastewater treatment systems meeting increasingly stringent treatment standards from a life cycle perspective. *Bioresource Technology* 126: 31–40.

Wells, A., M. Teria, and T. Eve. 2006. Green oxidations with laccase–mediator systems. *Biochemical Society Transactions* 34 (2): 304–308.

Wen, X., Y. Jia, and J. Li. 2010. Enzymatic degradation of tetracycline and oxytetracycline by crude manganese peroxidase prepared from Phanerochaete chrysosporium. *Journal of Hazardous Materials* 177 (1): 924–928.

Wenzel, H., H. F. Larsen, J. Clauson-Kaas, L. Høibye, and B. N. Jacobsen. 2008. Weighing environmental advantages and disadvantages of advanced wastewater treatment of micro-pollutants using environmental life cycle assessment. *Water Science and Technology* 57 (1): 27–32.

Wesenberg, D., I. Kyriakides, and S. N. Agathos. 2003. White-rot fungi and their enzymes for the treatment of industrial dye effluents. *Biotechnology Advances* 22 (1): 161–187.

Wong, D. W. S. 2009. Structure and action mechanism of ligninolytic enzymes. *Applied Biochemistry and Biotechnology* 157 (2): 174–209.

Yagub, M. T., T. K. Sen, S. N. Afroze, and H. M. Ang. 2014. Dye and its removal from aqueous solution by adsorption: A review. *Advances in Colloid and Interface Science* 209: 172–184.

Yalkowsky, S. H and R. M. Dannenfelser. 1992. *Aquasol Database of Aqueous Solubility.* Tucson, AZ: College of Pharmacy, University of Arizona.

Yang, J-S. 2004. Studies on extracellular enzyme of lignite degrading fungi—Penicillium sp. P6. *China Environmental Science—Chinese Edition* 24 (1): 24–27.

Yang, S., F. I. Hai, L. D. Nghiem, L. N. Nguyen, F. Roddick, and W. E. Price. 2013. Removal of bisphenol A and diclofenac by a novel fungal membrane bioreactor operated under non-sterile conditions. *International Biodeterioration & Biodegradation* 85: 483–490.

Yang, J., W. Li, T. B. Ng, X. Deng, J. Lin, and X. Ye. 2017. Laccases: Production, expression regulation, and applications in pharmaceutical biodegradation. *Frontiers in Microbiology* 8: 832.

Yıldırım, M. and B. Topkaya. 2012. Assessing environmental impacts of wastewater treatment alternatives for small-scale communities. *Clean–Soil, Air, Water* 40 (2): 171–178.

Yoshida, H., T. H. Christensen, and C. Scheutz. 2013. Life cycle assessment of sewage sludge management: A review. *Waste Management & Research* 31 (11): 1083–1101.

Zang, Y., Y. Li, C. Wang, W. Zhang, and W. Xiong. 2015. Towards more accurate life cycle assessment of biological wastewater treatment plants: A review. *Journal of Cleaner Production* 107: 676–692.

Zhang, Y. and S.-U. Geißen. 2012. Elimination of carbamazepine in a non-sterile fungal bioreactor. *Bioresource Technology* 112: 221–227.

Zouari-Mechichi, H., T. Mechichi, A. Dhouib, S. Sayadi, A. T. Martínez, and M. J. Martínez. 2006. Laccase purification and characterization from Trametes trogii isolated in Tunisia: Decolorization of textile dyes by the purified enzyme. *Enzyme and Microbial Technology* 39 (1): 141–148.

Zupanc, M., T. Kosjek, M. Petkovšek, M. Dular, B. Kompare, B. Širok, Ž. Blaževka, and E. Heath. 2013. Removal of pharmaceuticals from wastewater by biological processes, hydrodynamic cavitation and UV treatment. *Ultrasonics Sonochemistry* 20 (4): 1104–1112.

9 Determination of Pharmaceutical Compounds in Sewage Sludge from Municipal Wastewater Treatment Plants

Irene Aparicio, Julia Martín,
Juan Luis Santos, and Esteban Alonso

CONTENTS

9.1 INTRODUCTION

In the last few years, special attention has been paid to the presence of pharmaceutical compounds in the environment because of their bioactivity, widespread use, and potential health and ecological risks. The main routes for the appearance of pharmaceuticals in the environment are the release of wastewater effluents from wastewater treatment plants (WWTPs) and also land application of sewage sludge generated as a by-product of wastewater treatment. The European Union (EU) included for the first time in 2015 the monitoring of seven pharmaceutical compounds in surface water (EU, 2015). These compounds were included in the first watch list of substances for Union-wide monitoring in the aquatic environment, which was composed of 10 substances/groups of substances. The pharmaceutical compounds included in the watch list are the anti-inflammatory drug diclofenac, three macrolide antibiotics (erythromycin, clarithromycin, and azithromycin), and three hormones (17-α-ethinylestradiol, 17-β-estradiol, and estrone). In Europe, the implementation of a directive concerning urban wastewater treatment (EC, 1991) has increased the number of WWTPs, because it stated that by the end of 2005, secondary treatment had to be applied to wastewater coming from settlements of more than 10,000 population equivalent and to wastewater coming from settlements of more than 2,000 population equivalent when it is discharged to fresh waters and estuaries. As a result, the European Commission has reported that the amount of sewage sludge generated in the EU, due to the implementation of the wastewater treatment directive, has increased from 5.5 million tons dry matter (d.m.) in 1992 to 9 million tons by the end of its implementation. Sewage sludge generated in urban WWTPs is of great interest as a fertilizer because of its nitrogen, phosphorus, and organic matter content. Therefore, the application of sewage sludge generated in WWTPs to soils enables not only the disposal of this increasing by-product of wastewater treatment but also taking advantage of its agronomic properties as a fertilizer. That application is promoted in the EU by the sewage sludge directive 86/278/EEC (EC, 1986), which encourages the use of sewage sludge in agricultural soils and regulates its application to prevent harmful effects on soil, vegetation, animals, and man. Nevertheless, in that directive, concentration limits are fixed only for seven heavy metals (cadmium, copper, nickel, lead, zinc, mercury, and chromium) for sludge that is intended for application to agricultural soils. WWTPs are not designed to eliminate pharmaceutical compounds, but when removal occurs, it can be due to biodegradation or to sorption onto sludge, mainly in the case of compounds with high solid-water distribution coefficient (K_d) (Martínez-Alcalá et al., 2017). For some pharmaceuticals, biodegradation can result in the release of the parent compounds from their metabolites and conjugated forms, resulting in an increase, instead of a decrease, of the concentration of the pharmaceutical compound after wastewater treatment. Wastewater treatment can even affect enantiomeric compounds, causing their concentrations to deviate from the racemic proportions. For instance, Evans et al. (2015) determined the enantiomeric concentrations of nine pharmaceutical compounds in wastewater and in digested sludge, which had undergone activated sludge and anaerobic digestion. They reported that the nine compounds monitored had significantly different enantiomeric profiling in wastewater and in sludge. Moreover, the enantiomeric fractions of some compounds (alprenolol, norephedrine, and tramadol) significantly deviated from the racemic proportion. This fact was explained by biological processes occurring during activated sludge treatment or during anaerobic digestion.

A wide range of concentration levels have been reported for pharmaceutical compounds in sewage sludge. Verlicchi and Zambello (2015), in a review of 59 papers about the presence of 152 pharmaceuticals in different types of sewage sludge, reported that analgesics are in the range of 3–10,000 ng g^{-1} d.m. in primary sludge, 1–1000 ng g^{-1} d.m. in secondary sludge, 4–1000 ng g^{-1} d.m. in digested sludge, and 10–1000 ng g^{-1} d.m in compost; antibiotics are in the range of 5–4000 ng g^{-1} d.m. in primary sludge, 0.1–70,000 ng g^{-1} d.m. in secondary sludge, 1–8000 ng g^{-1} d.m. in digested sludge, and 0.8–200 ng g^{-1} d.m in compost; hormones are in the range of 4–400 ng g^{-1} d.m. in primary sludge, 0.1–300 ng g^{-1} d.m. in secondary sludge, 1–10,000 ng g^{-1} d.m. in digested sludge, and 20–200 ng g^{-1} d.m in compost; and psychiatric drugs are in the range of 5–2000 ng g^{-1} d.m. in primary sludge, 1–600 ng g^{-1} d.m. in secondary sludge, 0.1–3000 ng g^{-1} d.m. in digested sludge, and 0.1–900 ng g^{-1} d.m in compost.

To obtain information about the occurrence and fate of pharmaceutical compounds and their metabolites in sewage sludge, accurate and sensitive methods are needed. In this chapter, an overview of extraction and detection techniques applied to the determination of pharmaceutical compounds in sewage sludge from urban WWTPs is discussed. The application of analytical methodologies to obtain information about the evolution of their concentration in the sludge treatment line, their occurrence and fate in sewage sludge from different treatment technologies, and the ecotoxicological risk assessment of sewage sludge application to soils are also described.

9.2 METHODOLOGY FOR THE DETERMINATION OF PHARMACEUTICALS IN SEWAGE SLUDGE

The occurrence and fate of pharmaceutical compounds in sewage sludge have not been studied in such depth as has been done in wastewater. This fact can be explained by the difficulties of detecting and quantifying such low trace pollutants in the complex sludge matrix. The following section discusses the main extraction and determination techniques reported for the determination of pharmaceutical compounds in sewage sludge from urban WWTPs. An overview of analytical methods reported for the determination of pharmaceuticals in sewage sludge is shown in Table 9.1.

9.2.1 SAMPLE TREATMENT

The main goals of sludge sample treatment are to achieve good extraction recoveries for the target compounds, reducing interfering compounds to the lowest levels, which can involve a difficult and time-consuming step (Peysson and Vulliet, 2013). Moreover, after sample extraction, a clean-up step can still be required to reduce interfering compounds before analytical determination. Extraction methods should be simple and easy to perform, should require small volumes of organic solvents, should generate little waste, and should require cheap equipment to make the methodology

TABLE 9.1

Overview of Analytical Methods for the Determination of Pharmaceutical Compounds in Sewage Sludge

Therapeutic Group	Sample	Sample Amount	Extraction Technique	Extraction Solvents	Total Solvent Volume	Clean-up	Analytical Determination	Recoveries (%)	Limits of Detection (ng g^{-1} d.w.)	References
Anti-inflammatory drugs, anticancer agents, and β-blockers	Compost	0.05 g	UAE	MeOH:ethyl acetate (1:1, v/v)	4 mL	—	LC-MS/MS	87.3–113	0.66	López Zavala and Reynoso-Cuevas (2015)
148 pharmaceuticals (antibiotics, analgesic, anti-inflammatory drugs, antiepileptics, benzodiazepines, antipsychotics, antidepressants, and illicit drugs)	Sewage sludge	0.1 g	UAE	MeOH:water (pH2, EDTA 0.1%) (1:1, v/v)	2 × 3 mL	—	LC-MS/MS	50–110 (for more than 77% of the analytes)	<10 (for more than 91% of the analytes)	Gago-Ferrero et al. (2015)
Antibiotics										
Digested sludge	2 g	UAE	Phosphate buffer pH 6 (10mL); MeOH:water (75:25, v/v), 5% TEA (10 mL)	20 mL	SPE (ENV+)	LC-MS/MS	64–71	100–110[a]		Lindberg et al. (2005); Martín et al. (2010)
Primary sludge	1 g	UAE	MeOH (5 + 2 mL), acetone (2 mL)	9 mL	SPE (Oasis HLB)	HPLC-UV-FI	61–107	1.56–115		
Secondary sludge	1 g						47–97	1.55–112		
Digested sludge	1.5 g						41–110	0.45–79.8		
Compost	2 g	0.5 g	UAE	MeOH (4+2 mL), acetone (2+2 mL)	10 mL	SPE (Oasis MCX; RP C18; ENV+)	55–106	0.13–44.7	20–50[a]	Ternes et al. (2005)
Digested sludge								43–76		

(Continued)

TABLE 9.1 (CONTINUED)

Overview of Analytical Methods for the Determination of Pharmaceutical Compounds in Sewage Sludge

Therapeutic Group	Sample	Sample Amount	Extraction Technique	Extraction Solvents	Total Solvent Volume	Clean-up	Analytical Determination	Recoveries (%)	Limits of Detection (ng g−1 d.w.)	References
Estrogens	Secondary sludge	0.5 g	UAE	MeOH (4+3 mL), acetone (3+3 mL.)	13 mL	GPC	GC-MS/MS	83–119	2–4a	Ternes et al. (2002)
	Digested sludge	0.5 g						94–117	2–4a	
Anti-inflammatory drugs and estrogens	Sewage sludge	0.5 g	UAE	Phosphate buffer pH 2 (15 mL), ACN (20 mL.)	3 × 35 mL.	SPE (ENVI-18)	GC-MS	72–131	0.7–5.2	Zhang et al. (2016)
Antibiotics	Sewage sludge	0.5 g	UAE / MAE / PLE	MeOH:McIlvaine buffer (1:1; v/v, pH = 3)	10 mL / 10 mL / 15 mL	—	LC-MS/MS	97.0–103.6 / 97.2–104.8 / 97.3–104.2	2–5	Dorival-Garcia et al. (2013)
Anti-inflammatory drugs, antibacterial agents, anti-epileptics, and lipid regulators	Sewage sludge	1 g	UAE	MeOH (1% [v/v] formic acid)	12 mL	SPE (Envi-carb)	GC-MS	57.9–103.1	1.4–11	Yu and Wu (2012)
66 pharmaceuticals (anti-inflammatory drugs, antibiotics, anti-epileptics, β-blocker, nervous stimulant, and lipid regulators)	Primary sludge	20 mL	UAE	MeOH:water (pH 11) (9:1, v/v)	75 mL	SPE (Oasis HLB)	LC-MS/MS	70–115	—	Okuda et al. (2009)
	Digested sludge		PLE(1) / PLE(2)	Water at pH 2 / MeOH at pH 4 (PLE: 100 °C)	20 mL / 20 mL					
Anti-inflammatory drugs, antibiotics, anti-epileptics, β-blocker, nervous stimulant, and lipid regulators	Digested sludge	1 g	PLE	MeOH:water (1:1, v/v) (60 °C)	53 mL	SPE (Oasis HLB)	LC-MS	45–120 / 50–110	2–350 / 0.8–120	Barron et al. (2008)
Glucocorticoids	Digested sludge	1 g	PLE	MeOH:acetone (4:1, v/v) (40 °C)	n.d.	SPE (Oasis HLB)	LC-MS/MS	8–73	0.5–1	Herrero et al. (2013)

(Continued)

TABLE 9.1 (CONTINUED)

Overview of Analytical Methods for the Determination of Pharmaceutical Compounds in Sewage Sludge

Therapeutic Group	Sample	Sample Amount	Extraction Technique	Extraction Solvents	Total Solvent Volume	Clean-up	Analytical Determination	Recoveries (%)	Limits of Detection (ng g−1 d.w.)	References
Anti-inflammatory drugs, antibiotics, anti-epileptics, β-blocker, antidepressants, H2 receptor agonists, nervous stimulants, lipid regulators, and others (the top 30 dispensed in Scotland)	Sludge	1–10 g	PLE	MeOH or MeOH:formic acid (100:0.1, v/v) (70 °C)	n.d.	—	LC-MS/MS	>76	2–20	Langford et al. (2011)
Anti-inflammatory drugs, antibiotics, β-blocker, nervous stimulants, and lipid regulators	Primary sludge	1 g	PLE	MeOH:water (2:1, v/v) (100 °C)	22 mL	SPE (Oasis HLB)	LC-MS/MS	36.8–130	0.44–89.2[a]	Radjenović et al. (2009a)
	Secondary sludge							33.5–122	0.44–89.2[a]	
	Digested sludge							29.2–102	2.10–96.3[a]	
Anti-cancer drugs	Seven types of sludge	0.35 g	PLE	MeOH:water (65:35, v/v) (100 °C)	15 mL	SPE (Oasis MAX)	LC-MS/MS	92–110	2.5–74	Seira et al. (2013)
Antibiotics	Secondary sludge	0.2 g	PLE	MeOH:water (1:1, v/v) (100 °C)	22 mL	SPE (Oasis HLB)	LC-MS/MS	51–64	14–15[a]	Göbel et al. (2005)
	Digested sludge									
Anti-inflammatory drugs, antibiotics, β-blocker, nervous stimulants, and lipid regulators	Sewage sludge	5 g	PLE	MeOH:phosphoric acid 50 mM (1:1, v/v) (80 °C)	40 mL	—	LC/MS	72–109	14–32[a]	Nieto et al. (2007)

(Continued)

TABLE 9.1 (CONTINUED)

Overview of Analytical Methods for the Determination of Pharmaceutical Compounds in Sewage Sludge

Therapeutic Group	Sample	Sample Amount	Extraction Technique	Extraction Solvents	Total Solvent Volume	Clean-up	Analytical Determination	Recoveries (%)	Limits of Detection (ng g^{-1} d.w.)	References
Analgesics, antibacterials, anti-epileptics, β-blockers, lipid regulators, and non-steroidal anti-inflammatories	Sewage sludge	1 g	MAE	MeOH:water (3:2, v/v) (500 W, 6 min)	10 mL	SPE (Oasis HLB)	GC-MS	91–101	0.8–5.1	Azzouz and Ballesteros (2012)
Quinolone antibiotics	Compost	1 g	MAE	ACN:phosphoric acid (99:1, v/v) (120 °C, 5 min)		SALLEd-SPE (PSA)	LC-MS/MS	95–106	0.2–0.5	Dorival-García et al. (2015)
Enantiomers of antidepressants, β-blockers, and illicit drugs	Digested sludge	1 g	MAE	MeOH:water(120 °C, 30 min)	30 mL	SPE (Oasis MAX)	Chiral LC-MS/MS	65–140	0.03–80	Evans et al. (2015)
Non-steroidal anti-inflammatory drugs and estrogenics	Sewage sludge	5 g	MAE	CAN (110 °C, 15 min)	n.d.	SPE (Oasis HLB)	GC-MS	39–108	0.3–5.7	Kumirska et al. (2015)
50 pharmaceuticals (anesthetics, anti-inflammatory drugs, antibiotics, anticancer agents, antidepressants, antidiabetics, anti-epileptics, antihistamines, antihypertensives, β-blockers, H$_2$ receptor agonists, nervous stimulants, estrogens, lipid regulators, radiocontrast agents, and metabolites)	Digested sludge	0.5 g	MAE	MeOH:water (pH$_2$) (1:1, v/v) (110 °C, 30 min)	25 mL	SPE (Oasis MCX)	LC-MS/MS	40–152	0.1–24.1	Petrie et al. (2016)

(Continued)

TABLE 9.1 (CONTINUED)

Overview of Analytical Methods for the Determination of Pharmaceutical Compounds in Sewage Sludge

Therapeutic Group	Sample	Sample Amount	Extraction Technique	Extraction Solvents	Total Solvent Volume	Clean-up	Analytical Determination	Recoveries (%)	Limits of Detection (ng g−1 d.w.)	References
Fluoroquinolones	Compost	0.3 g	MAE	Water containing 40% w/v Mg(NO$_3$)2·6H$_2$O and 4% v/v NH$_3$) (135 °C, 20 min)	10 mL	SPE (Oasis HLB)	LC-MS	70–112	2.2–3.0	Speltini et al. (2015)
Antibiotics, anti-epileptic, anti-inflammatories, and a lipid regulator	Sewage sludge supernatant	5 mL	QuEChERS	ACN (0.1% acetic acid),Na2SO4.NaAc.EDTA	5 mL	Online SPE	LC-MS/MS	78–120	10–200 ng L−1	Bourdat-Deschamps et al. (2014)
119 pharmaceuticals and 17 hormonal steroids (antidepressants, anti-inflammatory, antihistamines, hormones, antibiotics, lipid regulators, antifungals, radio-contrast agents, anti-cancers, anticonvulsants, anthelmintic, anti-pyretics, β-blockers, antiarrhythmics, diuretics, antitussives, vasodilator, nervous stimulant, antiemetic, anti-diabetic, proton pump inhibitor, analgesics, anti-coagulant)	Limed sludge Digested sludge Dried sludge Compost	2 g	QuEChERS	ACN (1% acetic acid) (10mL); EDTA 0.1 M (10mL); Heptane (1 mL)	21 mL	d-SPE (MgSO4, PSA)	LC-TOF-MS	2–122	1–2500	Peysson et al. (2013)
Non-steroidal anti-inflammatories	Stabilized and non-stabilized sewage sludge	0.2 g	MSPD	Hexane:acetone (1:2, v/v)	15 mL	MSPD (KOH, Na$_2$SO$_4$, Florisil)	LC-QTOF-MS	86–105	0.005–0.05	Trifanes et al. (2016)

(Continued)

TABLE 9.1 (CONTINUED)
Overview of Analytical Methods for the Determination of Pharmaceutical Compounds in Sewage Sludge

Therapeutic Group	Sample	Sample Amount	Extraction Technique	Extraction Solvents	Total Solvent Volume	Clean-up	Analytical Determination	Recoveries (%)	Limits of Detection (ng g−1 d.w.)	References
45 pharmaceuticals (antibiotics, non-steroidal anti-inflammatory drugs, β-blockers, and antidepressants)	Sewage sludge	0.1 g	MSPD	MeOH (6 mL), ACN:oxalic acid 5% (8:2, v/v) (10 mL)	16 mL	MSPD (C18)	LC-MS/MS	50–107	0.117–5.55	Li et al. (2016)

a Limits of quantification.

b ACN: acetonitrile; d-SPE, dispersive solid-phase extraction; EDTA, ethylenediaminetetraacetic acid; GC-MS, gas chromatography–mass spectrometry; GC-MS/MS, gas chromatography–tandem mass spectrometry; GPC, gel permeation chromatography; HPLC-UV-Fl, high performance liquid chromatography with ultraviolet and fluorescence detectors; LC-MS/MS, liquid chromatography–tandem mass spectrometry; LC-QTOF-MS, liquid chromatography–time of flight-mass spectrometry; MAE, microwave-assisted extraction; MeOH, methanol; MSPD, matrix solid-phase dispersion; n.d., no data; PLE, pressurized liquid extraction; PSA, primary and secondary amine; QuEChERS, quick, easy, cheap, effective, rugged and safe; SALLE, salt-assisted liquid-liquid extraction; UAE, ultrasonic-assisted extraction.

affordable for routinely monitoring pharmaceuticals in sewage sludge. In the multiresidue determination of pharmaceutical compounds with different physicochemical properties, compromise solutions have to be applied, such as, for instance, not adjusting extraction solvent pH, which can result in good extraction recoveries for some compounds but poor recoveries for the most acidic pharmaceuticals. The most widely used extraction techniques for the determination of pharmaceuticals in sewage sludge are ultrasonic-assisted extraction (UAE), pressurized liquid extraction (PLE), and microwave-assisted extraction (MAE) (Table 9.1). Some authors have compared the efficiencies of several techniques for the extraction of pharmaceuticals from sewage sludge (Dorival-García et al., 2013; Okuda et al., 2009). Dorival-García et al. (2013) compared quinolone extraction from sewage sludge by UAE, MAE, and PLE. They concluded that MAE and PLE were better options for quinolone extraction than UAE, because they enabled higher extraction efficiencies, easy operation, a shorter analysis time, and a higher degree of automation than UAE. Okuda et al. (2009) compared UAE and PLE for the extraction of 66 pharmaceutical compounds and personal care products from sewage sludge, concluding that some compounds were better extracted by UAE, whereas others were better extracted by PLE. Recently, some easy and inexpensive methods have been reported for pharmaceutical determination in sewage sludge. These methods are the so-called quick, easy, cheap, effective, rugged, and safe extraction (QuEChERS) (Peysson and Vulliet, 2013; Bourdat-Deschamps et al., 2014) and matrix solid-phase extraction (MSPD) (Triñanes et al., 2016; Li et al., 2016). Regardless of the extraction technique applied, samples are first dried (to express the concentration per dry weight), homogenized, and sieved. To avoid the degradation or loss of pharmaceuticals at high temperatures, drying by lyophilization is preferred.

9.2.1.1 Ultrasound-Assisted Extraction (UAE)

In UAE, solid-liquid extraction is carried out in an ultrasound bath. UAE is an efficient and low solvent-consuming alternative to classical extraction techniques such as Soxhlet, is inexpensive in comparison with PLE and MAE equipment (Albero et al., 2015), and does not require the high temperatures applied in PLE, which could degrade some compounds (Martín et al., 2010; Herrero et al., 2013). This extraction technique is widely used for the extraction of organic pollutants from sewage sludge (Zuloaga et al., 2012). The main variables to optimize are the type and volume of extraction solvent and extraction cycles. Acetone, methanol (MeOH), and acetonitrile are the most commonly used extraction solvents for the determination of pharmaceutical compounds. In general, two or three extraction cycles with fresh solvent are applied with a total extraction time lower than 45 min (Albero et al., 2015). After extraction, clean-up by SPE is usually applied to remove interfering compounds. Some multiresidue methods have been reported using UAE. Gago-Ferrero et al. (2015) developed a multiresidue method for the extraction of 148 pharmaceuticals and illicit drugs from sewage sludge and determination by liquid chromatography-tandem mass spectrometry (LC-MS/MS). UAE was carried out with MeOH and acidified water containing ethylene diamine tetraacetic acid (EDTA). No clean-up was applied. Absolute recoveries were in the range of 50%–110% for more than 77% of studied compounds. Limits of detection (LODs) were below 0.01 ng kg^{-1} d.m.

for 91% of analytes. López Zavala and Reynoso-Cuevas (2015) reported the extraction of eight pharmaceuticals from compost using ethyl acetate:MeOH (1:1, v/v) as extraction solvent. Recoveries in the range 87%–113%, relative standard deviations lower than 11%, and limits of quantification (LOQ) in the order of 2 ng g^{-1} were obtained. Okuda et al. (2009) compared UAE and PLE extraction efficiencies for 66 pharmaceuticals and personal care products from sewage sludge. Fifty-two percent of the pharmaceuticals were better extracted by USE. They tested four MeOH:water mixtures as USE extraction solvents (0:10v/v, 1:9 v/v, 5:5 v/v, and 10:0 v/v) and four pH conditions (pH 2, 4, 7, and 11). The best UAE recoveries were obtained using MeOH:water pH 11 (1:9, v/v).

9.2.1.2 Pressurized Liquid Extraction (PLE)

In PLE, also known as accelerated solvent extraction (ASE), solid sample is introduced into an extraction cell together with an inert dispersant sorbent. The extraction is accelerated by applying high temperatures and pressures. Commercially available systems allow the simultaneous and automated extraction of up to 24 samples with low solvent consumption. PLE combines good recoveries and adequate precision with rapid extraction (Nieto et al., 2010). PLE has been applied to the determination of anticancer drugs (Seira et al., 2013), glucocorticoids (Herrero et al., 2013), antibiotics, anti-inflammatories, analgesics, antiepileptics, and hormones, among other pharmaceuticals in sewage sludge (Barron et al., 2008; Langford et al., 2011; Nieto et al., 2010; Okuda et al., 2009; Radjenović et al., 2009a; Zuloaga et al., 2012). The main variables to optimize are extraction solvent, temperature, pressure, extraction time, and number of cycles. Extraction is commonly carried out using MeOH:water mixtures at temperatures in the range from 70 to 100°C (Table 9.1), pressure 10–14 Mbar, and from one to three cycles of 5–15 min (Nieto et al., 2010). Afterward, the extract can be diluted with water and subjected to clean-up, commonly by SPE with Oasis HLB cartridges. The influence of the extraction temperature is an important factor to consider. An increase of extraction temperature can result in an increase of extraction recoveries for some compounds, but for others, can result in a decrease due to thermal decomposition. Herrero et al. (2013) reported an increase of extraction recoveries of nine glucocorticoids from sewage sludge when the temperature was increased from room temperature to 40 °C. Nevertheless, an increase of extraction temperature from 60 to 100 °C resulted in a decrease of recoveries, which were below 10% for the majority of the compounds at 100 °C. High recoveries in the range from 80 to 100% have been reported for some compounds, such as ketoprofen, norfloxacin, ciprofloxacin, bezafibrate, and caffeine (Nieto et al., 2010). The less polar compounds (i.e., log K_{ow} higher than 4) may require a larger number of cycles than polar compounds to be extracted (Radjenović et al., 2009). Nevertheless, extraction efficiency depends not only on the properties of the pharmaceutical but also on the type of sewage sludge. Radjenović et al. (2009) reported a multiresidue method for the determination of 31 pharmaceuticals from different types of sewage sludge with extraction recoveries in the range from 3.2% to 100%. They described different behavior not only from one pharmaceutical compound to another but also, for some pharmaceutical compounds, such as naproxen, azithromycin, or glibenclamide, from one type of sludge to another due to matrix components that can strongly influence the extraction of some compounds. Okuda et al. (2009) compared UAE and PLE for the extraction of 66 pharmaceuticals from sludge. Fifteen percent of the pharmaceuticals were better extracted by PLE using water conditioned at pH 2 as extraction solvent, whereas 33% of the pharmaceuticals were better extracted using methanol conditioned at pH 4. The others were better extracted by UAE. Barron et al. (2008) reported a multiresidue determination of 27 frequently prescribed and consumed pharmaceuticals in digested sludge. The optimized method involved two PLE cycles with MeOH:water (1:1, v/v), 60 °C, and 1500 p.s.i. The total extraction solvent volume was 53 mL for 1 g of sludge. Afterward, clean-up by SPE with Oasis HLB cartridges and LC-MS/MS determination was applied. Recoveries were higher than 60% for 75% of the pharmaceutical compounds. Seira et al. (2013) applied PLE to the determination of two widely administered anticancer drugs (ifosfamide and cyclophosphamide). They found that extraction solvent was the most decisive extraction factor and that interactions between some parameters also appeared very influential. Matrix effect was described as the most limiting analytical step for quantification, to different extents depending on the analyte and the type of sludge.

9.2.1.3 Microwave-Assisted Extraction (MAE)

In MAE, microwave energy is used to heat solvents in contact with solid or liquid samples and to promote partition of the analytes from the sample matrix into the extraction solvent. The main variables to optimize are type and volume of extraction solvent, temperature, extraction time, and microwave power (Sánchez-Prado et al., 2015; Dobor et al., 2010). Dorival-García (2015) applied MAE to the determination of 17 strongly sorbed quinolones in compost samples, achieving recoveries in the range from 72% to 96% and inter-day precision lower than 4% for most of the compounds. Speltini et al. (2015) applied a new low-pressurized microwave platform, improving the extraction recoveries of fluoroquinolone residues from compost in comparison with conventional MAE. Evans et al. (2015) reported an analytical method for multiresidue determination of chiral pharmaceutically active compounds, including β-blockers, antidepressants, and amphetamines, in wastewater and digested sludge. The method allowed the determination of the enantiomers of 11 pharmaceuticals in digested sludge. The method was validated for 1 and 3 g of sludge. Extraction was carried out with 20 mL of MeOH:water (1:1, v/v) heated to 120 °C for 30 min, and extract clean-up was carried out with OASIS MAX cartridges (Waters, UK). OASIS HLB cartridges were rejected, because interfering compounds were not successfully removed, resulting in an increase in the backpressure of the chiral LC column. Most of the recoveries were in the range from 65% to 140%. Intra- and inter-day precision, expressed as relative standard deviation (RSD%), were in the range 4.3%–44.1% and 5.1%–51.7%, respectively. Petrie et al. (2016) reported a multiresidue analytical method for the determination of pharmaceutical compounds (estrogens, antibiotics, anti-hypertensives, anti-inflammatories, lipid regulators, anti-histamines, anti-diabetics, β-blockers, H2 receptor agonists, X-ray contrast media, cytostatics, anesthetics, anti-depressants, anti-epileptics, analgesics, stimulants, and others) and some of their metabolites together with other emerging contaminants in wastewater, river water, and digested sludge. Digested sludge (0.5 g) was extracted with 25 mL of MeOH:water (pH: 2) (1:1,v/v) by MAE at 110 °C for 30 min, then clean-up by SPE with OASIS MCX cartridges (Waters, UK) was applied. Analytical determination was carried out by LC-MS/MS. Recoveries were in the range from 49% to 180%, but for the majority of compounds, they were in the range from 90% to 110%. Langford et al. (2011) applied MAE to the extraction of the 30 most dispensed pharmaceuticals in Scotland in 2007. Extraction was carried out with pure MeOH or MeOH:formic acid (100:0.1, v/v). Extracts were evaporated to 8–9 mL and centrifuged at 21,000 g, forming three layers. The middle layer was injected into the LC-MS/MS system without clean-up.

9.2.1.4   Quick, Easy, Cheap, Effective, Rugged and Safe Extraction (QuEChERS)

This method was developed by Anastassiades et al. (2003) for the determination of pesticides in fruits and vegetables. QuEChERS is based on sample extraction with acetonitrile, liquid-liquid partition by salting out with sodium chloride and magnesium sulfate, and simultaneous removal of residual water and clean-up by dispersive solid-phase extraction (d-SPE) with anhydrous magnesium sulfate and a disperser sorbent. The variables to optimize are the type and amount of salts, the presence and type of ligand (EDTA, citrate, etc.), the sample pH, acidification or not of acetonitrile (commonly with acetic acid), and the type and amount of the disperser sorbent (primary and secondary amine [PSA], C18, graphitized carbon, and their mixtures are most commonly used). The method is simple and rapid and does not require expensive equipment or large solvent volumes. Peysson and Vulliet (2013) reported a multiresidue method based on QuEChERS extraction and liquid chromatography/time-of-flight/mass spectrometry (LC-TOF-MS), which allowed the determination of 117 pharmaceuticals and hormone residues in sewage sludge. PSA, C18, graphitized carbon, and their mixtures were tested as disperser sorbents. The best recoveries were obtained with PSA. The major drawback of the method was the low recoveries of complexed molecules, tetracyclines, and fluoroquinolones. Bourdat-Deschamps et al. (2014) applied QuEChERS to the determination of 13 pharmaceuticals, including eight antibiotics (fluoroquinolones, tetracyclines, sulfonamides, and macrolide), an anti-epileptic (carbamazepine), anti-inflammatories (diclofenac and ibuprofen), and a lipid regulator (gemfibrozil) in sewage sludge supernatant from an urban municipal WWTP based on activated sludge. Sludge supernatant (5 mL) was extracted with acidified acetonitrile (1% acetic acid) after addition of sodium sulfate, sodium acetate, and Na_2EDTA. Clean-up was carried out with PSA. They reported that sodium sulfate, as QuEChERS salt, and Na_2EDTA are necessary for fluoroquinolone and tetracycline extraction. Recoveries were in the range from 78% to 120%; inter-day precision, expressed as RSD%, was 6%–30%; and LOQs were in the range of 10–200 ng L^{-1}.

9.2.1.5   Matrix Solid-Phase Dispersion (MSPD)

MSPD was developed in 1989 by Baker et al. (Baker et al., 1989) for the isolation of drug residues from tissues. Solid and semi-solid samples are mixed with an appropriate solid sorbent, introduced into an empty column, and eluted with the appropriate solvent. Extraction and clean-up are simultaneously carried out with low solvent consumption and without expensive instrumentation. The variables to optimize are the type of solid sorbent, elution solvent, and amount of additives. Recently, a few methods based on MSPD have been reported for the determination of pharmaceuticals in sewage sludge. Triñanes et al. (2016) reported a method for the determination of non-steroidal anti-inflammatory drugs. They proposed soaking 0.2 g of dried sludge sample with 100 μL of aqueous potassium hydroxide solution, mixing with anhydrous sodium sulfate, and adding 1 g of Florisil. The mixture was introduced into an empty column containing a 3 g layer of silica as clean-up sorbent. Elution was carried out with 15 mL of hexane:acetone (1:2, v/v). Analytical determination was carried out by liquid chromatography/quadrupole time-of-flight mass spectrometry (LC-QTOF-MS). Recoveries in the range of 86%–105% and LOQs in the range of 0.005–0.05 ng g^{-1} were obtained. Li et al. (2016) reported a method for the determination of 45 pharmaceuticals. Their best results were obtained by mixing sludge (0.1 g) with C18 and eluting with methanol and acetonitrile:oxalic acid 5% (8:2, v/v). Analytical determination was carried out by LC-MS/MS. Recoveries were in the range of 50%–107%, RSD% was lower than 15%, and LOQs were in the range of 0.11–5.5 ng g^{-1}.

9.2.2   EXTRACT CLEAN-UP

After sample extraction by UAE, MAE, and PLE, clean-up is usually applied to eliminate interfering compounds that can affect analytical determination, especially when LC-MS/MS with electrospray ionization source is employed (Table 9.1). The most frequently used technique for extract clean-up is SPE using Oasis HLB cartridges (Okuda et al., 2009; Barron et al., 2008; Dobor et al., 2010; Martín et al., 2010). Nevertheless, the use of other cartridges such as Oasis MAX (Seira et al., 2013) and Oasis MCX (Petrie et al., 2016), among others, has also been reported (Table 9.1). For instance, Ternes et al. (2005) reported the use of Oasis MCX cartridges for acidic pharmaceuticals, RP-C18 for neutral pharmaceuticals, and ENV+solute® /RP-C18 for iodinated contrast media.

9.2.3   ANALYTICAL DETERMINATION

LC-MS/MS is widely used for the determination of pharmaceuticals in sewage sludge (Zuloaga et al., 2012; Omar et al., 2016) (Table 9.1). Nevertheless, due to the high cost of LC-MS/MS, this equipment is unaffordable for many laboratories, making routine control of pharmaceuticals in sewage sludge difficult. In such cases, gas chromatography-mass spectrometry (GC-MS) can be an excellent alternative that provides comparable sensitivity and specificity to the LC-MS/MS system (Omar et al., 2016). Nevertheless, because most pharmaceuticals are polar and relatively non-volatile compounds, their analysis by GC-MS requires a previous derivatization step (Yu and Wu, 2012; Kumirska et al., 2015; Zhang et al, 2016). GC-MS has been commonly used for the determination of acidic pharmaceuticals (Dobor et al., 2010) and estrogens after derivatization. Kumirska et al. (2015) applied GC-MS to the determination of eight non-steroidal anti-inflammatory drugs and five estrogens. Derivatization was performed at 60 °C for 30 min by adding N,O-bis(trimethylsilyl)trifluoroacetamide (BSTFA) containing 1% of trimethylchlorosilane (TMCS) and pyridine (dried with solid KOH). Zhang et al. (2016) applied derivatization with BSTFA for 2 h at 40 °C to the determination of acidic pharmaceuticals and estrogenic hormones. Zuo et al. (2007) applied microwave irradiation to reduce the derivatization time of five estrogenic hormones. Derivatization was carried out with BSTFA and TMCS in pyridine solution by irradiation at 800 W for 60 s. Azzouz and Ballesteros (2012) applied derivatization with BSTFA + 1% TMCS for 20 min at 70 °C to the determination of eight pharmaceuticals (analgesics, antibacterials, anti-epileptics, β-blockers, lipid regulators, and non-steroidal anti-inflammatories) in sewage sludge. LC-MS/MS has been applied to the determination of a wider group of pharmaceutical compounds than GC-MS (Table 9.1). For instance, Gago-Ferrero et al. (2015) and Peysson and Vulliet (2013) developed LC-MS/MS methods for the determination of 148 and 136 pharmaceuticals, respectively. LC-MS/MS has also been applied to the determination of pharmaceutical enantiomers. The method reported by Evans et al. (2015) allows the determination of the enantiomers of 11 pharmaceutical compounds using a Chirobiotic V column (Sigma-Aldrich, UK) packed with Vancomycin. The main problem associated with LC-MS/MS determination of pharmaceuticals in sewage sludge is matrix effect. Matrix effect is due to co-extracted substances that cause signal suppression or enhancement, especially when electrospray source is used. Gago-Ferrero et al. (2015) reported signal suppression in the range from −92% to −3% for 136 out of the 148 target compounds and signal enhancement in the range of 11%–90% for the others. Matrix-matched calibration curves are commonly applied for quantification when matrix effect cannot be reduced by clean-up (Dorival-García et al., 2015). High performance liquid chromatography (HPLC) with diode-array (DAD) and fluorescence (Fl) detectors online has also been reported for the determination of pharmaceuticals in sewage sludge (Martín et al., 2010). The main advantage of HPLC-DAD-Fl is the use of relatively inexpensive detectors with adequate selectivity and sensitivity (Ribeiro et al., 2013), which makes HPLC-DAD(Fl) significantly more affordable for routine laboratories than LC-MS/MS (Martín et al., 2010). Moreover, the derivatization required for GC-MS determination is not needed. Martín et al. (2012a) employed HPLC-DAD-Fl to obtain information about the occurrence and fate of pharmaceuticals in wastewater and sludge from WWTPs, their removal, the ecotoxicological impact of wastewater discharges and sludge disposal (Martín et al., 2012a,b), and their distribution and temporal evolution alongside sewage treatment (Martín et al., 2012b). Ribeiro et al. (2013) applied HPLC-Fl to evaluate the enantioselectivity of the biodegradation of alpreonolol and propranolol using a Vancomycin-based chiral column. The results indicated slightly higher biodegradation rates for the S-enantiomeric forms of both β-blockers.

9.3 APPLICABILITY OF ANALYTICAL DETERMINATION OF PHARMACEUTICALS IN SEWAGE SLUDGE: CASE STUDIES

Pharmaceutical compounds can be degraded or accumulated onto sludge depending not only on their different physicochemical properties but also on the different composition of each type of sludge, which can result in different retention mechanisms. The sorption onto sludge is strongly affected by many factors, including the characteristics of the compound (molecular structure, in particular the presence of amino or carboxylic acid groups in the molecule, and chemical properties, including K_{ow}, pK_a, and K_d) and the sludge (organic compound fraction, cation exchange capacity, suspended solid size) and operating conditions (pH and sludge retention time) (Verlicchi and Zambello, 2015). For instance, Martínez-Alcalá et al. (2017) reported that the mechanism of elimination of carbamazepine, naproxen, and diclofenac from wastewater is due to sorption onto sludge, whereas the elimination of ibuprofen and ketoprofen is due to biodegradation (54.3% and 99.7%, respectively), and to a lower extent to sorption onto sludge (45.7% and 0.3%, respectively).

In this section, we describe the application of the analytical determination of pharmaceuticals in sewage sludge to obtain information about their evolution in the sludge treatment line and their occurrence and fate in sewage sludge from different treatment technologies, and assess the ecotoxicological risk of sewage sludge application to agricultural soils. Sixteen pharmaceutical compounds from seven therapeutic groups (five anti-inflammatories, two antibiotics, an antiepileptic drug, a β-blocker, a nervous stimulant, four hormones, and two lipid regulators) were selected based on their high consumption and environmental relevance. The physicochemical properties of the monitored pharmaceuticals are shown in Table 9.2. The analytical method applied was the one described in Martín et al. (2010).

9.3.1 Evolution of Concentrations of Pharmaceuticals in Sludge Treatment Line

To obtain information about the concentrations of pharmaceutical compounds in the sludge treatment line, sewage sludge from different treatment stages was sampled from four urban WWTPs based on activated sludge technology and from a composting plant. Wastewater treatments involved pretreatment and primary (settling) and secondary (activated sludge) treatments. Primary sludge is produced by settling pretreated wastewater in primary clarifiers. Secondary sludge is produced after activated sludge processes in a biological reactor and settling in secondary clarifiers. Digested sludge is produced by the anaerobic digestion of primary and secondary sludge mixtures. Compost sludge is obtained in a composting plant where anaerobically digested and dehydrated sludge from the above-mentioned WWTPs is treated in dynamic batteries thermally controlled with aeration facilitated by turning. Nine samples of each type of sludge were sampled from each WWTP and from the composting plant in a 1-year monitoring campaign. The characteristics of each type of sludge are shown in Table 9.3.

In Figures 9.1 and 9.2, the concentrations of the most concentrated pharmaceuticals (Martín et al., 2012b) are shown as box-and-whisker plots. Lines in each box show the lower (≤25%), median (≤50%), and upper quartile (≤75%). Lines from each box show the highest and lowest concentrations. The concentrations of some compounds (salicylic acid, ibuprofen, and caffeine) decreased from primary to secondary sludge, but the concentrations of others (naproxen, 17β-estradiol, and estrone) remained constant or even increased (carbamazepine, propranolol, and 17α-ethinylestradiol) from primary to secondary sludge. This fact can be explained not only by the different physicochemical properties of the pharmaceutical compounds (log K_{ow} and

TABLE 9.2

Physicochemical Properties of the Selected Pharmaceutical Compounds

Therapeutic Group	Pharmaceutical Compound	S_w (mg L^{-1})	Log K_{ow}	pK_a	Log K_d			
					Primary Sludge	Secondary Sludge	Digested Sludge	Soil
Anti-inflammatories	Diclofenac	2.4	4.5	4.2	2.29[a]	2.07[a]	2.70[b]	2.21[c]
	Ibuprofen	21	4	4.9	2.11[d]	2.37[d]	2.10[b]	1.45[e]
	Ketoprofen	300	3.1	4.5	1.50[b]	1.24[d]	2.40[b]	0.95[f]
	Naproxen	16	3.2	4.2	2.11[d]	1.53[d]	1.55[f]	0.95[f]
	Salicylic acid	2000	1.2	3.5	1.31[d]	2.25[d]	1.36[b]	1.91[f]
Antibiotics	Sulfamethoxazole	610	0.9	1.7, 5.6	0.51[a]	1.89[a]	1.04[b]	0.90[f]
	Trimethoprim	171	1.4	6.6	2.63[a]	2.40[a]	1.83[b]	1.41[f]
Antiepileptic	Carbamazepine	17.7	2.5	13.9	2.50[a]	2.13[a]	1.15[b]	1.40[f]
β-Blocker	Propranolol	3009	0.7	9.5	2.81[a]	2.56[a]	2.52[b]	1.76[f]
Nervous stimulant	Caffeine	21,700	−0.1	14	1.00[b]	—	2.30[b]	1.40[f]
Estrogens	17α-Ethinylestradiol	11.3	3.7	10.4	2.44[g]	2.54[g]	2.46[h]	1.75[i]
	17β-Estradiol	3.6	3.9	10.4	4.02[j]	4.54[j]	2.56[a]	1.85[i]
	Estriol	441	2.8	9.8	2.27[d]	2.21[d]	2.67[k]	2.67[k]
	Estrone	30	3.4	10.4	2.77[j]	2.90[j]	2.45[h]	1.40[i]
Lipid regulators	Clofibric acid	582	2.6	3.2	—	0.68[g]	0.70[b]	0.95[f]
	Gemfibrozil	10.9	4.8	4.7	1.36[a]	1.29[a]	—	0.11[i]

K_a, acid ionization constant; K_d, solid–water distribution coefficient; K_{ow}, octanol–water distribution coefficient; S_w, solubility in water.

[a] Radjenović et al. 2009a; [b] Okuda et al. 2009; [c] Drillia et al., 2005; [d] Unpublished experimental data; [e] Stuer-Lauridsen et al. 2000; [f] Barron et al. 2009; [g] Ternes et al. 2004; [h] Carballa et al. 2008; [i] Sarmah et al. 2008; [j] Carballa et al., 2007; [k] López de Alda et al. 2002; [l] Krascsenits et al. 2008.

TABLE 9.3

Main Values of Sludge Characterization Parameters

Parameter	Primary Sludge	Secondary Sludge	Anaerobically Digested Sludge	Compost
pH	6.48–7.15	6.66–7.20	8.40–8.61	6.55–8.29
Moisture content (%)	95.7–96.8	96.2–96.8	75.3–80.2	28.6–36.3
Dryness (%)	3.23–4.28	3.17–3.81	19.8–24.7	63.7–71.4
Organic matter (%)	62.6–74.4	73.3–78.7	48.6–52.1	20.2–26.6
Total nitrogen (%)	2.01–2.48	4.22–6.28	3.27–4.04	2.54–4.52
Total phosphorus (%)	0.13–0.28	0.28–0.39	0.26–0.38	0.06–0.10

pK_a values) (Table 9.2) but also by the different composition of primary and secondary sludge (Table 9.3) resulting in different retention mechanisms. The higher concentrations of some compounds in primary sludge could be due to retention by electrostatic interactions, whereas the higher concentrations of other compounds in secondary sludge may be due to the hydrolysis of conjugates, releasing the pattern compound, and to sorption enhancement of non-ionic compounds, due to the higher organic matter content of secondary sludge (Table 9.3), by non-ionic van der Waals interactions (Martín et al., 2012a; Ternes et al. 2004; Verlicchi and Zambello, 2015). After anaerobic digestion, the concentration levels of most of the pharmaceuticals decreased by biodegradation (Figures 9.1 and 9.2). Nevertheless, the concentration of some compounds, such as caffeine and 17β-estradiol (Figure 9.2), increased.

9.3.2 Occurrence and Fate of Pharmaceuticals in Sewage Sludge from Different Treatment Technologies

Sludge stabilization can be carried out by anaerobic digestion (in anaerobic treatment plants [AnTP] and in anaerobic wastewater stabilization ponds [AnWSPs]) or by aerobic digestion (in aerobic treatment plants [AeTP] and in composting plants [CP]). AnWSPs are low-cost urban wastewater treatments based on shallow ponds typically used to treat wastewater from small towns. Aerobic digestion is a bacterial process in the presence of oxygen. Composting is a natural process of microbial decomposition of organic matter enhanced by sludge aeration by regularly turning. In previous work by our group, the 16 above-mentioned pharmaceuticals (Table 9.2) together with the anti-inflammatory acetaminophen; the antibiotics ciprofloxacin, norfloxacin, and ofloxacin; the β-blocker atenolol; and the lipid regulator bezafibrate were analyzed in sewage sludge from four AnTPs, three AeTPs, three AnWSPs, and a CP (Martín et al., 2015). Four sampling campaigns in a 1-year period were carried out. The most contaminated samples were primary sludge from AnTPs (mean concentration: 179 ng g^{-1} d.m.), mixed sludge (mixture of primary and secondary sludge) from AeTPs (mean concentration: 310 ng g^{-1} d.m.), and lagoon sludge from AnWSPs (mean concentration: 142 ng g^{-1} d.m.). A higher decrease was observed after anaerobic digestion than after aerobic digestion, which can be explained by

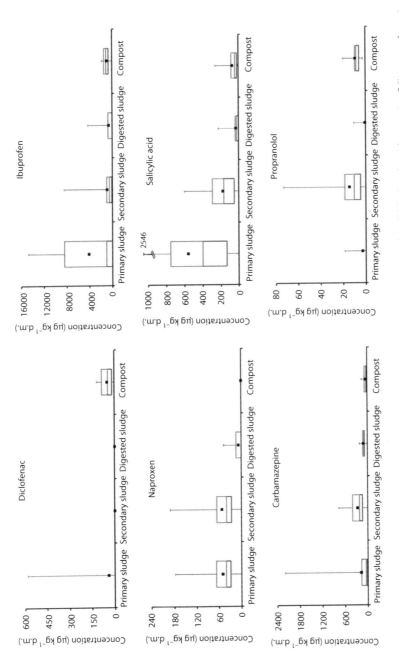

FIGURE 9.1 Box-and-whisker plots of anti-inflammatory drugs, an antiepileptic drug, and a β-blocker in primary (n=36), secondary (n=36), anaerobically digested (n=36), and compost (n=9) sludge.

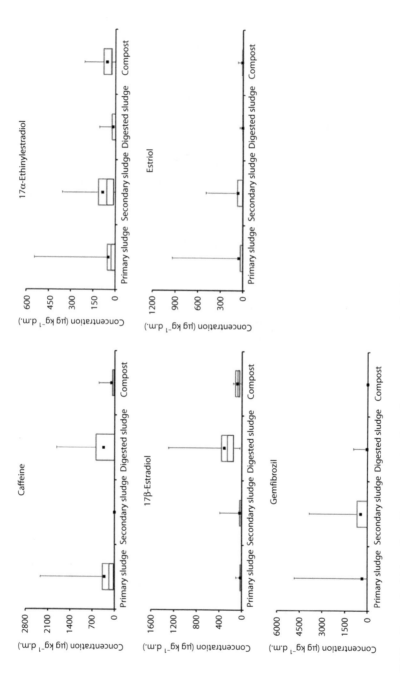

FIGURE 9.2 Box-and-whisker plots of a nervous stimulant, three estrogens, and a lipid regulator in primary ($n = 36$), secondary ($n = 36$), anaerobically digested ($n = 36$), and compost ($n = 9$) sludge.

higher degradation under anaerobic conditions. In contrast, lagooning treatment caused lower elimination of all the compounds detected (Martín et al., 2015).

9.3.3   ECOTOXICOLOGICAL RISK ASSESSMENT OF SEWAGE SLUDGE APPLICATION ONTO SOILS

Ecotoxicological risk assessment can be evaluated by means of risk quotient values (RQ). RQ values were calculated for digested-sludge and compost organisms and for soil organisms after soil application of digested sludge or compost. RQ values for sludge organisms are calculated as the ratios between the measured environmental concentration (MEC) in sludge and the highest concentration of the pharmaceutical below which no adverse effect is expected to occur (predicted no-effect concentration [PNEC]). RQ values for soil organisms are calculated as the ratio between the predicted environmental concentration (PEC) in soil after sludge application and the PNEC of the pharmaceutical compound. The criterion usually applied to evaluate the ecotoxicological risk is that proposed by Hernando (2006), which considers there to be a high risk if $RQ \geq 1$, a medium risk if $0.1 \leq RQ \leq 1$, and a low risk if $RQ \leq 0.1$ (Hernando et al., 2006; Verlicchi and Zambello, 2015). According to the European Technical Guidance Document on Risk Assessment EUR 20418 EN/2 (EC-TGD, 2003), the concentration of a pollutant in soil after the first year of sludge application is given by the equation

$$PEC = \frac{C_{sludge} \times APPL_{sludge}}{DEPTH_{soil} \times RHO_{soil}}$$

where:

C_{sludge}	is the concentration in dry sewage sludge
$APPL_{sludge}$	is the dry sludge application rate
$DEPTH_{soil}$	is the mixing depth of soil
RHO_{soil}	is the bulk density of wet soil
$APPL_{sludge}$, $DEPTH_{soil}$, and RHO_{soil}	values applied to agricultural soils are 0.5 kg_{dw} m^2 year^{-1}, 0.20 m, and 1700 kg m^{-3}, respectively (EC-TGD, 2003). Due to the lack of toxicity data for soil and sludge organisms, PNEC concentrations for soil ($PNEC_{soil}$) and sludge ($PNEC_{sludge}$) organisms are evaluated from PNEC data for aquatic organisms ($PNEC_{water}$) and the sludge ($K_{d\ sludge}$) or soil ($K_{d\ soil}$) solid-water partition coefficient of the pharmaceutical compound (Martín et al., 2012b; Verlicchi and Zambello, 2015):

$$PNEC_{sludge,\ soil} = PNEC_{water} \times K_{d\ sludge,\ soil}$$

$PNEC_{water}$ is calculated from the available toxicity data, applying assessment factors depending on the availability of long-term no observed effect concentration (NOEC) values for species from three, two, or just one trophic level (assessment factors of 10, 50, and 100, respectively) or at least one short-term $L(E)C_{50}$ from each of three trophic levels (fish, Daphnia, and algae) (assessment factor of 1000) (EC-TGD, 2003).

In Table 9.4, PEC, PNEC, and RQ values for the pharmaceutical compounds are shown. PEC values were calculated from the highest concentrations measured in anaerobically digested sludge from four WWTPs and in compost samples from a composting plant where the digested sludge from the above-mentioned WWTP is composted. Nine digested sludge samples were sampled from each WWTP and from the composting plant. $PNEC_{soil}$ values were calculated from short-term $L(E)C_{50}$ data available in literature (Martín et al., 2012b), applying an assessment factor of 1000. RQ values were higher than 1 for ibuprofen, 17α-ethinylestradiol, and 17β-estradiol. Nevertheless, after sludge application to soils, RQ values were lower than 1, except for 17β-estradiol in digested sludge–amended soils; therefore, no significant ecotoxicological risk is expected to occur. The RQ for 17β-estradiol is still slightly higher than 1 (RQ: 1.08) in digested sludge–amended soils, but this value corresponds to the worst-case scenario. It was calculated from the highest concentration measured in the digested sludge samples analyzed.

9.4 CONCLUSIONS AND FUTURE TRENDS

Several multiresidue methods have been reported for the determination of a wide group of pharmaceuticals belonging to different therapeutic groups and with different physico-chemical properties. Reported multiresidue methods allow the determination of up to 148 pharmaceuticals (Gago-Ferrero et al., 2015) by UAE and LC-MS/MS or 119 pharmaceuticals (Peysson et al., 2013) by QuEChERS and LC-TOF-MS determination. The most widely reported techniques for sample extraction are UAE, PLE, and MAE, and afterward, clean-up by SPE using Oasis HLB cartridges is commonly applied. Recently, QuEChERS (Bourdat-Deschamps et al., 2014; Peysson et al., 2013) and MSPD (Li et al, 2016; Triñanes et al., 2016) have been applied to the determination of pharmaceuticals in sewage sludge. In both methods, extraction and clean-up are carried out simultaneously, reducing analysis time. Analytical determination is carried out by LC-MS/MS and to a lower extent, by GC-MS/MS and HPLC-DAD-(FI).

Future investigations should focus on the occurrence of metabolites, the stereoselective degradation of chiral pharmaceuticals, and pharmaceuticals with a high sorption potential in treated sludge. More investigation is needed about the fate of pharmaceuticals in different types of sludge and in sludge-amended soils. The main drawbacks of the most commonly extraction techniques reported (UAE, PLE, and MAE) are the need for a further clean-up step to remove interfering compounds and the use of high-cost equipment (PLE and MAE). QuEChERS and MSPD are promising low-cost extraction techniques requiring low solvent volumes, which have been reported for multiresidue extraction of up to 136 pharmaceuticals and hormonal steroids (Peysson et al., 2013) and 45 pharmaceuticals (Li et al., 2016), respectively. The

TABLE 9.4
Predicted Environmental Concentrations (PEC), Predicted Non-Effects Concentrations in Soils (PNEC$_{soil}$), and Risk Quotients for Sludge Organisms and Sludge-Amended Soil Organisms

Pharmaceutical Compound	PEC-PNEC (ng g^{-1})			Risk Quotients			
	PEC$_{digested}$[a]	PEC$_{compost}$[a]	PNEC$_{soil}$[b]	Digested Sludge	Compost	Digested Sludge–Amended soil	Compost-Amended Soil
Diclofenac	—	0.18	1595	—	0.02	—	0.00011
Ibuprofen	6.2	2.64	256	2.35	1.57	0.01532	0.01033
Ketoprofen	—	—	140	—	—	—	—
Naproxen	0.07	0.002	189	0.04	0.01	0.00028	0.00001
Salicylic acid	0.33	0.38	3520	0.1	0.26	0.00004	0.00011
Sulfamethoxazole	—	0.09	1.19	—	37.8	—	0.07665
Trimethoprim	—	—	3102	—	—	—	—
Carbamazepine	0.31	0.23	347	0.91	0.8	0.00075	0.00066
Propranolol	0.02	0.03	40.3	0.01	0.09	0.00024	0.00073
Caffeine	2.67	0.73	2185	0.04	0.03	0.00049	0.00033
17α-Ethinylestradiol	0.16	0.3	1.69	7.62	23.3	0.05814	0.17776
17β-Estradiol	1.9	0.23	0.99	142	30.7	1.08247	0.23437
Estriol	0.06	0.1	711	0.02	0.09	0.00003	0.00013
Estrone	—	—	2.51	—	—	—	—
Clofibric acid	—	—	642	—	—	—	—
Gemfibrozil	—	—	13.3	—	—	—	—

—, not detected in sewage sludge.

[a] Martin et al. 2012a.

[b] Martin et al. 2012b.

main barrier to extending the monitoring of pharmaceuticals to routine laboratories is the high cost of the LC-MS/MS instruments commonly reported for pharmaceutical determination in sludge. Therefore, more investigation is needed to promote the determination of at least some of the most environmentally relevant pharmaceutical compounds by means of the development of analytical methods affordable for a larger number of routine laboratories.

ACKNOWLEDGMENTS

The authors would like to thank the Ministerio de Educación y Ciencia, Spain (Project no. CGL2007-62281) and the Programa de Becas de Formación de Profesorado Universitario (Ministerio de Educación, Spain) for financial support to carry out the case studies reported in Section 9.3.

REFERENCES

Albero, B., Sánchez-Brunete, C., García-Valcárcel, A.I., Pérez, R.A., and Tadeo, J.T. 2015. Ultrasound-assisted extraction of emerging contaminants from environmental samples. *Trends in Analytical Chemistry* 71: 110–118.

Anastassiades, M., Lehotay, S.J., Štajnbaher D., and Schenck F.J. 2003. Fast and easy multiresidue method employing acetonitrile extraction/partitioning and "dispersive solid-phase extraction" for the determination of pesticide residues in produce. *Journal of AOAC International* 86: 412–431.

Azzouz, A. and Ballesteros, E. 2012. Combined microwave-assisted extraction and continuous solid-phase extraction prior to gas chromatography–mass spectrometry determination of pharmaceuticals, personal care products and hormones in soils, sediments and sludge. *Science of the Total Environment* 419: 208–215.

Baker, S.A., Long, A.R., and Short, C.R. 1989. Isolation of drug residues from tissues by solid phase dispersion. *Journal of Chromatography A* 475: 353–361.

Barron, L., Purcell, M., Havel, J., Thomas, K., Tobin, J., and Paull, B. 2009. Occurrence and fate of pharmaceuticals and personal care products within sewage sludge and sludge-enriched soils, 2005-FS-30-M2 STRIVE Report. Environmental Protection Agency, Dublin, Ireland, 73 pp.

Barron, L., Tobin, J., and Paull, B. 2008. Multi-residue determination of pharmaceuticals in sludge and sludge enriched soils using pressurized liquid extraction, solid phase extraction and liquid chromatography with tandem mass spectrometry. *Journal of Environmental Monitoring* 10(3): 353–361.

Bourdat-Deschamps, M., Leang, S., Bernet, N., Daudin, J.J., and Nélieu, S. 2014. Multi-residue analysis of pharmaceuticals in aqueous environmental samples by online solid-phase extraction–ultra-high-performance liquid chromatography-tandem mass spectrometry: Optimisation and matrix effects reduction by quick, easy, cheap, effective, rugged and safe extraction. *Journal of Chromatography A*, 1349: 11–23.

Carballa, M., Omil, F., and Lema, J.M. 2007. Calculation methods to perform mass balances of micropollutants in sewage treatment plants. Application to pharmaceutical and personal care products (PPCPs). *Environmental Science and Technology* 41(3): 884–890.

Carballa, M., Omil, F., and Lema, J., 2008. Comparison of predicted and measured concentrations of selected pharmaceuticals, fragrances and hormones in Spanish sewage. *Chemosphere* 72: 1118–1123.

Dobor, J., Varga, M., Yao, J., Chen, H., Palkó, G., and Záray, G. 2010. A new sample preparation method for determination of acidic drugs in sewage sludge applying microwave assisted solvent extraction followed by gas chromatography-mass spectrometry. *Microchemical Journal* 94: 36–41.

Dorival-García, N., Labajo-Recio, C., Zafra-Gómez, A., Juárez-Jiménez, B., and Vílchez, J.L. 2015. Improved sample treatment for the determination of 17 strong sorbed quinolone antibiotics from compost by ultra high performance liquid chromatography tandem mass spectrometry. *Talanta* 138: 247–257.

Dorival-García, N., Zafra-Gómez, A., Camino-Sánchez, F.J., Navalón, A., and Vílchez, J.L. 2013. Analysis of quinolone antibiotic derivatives in sewage sludge samples by liquid chromatography-tandem mass spectrometry: Comparison of the efficiency of three extraction techniques. *Talanta* 106: 104–118.

Drillia, P., Stamatelatou, K., and Lyberatos, G. 2005. Fate and mobility of pharmaceuticals in solid matrices. *Chemosphere* 60: 1034–1044.

EC, 1986. Council Directive 86/278/EEC of 12 June 1986 on the protection of the environment, and in particular of the soil, when sewage sludge is used in agriculture. *Official Journal of European Communities No L* 181: 6–12.

EC, 1991. Council Directive 91/271/EEC of 21 May 1991 concerning urban waste water treatment. *Official Journal of European Communities No L* 135: 40–52.

EC-TGD, 2003. Technical Guidance Document on Risk Assessment, Part II, EUR 20418 EN/2. *European Commission*, Joint Research Centre.

EU, 2015. Commission implementing decision (EU) 2015/495 of 20 March 2015 establishing a watch list of substances for Union-wide monitoring in the field of water policy pursuant to Directive 2008/105/EC of the European Parliament and of the Council. *Official Journal of European Communities No L* 78: 40–42.

Evans, S.E., Davies, P., Lubben, A., and Kasprzyk-Hordern, B. 2015. Determination of chiral pharmaceuticals and illicit drugs in wastewater and sludge using microwave assisted extraction, solid-phase extraction and chiral liquid chromatography coupled with tandem mass spectrometry, *Analytica Chimica Acta* 882: 112–126.

Gago-Ferrero, P., Borova, V., Dasenaki M.E., and Thomaidis, N.E. 2015. Simultaneous determination of 148 pharmaceuticals and illicit drugs in sewage sludge based on ultrasound-assisted extraction and liquid chromatography–tandem mass spectrometry. *Analytical and Bioanalytical Chemistry* 15: 4287–4297.

Göbel, A., Thomsen, A., McArdell, C.S., Alder, A.C., Giger, W., Theiß, N., Löffler, D., and Ternes, T.A. 2005. Extraction and determination of sulfonamides, macrolides, and trimethoprim in sewage sludge. *Journal of Chromatography A* 1085(2): 179–189.

Hernando, M.D., Mezcua, M., Fernández-Alba, A.R., and Barceló, D., 2006. Environmental risk assessment of pharmaceutical residues in wastewater effluents, surface waters and sediments. *Talanta* 69: 334–342.

Herrero, P., Borrull, F., Marcé, R.M., and Pocurull, E. 2013. Pressurised liquid extraction and ultra-high performance liquid chromatography-tandem mass spectrometry to determine endogenous and synthetic glucocorticoids in sewage sludge. *Talanta* 103: 186–193.

Krascsenits, Z., Hiller, E., and Bartal, M. 2008. Distribution of four human pharmaceuticals, carbamazepine, diclofenac, gemfibrozil, and ibuprofen between sediment and water. *Journal of Hydrology and Hydromechanics* 56 (4): 237–246.

Kumirska, J., Migowska, N., Caban, M., Łukaszewicz, P., and Stepnowski, P. 2015. Simultaneous determination of non-steroidal anti-inflammatory drugs and oestrogenic hormones in environmental solid samples. *Science of the Total Environment* 508: 498–505.

Langford, K.H., Reid, M., and Thomas, K.V. 2011. Multi-residue screening of prioritised human pharmaceuticals, illicit drugs and bactericides in sediments and sludge. *Journal of Environmental Monitoring* 13: 2284–2291.

Li, M., Sun, Q., Li, Y., Lv, M., Lin, L., Wu, Y., Ashfaq, M., and Yu, C.P. 2016. Simultaneous analysis of 45 pharmaceuticals and personal care products in sludge by matrix solid-phase dispersion and liquid chromatography-tandem mass spectrometry. *Analytical and Bioanalytical Chemistry* 18: 4953–4964.

Lindberg, R.H., Wennberg, P., Johansson, M.I., Tysklind, M., and Andersson, B.A.V. 2005. Screening of human antibiotic substances and determination of weekly mass flows in five sewage treatment plants in Sweden. *Environmental Science and Technology* 39(10): 3421–3429.

López de Alda, M.J., Gil, A., Paz, E., and Barceló, D., 2002. Occurrence and analysis of estrogens and progestogens in river sediments by liquid chromatography-electrospray-mass-spectrometry. *Analyst* 127: 1299–1304.

López Zavala, A.M. and Reynoso-Cuevas, L. 2015. Simultaneous extraction and determination of four different groups of pharmaceuticals in compost using optimized ultrasonic extraction and ultrahigh pressure liquid chromatography – mass spectrometry. *Journal of Chromatography A* 1423: 9–18.

Martín, J., Camacho-Muñoz, D., Santos, J.L., Aparicio, I., and Alonso, E. 2012a. Distribution and temporal evolution of pharmaceutically active compounds alongside sewage sludge treatment. Risk assessment of sludge application onto soils. *Journal of Environmental Management* 102: 18–25.

Martín, J., Camacho-Muñoz, D., Santos, J.L., Aparicio, I., and Alonso, E. 2012b. Occurrence of pharmaceutical compounds in wastewater and sludge from wastewater treatment plants: Removal and ecotoxicological impact of wastewater discharges and sludge disposal. *Journal of Hazardous Materials* 239–240: 40–47.

Martín, J., Santos, J.L., Aparicio, I., and Alonso, E. 2010. Multi-residue method for the analysis of pharmaceutical compounds in sewage sludge, compost and sediments by sonication-assisted extraction and LC determination. *Journal of Separation Science* 33(12): 1760–1766.

Martín, J., Santos, J.L., Aparicio, I., and Alonso, E. 2015. Pharmaceutically active compounds in sludge stabilization treatments: Anaerobic and aerobic digestion, wastewater stabilization ponds and composting. *Science of the Total Environment* 503–504: 97–104.

Martínez-Alcalá, I., Guillén-Navarro, J.M., and Fernández-López, C. 2017. Pharmaceutical biological degradation, sorption and mass balance determination in a conventional activated-sludge wastewater treatment plant from Murcia, Spain. *Chemical Engineering Journal* 316: 332–340.

Nieto, A., Borrull, F., Marcé, R.M., and Pocurull, E. 2007. Selective extraction of sulfonamides, macrolides and other pharmaceuticals from sewage sludge by pressurized liquid extraction. *Journal of Chromatography A* 1174(1–2): 125–131.

Nieto, A., Borrull, F., Pocurull, E., and Marce, R.M. 2010. Pressurized liquid extraction: A useful technique to extract pharmaceuticals and personal-care products from sewage sludge. *Trends in Analytical Chemistry* 29: 752–764.

Okuda, T., Yamashita, N., Tanaka, H., Matsukawa, H., and Tanabe, K., 2009. Development of extraction method of pharmaceuticals and their occurrences found in Japanese wastewater treatment plants. *Environment International* 35: 815–820.

Omar, T.F.T., Ahmad, A., Aris, A.Z., and Yusoff, F.M. 2016. Endocrine disrupting compounds (EDCs) in environmental matrices: Review of analytical strategies for pharmaceuticals, estrogenic hormones, and alkylphenol compounds. *Trends in Analytical Chemistry* 85: 241–259.

Petrie, B., Youdan, J., Barden, R., and Kasprzyk-Hordern, B. 2016. Multi-residue analysis of 90 emerging contaminants in liquid and solid environmental matrices by ultra-high-performance liquid chromatography tandem mass spectrometry. *Journal of Chromatography A* 1431: 64–78.

Peysson, W. and Vulliet, E. 2013. Determination of 136 pharmaceuticals and hormones in sewage sludge using quick, easy, cheap, effective, rugged and safe extraction followed by analysis with liquid chromatography-time-of-flight-mass spectrometry. *Journal of Chromatography A* 1290: 46–61.

Radjenović, J., Jelić, A., Petrović, M., and Barceló, D. 2009a. Determination of pharmaceuticals in sewage sludge by pressurized liquid extraction (PLE) coupled to liquid chromatography-tandem mass spectrometry (LC-MS/MS). *Analytical and Bioanalytical Chemistry* 393: 1685–1695.

Radjenovic, J., Petrovic, M., and Barcelo, D., 2009b. Fate and distribution of pharmaceuticals in wastewater and sewage sludge of the conventional activated sludge (CAS) and advanced membrane bioreactor (MBR) treatment. *Water Research* 43: 831–841.

Ribeiro, A.R., Afonso, C.M., Castro, P.M.L., and Tiritan, M.E. 2013. Enantioselective biodegradation of pharmaceuticals, alprenolol and propranolol, by an activated sludge inoculum. *Ecotoxicology and Environmental Safety* 87: 108–114.

Sánchez-Prado, L., García-Jares, C., Dagnac, T., and Llompart, M. 2015. Microwave-assisted extraction of emerging pollutants in environmental and biological samples before chromatographic determination. *Trends in Analytical Chemistry* 71: 119–143.

Sarmah, A.K., Northcott, G.L., and Scherr, F.F. 2008. Retention of estrogenic steroid hormones by selected New Zealand soils. *Environment International* 34 (6): 749–755.

Seira, J., Claparols, C., Joannis-Cassan, C., Albasi, C., Montréjaud-Vignoles, M., and Sablayrolles, C. 2013. Optimization of pressurized liquid extraction using a multivariate chemometric approach for the determination of anticancer drugs in sludge by ultra high performance liquid chromatography–tandem mass spectrometry. *Journal of Chromatography A*, 1283: 27–38.

Speltini, A., Sturini, M., Maraschi, F., Viti, S., Sbarbada, D., and Profumo, A. 2015. Fluoroquinolone residues in compost by green enhanced microwave-assisted extraction followed by ultra performance liquid chromatography tandem mass spectrometry. *Journal of Chromatography A* 1410: 44–50.

Stuer-Lauridsen, F., Birkved, M., Hansen, L.P., Holten Lützhøft, H.C., and Halling- Sørensen, B., 2000. Environmental risk assessment of human pharmaceuticals in Denmark after normal therapeutic use. *Chemosphere* 40(7): 783–793.

Ternes, T.A., Andersen, H., Gilberg, D., and Bonerz, M. 2002. Determination of estrogens in sludge and sediments by liquid extraction and GC/MS/MS. *Analytical Chemistry* 74: 3498–3504.

Ternes, T.A., Bonerz, M., Herrmann, N., Löffler, D., Keller, E., Lacida, B.B., and Alder, A.C. 2005. Determination of pharmaceuticals, iodinated contrast media and musk fragrances in sludge by LC tandem MS and GC/MS. *Journal of Chromatography A* 1067(1–2): 213–223.

Ternes, T.A., Herrmann, N., Bonerz, M., Knacker, T., Siegrist, H., and Joss A. 2004. A rapid method to measure the solid–water distribution coefficient (Kd) for pharmaceuticals and musk fragrances in sewage sludge. *Water Research* 38: 4075–4084.

Triñanes, S., Casais, M.C., Mejuto, M.C., and Cela, R. 2016. Matrix solid-phase dispersion followed by liquid chromatography tandem mass spectrometry for the determination of selective ciclooxygenase-2 inhibitors in sewage sludge samples. *Journal of Chromatography A*, 1462: 35–43.

Verlicchi, P., and Zambello, E. 2015. Pharmaceuticals and personal care products in untreated and treated sewage sludge: Occurrence and environmental risk in the case of application on soil—A critical review. *Science of the Total Environment* 538: 750–767.

Yu, Y. and Wu, L. 2012. Analysis of endocrine disrupting compounds, pharmaceuticals and personal care products in sewage sludge by gas chromatography–mass spectrometry. *Talanta* 89: 258–263.

Zhang, M., Mao, Q., Feng, J., Yuan, S., Wang, Q., Huang, D., and Zhang, J. 2016. Validation and application of an analytical method for the determination of selected acidic pharmaceuticals and estrogenic hormones in wastewater and sludge. *Journal of Environmental Science and Health – Part A Toxic/Hazardous Substances and Environmental Engineering* 51(11): 914–920.

Zuloaga, O., Navarro, P., Bizkarguenaga, E., Iparraguirre, A., Vallejo, A., Olivares, M., and Prieto, A. 2012. Overview of extraction, clean-up and detection techniques for the determination of organic pollutants in sewage sludge: A review. *Analytica Chimica Acta* 736: 7–29.

Zuo, Y., Zhang, K., and Lin, Y. 2007. Microwave-accelerated derivatization for the simultaneous gas chromatographic–mass spectrometric analysis of natural and synthetic estrogenic steroids. *Journal of Chromatography A* 1148: 211–218.

10 Green and Eco-Friendly Materials for the Removal of Phosphorus from Wastewater

Shraddha Khamparia, Dipika Jaspal, and Arti Malviya

CONTENTS

10.1 INTRODUCTION

The prodigious growth of various industries during industrialization and urbanization has degraded the quality of water, which is one of the most vital substances required for the sustenance of living species. The perpetual unloading of harmful industrial and municipal wastes into water resources detrimentally affects living organisms directly or indirectly. However, aquatic pollution of any kind disturbs the exquisite balance maintained by nature (Masters and Ela, 1991; Gupta et al., 2015; Malviya et al., 2015). About 97% of the total water in the planet lies in the oceans and is accompanied by a high concentration of salts, making it unfit for consumption. Besides acting as a sink for industrial and sewage wastes, sea water is also used as a coolant in thermal power stations (Hammer, 1986). Although desalination techniques are widely available, they are economically unfeasible due to the need for a large capital investment. Thus, the only fresh water available for different purposes is the 3% of water in the form of glaciers (68.7%), ground water (30.1%), and surface water (1.2%). Water for different purposes is withdrawn from the limited sources of fresh water such as lakes, rivers, and ponds.

Hence, the burden on water reservoirs is increasing day by day because of increased population and pollution. Surface water and ground water are contaminated to a large extent due to various anthropogenic activities (Manivasakam, 2005). Different industries such as textiles, chemical, paper, food, and plastic consume enormous amounts of water during various processing steps (Mittal et al., 2008, 2009a; Khamparia and Jaspal, 2016a, 2017a,b). Raw sewage and industrial wastes are considered as point source discharges, which are generally treated before dumping, thus avoiding contamination reaching water resources, while non-point sources of pollution often include runoff from agricultural lands and synthetic wastes from urban areas, which are difficult to eliminate by simple techniques (Todd and Mays, 2005). The versatile nature of water enables the effortless amalgamation of this indispensable source with any organic or inorganic pollutant (Albadarin et al., 2017; Daneshvar et al., 2017; Thakur et al., 2017). The wastewater discharged from both point and non-point sources grievously perturbs the functioning of different life-sustaining cycles. Hence, there is an urgent need to treat wastewater generated by various industries by following a sustainable approach. Certain substances, such as phosphorus, fluorine, chlorine, iron, nitrate, chromium, arsenic, and so on, can produce undesirable effects in the water bodies if present in excessive amounts. This chapter addresses harmful effects due to the excessive presence of phosphorus and discusses the available treatment technologies for its removal along with its recovery and recycling. Additionally, in the era of sustainability, this chapter revolves around the green and eco-friendly methods adopted for removal of phosphorus from wastewater.

10.1.1 TOXIC MATERIALS IN WATER

Surface, ground, domestic, and commercial wastes often include a variety of pollutants. Both inorganic and organic compounds are present in wastewater and have inimical effects on humans, animals, and plants. Heavy metals, synthetic dyes, pigments, pesticides, insecticides, and by-products from the textile, paper, printing, leather, and pharmaceuticals industries are considered to be the most hazardous substances present in the effluent (Bailey et al., 1999; Forgacs et al., 2004; Banat et al., 1996; Wauchope, 1978; Dunier and Siwicki, 1993; Heberer, 2002; Mittal et al., 2009b; Khamparia and Jaspal, 2016b). Substances and elements such as ammonia, nitrogen, phosphorus, and so on are widely found in effluents from domestic sewage and municipal wastes (Sharma, 2010; Cordell et al., 2009; Correll, 1998). Inorganic anions such as nitrates, sulfates, phosphates, fluoride, chlorates, and cyanides present in high concentrations induce toxicity in the receiving reservoirs (Kapoor and Viraraghavan, 1997; Postgate, 1959; Carpenter et al., 1998; Bhatnagar et al., 2011; Vanwijh and Hutchinson, 1995; Abel, 1996; Naushad et al., 2014). Organic compounds in wastewater often include phenolic compounds, surfactants, polynuclear aromatic hydrocarbons, polychlorinated biphenyls, polybrominated diphenyl ethers, and humic substances, which have detrimental effects (Kobayashi and Rittman, 1982; Scott and Ollis, 1995). Wastewater generated from sewage, agricultural activities, and fertilizer industries is high in ammonium content. The increased dosage of ammonium leads to eutrophication in water bodies with depletion of dissolved oxygen and toxicity in fish. The existence of heavy metals such as arsenic, mercury, lead, cadmium, copper, chromium, manganese, nickel, and zinc causes extreme health issues, and damages aquatic flora and fauna (Wang and Peng, 2010). Some of the organic pollutants are non-biodegradable and persist in nature, thus affecting the environment for decades (La Farre et al., 2008). Also, most of them are highly toxic and carcinogenic in nature.

10.1.2 HARMFUL EFFECTS OF PHOSPHORUS

Phosphorus is a non-renewable resource, necessary for the growth of different organisms in the ecosystem. Elemental phosphorus is toxic and exists in ortho, tri-poly, pyro, and organic states. Phosphorus is used in nature in very limited and small quantities (Weikard and Seyhan, 2009; Frossard et al., 2000). It is found in the Earth's crust in igneous and phosphorus rocks in relatively large proportions, while small quantities are available in sea water (Nash, 1984). Phosphorus enters into the environment by natural biogeochemical processes such as weathering. It plays an important role in different cycles of nature, such as the carbon and phosphorus cycle (Mackenzie et al., 2002). This delicate balance of phosphorus in the environment is maintained adequately by nature's forces (Schroder et al., 2011). Weathering is a lengthy process, through which phosphorus enters naturally into the system in a calculated manner, but interference through manual weathering leads to an increase in the rate of extraction, resulting in global changes disturbing the stability. Most importantly, excessive availability of phosphorus disturbs the ecological balance by triggering the eutrophication of water bodies, thereby resulting in the uncontrolled growth of aquatic plants and algae. Thus, excessive growth of water weeds and algae is the result of high levels of phosphorus in the aquatic system. Blooming of algae increases the chlorine dosage that needs to be supplied for treating water to be made palatable (Fisher et al., 2004), leading to the formation of high levels of cancer-causing disinfectant by-products (Wang et al., 2007). Furthermore, the excessive level of phosphorus stimulates the activity of noxious microbes (Nygaard and Tobiesen, 1993). Noxious *Microcystis* cyanobacteria produces a toxic substance (microcystin), which is poisonous to aquatic species and may cause hepatocellular carcinoma in humans (Vezie et al., 2002; Chong et al., 2000). Quantitatively, the addition of even 0.1–5.6 µg L^{-1} triggers the process of algal blooming (Bowman et al., 2007). Moreover, depletion of dissolved oxygen is one of the serious consequences of enhanced levels of phosphorus in the aquatic system (Tiessen, 1995). Another source of phosphorus is the fertilizers used in agriculture and treated and untreated sewage wastewater (Kirkham, 1982). Phosphorus occurs in nature in an insoluble form, which often controls the proliferation of the soluble form, with major implications for agriculture (Bennett et al., 2001). Phosphorus is considered to be one of the main constituents of domestic sewage, which includes phosphorus from detergents and human feces. Sewage sludge ranks second on the basis of phosphorus content and is a promising source for its recovery (Cordell et al., 2011; Havukainen et al., 2016). Phosphorus is widely used in detergents. Soaps and detergents contain large amounts of phosphorus as their basic ingredient; it emerges into the aquatic system through washing. During World War II, soaps were replaced by biodegradable detergents, which were modified in 1965, removing the demerits of biodegradable detergents (Devey and Harkness, 1973) with the development of soluble mono-calcium and ammonium phosphates. The sodium salt of triphosphoric acid plays a main role in detergents by sequestering the magnesium and calcium ions and working as a softening agent in water. But the problem with the modified detergents lay in their non-biodegradable nature and large quantity of phosphorus, which when released into the water system, stimulated enormous blooming of algae. Additionally, excessive phosphates cause a deadly threat to the coral reef ecosystem (Pastorok and Bilyard, 1985). Phosphorus is also used as a dispersing agent for clays in oil well drilling and cement technology, as a fertilizer, and as a flame retardant (Valsami-Jones, 2004). It has wide applications in the production of baking powder, food additives, matchsticks, and toothpastes. Phosphorus is also used in industries as corrosion inhibitors for protection of metal surfaces. (Corbridge, 2013). Hence, the effluents from these industries contain phosphorus.

Inorganic phosphates act as key substrates in the metabolism and biosynthesis of nucleic acids (Mudd et al., 1958). They also assist in the formation of bones and perform as a buffer during blood purification. Organic phosphorus in the form of esters is involved in enzymatic action in biochemical reactions as coenzymes. A slight increase or decrease in phosphorus affects the functioning of the body to a great extent and can be lethal (Takeda et al., 2004). Phosphorus is found in milk products and meat. Intake of phosphorus beyond the permissible limit (1.3 g per day) leads to a shortage of calcium in the body, causing hypocalcemia (Simesen et al., 1970) and digestive problems (Kumar and Puri, 2012). The World Health Organization recommends a maximal discharge limit of 0.5–1.0 mg L^{-1}. Moreover, it is known that an increase in phosphorus in nature of even 1 mg L^{-1} leads to excessive eutrophication, which detrimentally affects the quality of water reservoirs.

Thus, it becomes essential to design effective methods for the treatment of phosphorus-containing wastewater to maintain an optimal balance. Several methods, such as biological treatment, chemical precipitation, ion exchange, membrane filtration, and adsorption have been effectively explored to date for this purpose.

10.2   AVAILABLE TECHNOLOGIES FOR PHOSPHORUS REMOVAL

Phosphorus in water is present in dissolved and particulate forms. Different methods have been adopted for the removal of phosphorus from wastewater, such as sedimentation, chemical precipitation, ion exchange, membrane filtration, reverse osmosis, electro-dialysis, and biological methods.

A sedimentation method followed by filtration, or membrane technology, is required to convert soluble phosphorus into particulate matter (Painter Omoike, 1999). Chemical precipitation using alum, lime, or iron salts has been widely employed for the removal of orthophosphate. It requires a large amount of chemicals and results in excessive production of chemical sludge (Clark et al., 1997). The ion exchange method encounters a decrease in removal ability after regeneration. The membrane filtration method undergoes fouling and hence, is uneconomical (Blaney et al., 2007). Physical processes such as reverse osmosis (Bohdziewicz and Sroka, 2005) and electro-dialysis (Simons, 1979) suffer from the drawback of high cost, whereas biological methods suffer from the drawbacks of limitation to low concentrations and requirement for skilled labor for operation (Smolders et al., 1994). The adsorption process is considered to be the most prominent among all methods for removing phosphorus (Karageorgiou et al., 2007). Even at low concentrations, adsorption is workable and exhibits a high efficiency for phosphorus removal. Adsorbents such as bentonite, zeolite, red mud, lanthanum hydroxide, and iron-related adsorbents have been successfully employed for the removal of phosphorus by the adsorption technique (Sakadevan and Bavor, 1998; Boujelben et al., 2008; Drizo et al., 2006). Still, there remains a huge demand for cost-effective approaches and methods for the maintenance of phosphorus balance. Among several methods studied, the adsorption technique surpasses the rest, since it can be used for reduction, elimination, and reclamation by using suitable adsorbents (Ali and Gupta, 2006; Wahab et al., 2011a). The adsorption technique using solid adsorbents has been successfully applied for the removal of phosphorus. It is found to be highly effective and inexpensive and has the ability to recycle phosphorus, thus fulfilling all the criteria for an efficient treatment method.

10.2.1 REMOVAL FROM VARIOUS SOURCES

Phosphorus is present in several sources other than wastewater, such as in sediments, lakes, waste stabilization ponds, swine lagoon liquid, and so on, which have been treated by different methods. This subsection demonstrates the removal of phosphorus from different sources other than wastewater. Activated alumina treated with aluminum sulfate was adopted for the treatment of river and lake water with a low concentration of phosphorus. The adsorption ability of modified adsorbent was enhanced by 1.7 times when compared with the untreated one (Hano et al., 1997). Further, excessive phosphorus present in a lake in Germany was eradicated using Periphyton submerged on artificial substrata (Jobgen et al., 2004). In 2005, biological phosphorus present in sediments and causing eutrophication in lakes and ponds was treated with a microorganism having the ability to solubilize phosphorus. *Burkholderia glathei*, isolated from rhizospheric soil, was investigated during the treatment of phosphorus-contaminated sediments in a bioslurry reactor (Kim et al., 2005).

The uptake of phosphorus in a waste stabilization pond was examined using microalgae. Continuous culture bioreactors were used for the experiment to find the effects of different parameters such as phosphate concentration, intensity of light, and reaction temperature. Powell and co-workers advocated the successful biological removal of phosphorus in a waste stabilization pond using microalgal biomass (Powell et al., 2008).

10.2.2 REMOVAL FROM WASTEWATER

Urano and Tachikawa (1991) investigated activated alumina for the removal of phosphorus from domestic and industrial wastewaters. Simultaneous removal and recovery were attained after the experiment, suggesting that adsorption was a suitable method for the effective removal of phosphorus from wastewater. It was noted that no harmful by-products were formed, and moreover, it was a low-cost material. Other than adsorption, phycoremediation using algae was examined for the removal of phosphorus from wastewater generated from high–organic content wastewater. A filamentous microalga having high autoflocculation competence along with immobilized cells was studied for the removal and recycling of phosphorus in this integrated system (Olgui, 2003). Phosphorus-contaminated runoff water from a simulated plant nursery in Australia was treated using subsurface horizontal flow reed beds (Huett et al., 2005). Pratt and his team attempted to remove phosphorus from effluent generated from a waste stabilization pond using melter slag filter. The results indicated that adsorption of phosphorus was taking place on the surface of the slag, which contained metal oxides and oxyhydroxides (Pratt et al., 2007). Szogi and researchers developed a novel process for removal of the phosphorus from livestock effluent generated from a swine lagoon. The new treatment included bionitrification followed by the precipitation of phosphorus by increasing the pH of the contaminated effluent. Along with this, solid residual calcium phosphate was generated from the process, which could be reused as fertilizer in the farms (Szogi et al., 2009). A chemical precipitation method was adopted by Naseef (2012) during the removal of phosphorus from industrial wastewater. Lanthanum compounds doped with natural materials such as red mud, bentonite, zeolite, and montmorillonite have been used by several researchers as highly activated species for phosphorus adsorption. But still, these are not widely available and are not economically viable. An efficient adsorbent was developed using a by-product (lithium silica fume) from the silicon industry, which was found to be highly effective in removing phosphorus from wastewater (Xie et al., 2016).

Wetlands are often constructed for the treatment of different phosphorus-containing industrial effluents, employing several natural and modified filters. Phosphorus from a subsurface flow constructed wetland was adsorbed by bed-sand medium. The choice of the sand best suited for effective removal of phosphorus was governed by the calcium content. After a variety of sands, clay aggregates, powdered marbles, and so on were examined, calcite and marble were considered to have high phosphorus-binding abilities. The authors suggested that the addition of calcite or marble to sand may considerably increase the uptake of phosphorus, thereby removing phosphorus from constructed wetlands (Brix et al., 2001).

Both serpentinite and modified and unmodified steel slag obtained from an electric arc furnace were investigated for their adsorption capacity for the adsorption of phosphorus from wastewater. Steel slag showed 100% efficiency in adsorbing phosphorus due to the presence of metal hydroxides and precipitation of hydroxyapatite. It was found that lime-doped steel slag and serpentinite had lower efficacy. Steel slag was concluded to be the promising solution for removal of phosphorus from constructed wetlands and filter beds using adsorption followed by a precipitation mechanism (Drizo et al., 2006). Vymazal (2007) suggested harvesting of aboveground biomass of emergent vegetation to be the best solution for the elimination of phosphorus from wetlands at low concentrations. A hydrated oil-shale ash sediment obtained from a thermal power plant was employed to remove phosphorus from constructed wetlands in Estonia. Batch experiments resulted in a high phosphorus uptake capacity of about 67%–85%. The adsorption capacity of the material was due to the presence of reactive calcium minerals (Kaasik et al., 2008). Phosphate is considered to be one of the major pollutants in wetlands (Naushad et al., 2018).

Babatunde and co-researchers used residue generated from an aluminum-based water treatment plant for removing phosphorus in the engineering wetlands (Babatunde et al., 2009). Dunets and team examined concrete waste and basic oxygen steel slag from southern Ontario. Almost 99% phosphorus adsorption capacity was observed in this case. The authors suggested taking proper care while designing the filter and provided an effective design that could be implemented on a large scale (Dunets et al., 2015). Other industrial wastes, such as fly ash (Drizo et al., 1999); skin split, a waste from the leather industry (Farahbakhshazad and Morrison, 2003); electric arc furnace steel slag (Drizo et al., 2006); and lithium silica fume (Xie et al., 2016) have been employed for the treatment of phosphorus-contaminated water.

10.2.3 Use of Green and Eco-Friendly Materials

The eradication and recycling of phosphorus can be increased by the adoption of green and environmental friendly materials. Several researchers are in pursuit of finding the most capable low-cost natural material for the removal of phosphorus. Mineral-based naturally occurring materials, which include rocks, soils, sand, clay, and aggregates, have been explored for their efficiency in removing phosphorus from wastewater. Calcite, a natural adsorbent, was employed for the elimination of phosphorus from synthetic wastewater. The best results were obtained at a high concentration of calcite. The study also revealed the loaded adsorbent to be a fertilizer for acidic soils (Karageorgiou et al., 2007). Boujelben and group examined the phosphorus removal potential of synthetic iron oxide–coated sand, natural iron oxide–coated sand, and iron oxide–coated crushed brick. Comparative analysis revealed the last option to be the best (Boujelben et al., 2008). Similarly, the removal efficiency of several natural materials has been investigated by different researchers for phosphorus removal. Table 10.1 shows a number of naturally occurring materials used for the treatment of phosphorus-contaminated water.

Abundantly found, renewable, eco-friendly materials are endorsed by the scientific community as the most attractive and reliable solutions for phosphorus treatment. Also, as natural materials are biodegradable in nature, after uptake in the adsorption process, phosphorus-loaded adsorbents can be easily used on agricultural fields in the form of compost and fertilizer. Other than naturally occurring materials of mineral origin, lignocellulosic materials have been identified as potential candidates for the eradication of phosphorus from wastewater.

A green material, algal turf scrubber, was used during the treatment of runoff water from agricultural fields and eutrophic lake water. The biomass consisted of a natural mixed assembly of periphyton, microalgae, and bacteria. A 152.4 m long and 6.5 m wide scrubber was built with UV disinfection and used experimentally to treat the secondary effluent from an evaporation pond. The results suggested that the employed biomass had a great potential to treat wastewater, which could be controlled by varying the hydraulic loading rate (Craggs et al., 1996). Different sources of wastewater (sewage water, aquaculture water, well water, and mineral growth medium) were tested with water fern, a natural material, for the removal of phosphorus (Forni et al., 2001). Jayaweera and Kasturiarachchi (2004) investigated the efficiency of water hyacinth in phosphorus removal from industrial wastewater. The

TABLE 10.1
Natural Adsorbents for Phosphorus Removal

Adsorbents	P Conc.(mg L⁻¹)	Adsorption Capacity(mg g⁻¹)	Reference
Gravel	–	25.8–47.5	Mann and Bavor (1993)
Peat	70–258	12	Talbot et al. (1996)
Laterite	5–50	0.99	Wood and McAtamney (1996)
Soils	500–10000	0.005	Sakadevan and Bavor (1998)
Zeolite	500–10000	2.15	Sakadevan and Bavor (1998)
Bauxite	2.5–40	0.612	Drizo et al. (1999)
Limestone	5–25	0.3	Johansson (1999)
Limestone	2.5–40	0.682	Drizo et al. (1999)
Opoka	5–25	0.1	Johansson (1999)
Spodosol	5–25	1	Johansson (1999)
Shale	2.5–40	0.065–0.730	Drizo et al. (1999)
Burnt oil shale	2.5–40	0.065–0.730	Drizo et al. (1999)
Zeolite	2.5–40	0.065–0.730	Drizo et al. (1999)
Gravel	15	184–296	Tanner et al. (1999)
Limestone	2.5–40	0.682 g kg⁻¹	Drizo et al. (1999)
Shellsand	5–1000	4	Roseth (2000)
Opoka	5–25	0–0.1	Johansson and Gustafsson (2000)
Sand	0–100	0.417	Pant et al. (2001)
Sand	0–320	3.941	Arias et al. (2001)
Danish sand	10	0.052–0.165	Arias et al. (2001)
Sand	330	0.7–0.8	Pant et al. (2001)
Dolomite	–	0.2	Pant et al. (2002)
Shale	–	–	Pant and Reddy (2003)
Bauxite	0–100	2.95	Altundogan and Tumen (2002)
Sand	0.5–2.0	7.2–86.4	Farahbakhshazad and Morrison (2003)
Dolomite	–	7.34–52.02	Karaca et al. (2004)
Limerock	–	–	DeBusk et al. (1995)
Apatite	0–500	4.76	Molle et al. (2005)
Gravels	5–1000	3.6	Vohla et al. (2005)
Sand	5–1000	2.45	Vohla et al. (2005)
Shellsand	5–1000	8	Sovik and Klove (2005)
Sedimentary Apatite	0.5	13.9	Molle et al. (2005)
Apatite	5–150	1.09	Bellier et al. (2006)
Wollastonite	0.8–1700	12	Hedstrom and Rastas (2006)
Calcinated Alunite	25–150	0.8	Özacar (2006)
Dolomite and sand	0–100	0.168 g kg⁻¹	Prochaska and Zouboulis (2006)
Modified palygorskites		8.31	Ye et al. (2006)

(Continued)

TABLE 10.1 (CONTINUED)
Natural Adsorbents for Phosphorus Removal

Adsorbents	P Conc.(mg L^{-1})	Adsorption Capacity(mg g^{-1})	Reference
Shellsand	0–480	9.6	Adam et al. (2007)
Wollastonite	–	0.467–1.8	Gustafsson et al. (2008)
Polonite	5.30 mg dm^{-3}	–	Gustafsson et al. (2008)
Peat	4.9	0.081	Koiv et al. (2010)
Phoslock	–	9.5–10.5	Haghseresht et al. (2009)
Lanthanum(III) modified bentonite	5	14	Kuroki et al. (2014)
Red soil	20	0.998	Rout et al. (2015)

exact mechanism during the phytoremediation was assimilation and sorption during the removal of phosphorus. Zirconium(IV) immobilized orange waste gel from a widely available biomass was investigated for its possible use in the removal of phosphorus from aqueous solution. When the adsorption capacity was compared with that of synthetic materials, it was four times higher than for zirconium ferrite. At low pH, about 85% removal was attained in fixed bed column studies (Biswas et al., 2008). *Posidonia oceanica* fibers, found abundantly in the coastal zones of the Mediterranean basin, were investigated for the removal of phosphorus from wastewater by Wahab and his research team, who proposed a sustainable approach for using the plant waste in agricultural fields as a substitute for chemical fertilizers. A dual strategy was chosen, in which the plant assisted in the removal of orthophosphate from wastewater and phosphorus-loaded plant species were used as fertilizer and compost (Wahab et al., 2011a). Batch experimentation was performed on synthetic and secondary effluent using Posidonia oceanic fibers, and the effects of several parameters were studied. Eighty percent phosphorus removal was obtained from synthetic solution. An important phenomenon was noted with secondary effluent, in which competition was seen between sulfate and phosphorus for adsorption sites. The authors found that a continuous reactor was more beneficial than a column (Wahab et al., 2011b). A green alga, *Neochloris oleoabundans*, was used for the treatment of phosphorus from synthetic and real municipal wastewater. Additionally, the biomass had great potential to be used as biofuel (Wang and Lan, 2011). Greenhouse effluent also contained a large amount of phosphorus, which was a big challenge for researchers.

Gastropod shell, a green and eco-friendly natural material, was employed in a column reactor to decontaminate wastewater from aquaculture. It also assisted in recovering phosphorus and can be reused as a soil conditioner, which is better than a fertilizer. Chemisorption was found to be the dominating driving force while eliminating phosphorus from waste water. About 73.32% phosphorus was removed with a maximum uptake of 185.8 mg g^{-1}. Phosphate was precipitated by a thermally treated calcium mineral derived from gastropod shell, which resulted in high phosphate recovery (Oladoja et al., 2015). African land snail shell was tested for its ability to capture phosphorus from waste water with the aim of waste minimization and economic value. Batch adsorption studies were performed in comparison with other sorbents, which proved that land snail shell was a promising adsorbent for phosphorus removal and recovery as fertilizer (Oladoja et al., 2017). With similar perspectives, many more plant materials, weeds, and agricultural wastes have been exploited for their removal capacity for the elimination of phosphorus. Table 10.2 lists different biosorbents used for the removal of phosphorus.

TABLE 10.2
Biosorbents for Phosphorus Removal

Biosorbents	P Conc.(mg L^{-1})	Adsorption Capacity(mg g^{-1})	Reference
Banana stem (lignocellulosic residue)	–	72.46	Anirudhan et al. (2006)
Giant reed quaternary amino anion exchanger	–	0.821	Wang et al. (2010)
Giant reed	–	0.836	Xu et al. (2011)
Green algae *Neochloris oleoabundans*	–		Wang and Lan (2011)
Green alga *Chlorella* sp.	–	62.43	Wang et al. (2013)
Marine microalgae (Biochar)	–	157.7	Jung et al. (2016)
Algal turf scrubber	–	0.73 ± 0.28g m^{-2} d^{-1}	Craggs et al. (1996)
Maerl	0–5000	7.49	Gray et al. (2000)
Water fern (*Azolla filiculoides* Lam.)	–	–	Forni et al. (2001)
Water hyacinth	–	–	Jayaweera and Kasturiarachchi (2004)
Oyster shell	50	0.180–7.925	Seo et al. (2005)
Alum sludges	1.68	2.66	Babatunde et al. (2008)
Orange waste gel loaded with zirconium	–	57	Biswas et al. (2008)
Date palm fibers	50	4.35	Riahi et al. (2009)
Scallop shells	100	23	Yeom and Jung (2009)
Iron hydroxide-eggshell waste	2.8–110	14.49	Mezenner and Bensmaili (2009)
Coir pith activated carbon	–	–	Kumar et al. (2010)
Modified oyster shell	–	0.92	Yu et al. (2010)
Posidonia oceanica fibers (Plant waste)	15–100	7.45	Wahab et al. (2011a)
Modified sugarcane bagasse	50	21.3	Zhang et al. (2011)
Chemically modified sawdust of Aleppo pine	–	116.25	Benyoucef and Amrani (2011)

(Continued)

TABLE 10.2 (CONTINUED)
Biosorbents for Phosphorus Removal

Biosorbents	P Conc. (mg L^{-1})	Adsorption Capacity(mg g^{-1})	Reference
Shell of African land snail	25–300	222.22	Oladoja et al. (2012)
Palm surface fibers	50	26.05	Ismail (2012)
Granular date stones	50	26.66	Ismail (2012)
Modified coconut shell fibers	–	200	De Lima et al. (2012)
Iron loaded soybean milk residues (okara)	25	4.785	Nguyen et al. (2013)
Apple peels	–	20.35	Mallampati and Valiyaveettil (2013)
Peanut shell biochar	–	6.79	Jung et al. (2015)
Fruit (*Citrus limetta*) juice residue	–	0.9585	Yadav et al. (2015)
Ferric oxide loaded cotton stalk biochar	–	0.963	Ren et al. (2015)
Wheat straw biochar	–	2.47–16.58	Li et al. (2016)
Modified rice husk composite	–	21	Seliem et al. (2016)
Activated rice husk ash	–	0.89	Mor et al. (2016)
Engineering waste egg shell	–	248.73	Chen et al. (2016)
Water hyacinth magnetic biochar	–	5.07	Cai et al. (2017)

10.3 PHOSPHORUS RECYCLING

The foremost step during the recycling of phosphorus is the recovery of phosphorus from the adsorbed material. In some treatments, phosphorus was recovered from adsorbent material by a thermal method, which involved a large investment, thereby increasing the treatment cost. Hence, for easier recovery and recycling of phosphorus, natural biodegradable materials have attracted attention. Moreover, the loaded materials can be used as fertilizers, compost, or soil conditioners. Recycling lowers the environmental risk by reducing the burden on the limited sources of phosphorus. The depletion of phosphorus ores is occurring at an exponential rate, and recycling and recovery are in the limelight for the scientific community. Also, the recycling and recovery of phosphorus from sludge further fulfills the concept of zero generation of solid waste and lowers the burden on available natural resources. Several technologies are employed for recovery, which have certain limitations due to the cost of recovery and the difficulty of the processes (Hukari et al., 2016). Sewage sludge from municipal wastes contains large amounts of phosphorus. But generally, along with phosphorus, it also includes heavy metals and other pollutants. A thermochemical treatment is adopted for recovering phosphorus from sewage sludge

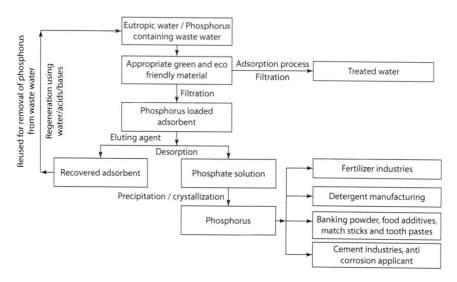

FIGURE 10.1   Schematic diagram of simultaneous removal and recovery of phosphorus from waste water via adsorption process

while removing heavy metals. The phosphorus can be used on agricultural fields. A significant increase in the bioavailability of phosphorus was noticed with the introduction of new mineral phases such as chlorapatite, farringtonite, and stanfieldite during the treatment process (Adam et al., 2009). Liao et al. (2005) developed a novel microwave treatment for sewage sludge to recover phosphorus discharged from municipal wastes. About 76% of total phosphorus was recovered from the sludge and could be reused on agricultural fields. The rapid disappearance of phosphorus from the environment motivated scientists to search for the recovery and recycling of phosphorus compounds from waste materials. Blocher and his research team proposed a novel concept of low-pressure wet oxidation for the decontamination of sewage sludge followed by phosphorus dissolution. Additionally, heavy metals from the sludge were removed using a nanofiltration method (Blocher et al., 2012). In a different study, phosphorus was recovered from sludge ash through struvite precipitation. Ninety-five percent of phosphorus was extracted, accompanied by leaching of heavy metals, which was overcome by employing a cation exchange resin (Xu et al., 2012).

In the past half century, humans have massively intervened in the global phosphorus cycle, resulting in the mobilization of about half a billion tonnes of limited phosphorus into the hydrosphere. Moreover, enhanced pollution due to excessive phosphorus content has been the motivating factor for the sustainable use of phosphorus, including recovery and recycling approaches. This challenge has been tackled by the adoption of green and eco-friendly materials for phosphorus-contaminated waste. Cordell in his extensive work demonstrated 30 different kinds of methods for the recovery of phosphorus (Cordell et al., 2011). The drawback of the old strategies for phosphorus recovery was their inability to manage the complex waste materials and residues produced during the treatment method, while some had inappropriate designs along with a high cost.

De-Bashan and group emphasized the effective removal of toxic pollutants from waste while simultaneously recovering phosphorus. Phosphorus was mainly recovered in the form of insoluble salt precipitates, microbial mass present in activated sludge, or plant biomass during the removal process (De-Bashan and Bashan, 2004). The major drawback of the available recycling techniques is that the loaded material contains not only phosphorus but also other pollutants, which may be toxic. Research is currently required to find a process that can easily separate toxic pollutants from phosphorus-loaded materials so that the latter can be used in fertilizers. From this viewpoint, enhanced biological phosphorus removal encompassing struvite and hydroxyapatite is an emergent technology for this purpose. It is the most cost-effective method, since it requires fewer chemicals during precipitation and reduces the size of the treatment facility, thereby decreasing the volume of effluent during primary or secondary treatment (Stratful et al., 1999). Figure 10.1 shows a block diagram for the recovery and recycling of phosphorus.

10.4 CONCLUSION

The adsorption of phosphorus in wastewater via natural, green, eco-friendly materials provides a sustainable solution to the management of phosphorus levels. Also, several are cost effective. The recovered phosphorus could be used in varied applications. Natural materials of plant origin are biodegradable in nature and could be directly used as fertilizer, thereby curtailing the excessive demand for phosphorus in fertilizers. Taking into account the fact that phosphorus is a vital limiting nutrient present in the ecosystem, whose delicate balance is perturbed by human activities, simultaneous removal and recovery should be an integral part of an efficient treatment system.

REFERENCES

Abel, P.D. 1996. *Water Pollution Biology.* CRC Press, Boca Raton, FL.

Adam, C., B. Peplinski, M. Michaelis, G. Kley, and F.G. Simon. 2009. Thermochemical treatment of sewage sludge ashes for phosphorus recovery. *Waste Management* 29: 1122–8.

Adam, K., T. Krogstad, L. Vrale, A.K. Sovik, and P.D. Jenssen. 2007. Phosphorus retention in the filter materials shellsand and Filtralite P—Batch and column experiment with synthetic P solution and secondary wastewater. *Ecological Engineering* 29: 200–8.

Albadarin, A.B., M.N. Collins, M. Naushad, and S. Shirazian. 2017. Activated lignin–chitosan extruded blends for efficient adsorption of methylene blue. *Chemical Engineering Journal* 307: 264–72.

Ali, I. and V.K. Gupta. 2006. Advances in water treatment by adsorption technology. *Nature Protocols* 1: 2661–7.

Altundogan, H.S. and F. Tumen. 2002. Removal of phosphates from aqueous solutions by using bauxite. I: Effect of pH on the adsorption of various phosphates. *Journal of Chemical Technology and Biotechnology* 77: 77–85.

Anirudhan, T.S., B.F. Noeline, and D.M. Manohar. 2006. Phosphate removal from wastewaters using a weak anion exchanger prepared from a lignocellulosic residue. *Environmental Science and Technology* 40: 2740–5.

Arias, C.A., M. Del Bubba, and H. Brix. 2001. Phosphorus removal by sands for use as media in subsurface flow constructed reed beds. *Water Research* 35: 1159–68.

Babatunde, A.O., Y.Q. Zhao, A.M. Burke, M.A. Morris, and J.P. Hanrahan. 2009. Characterization of aluminium-based water treatment residual for potential phosphorus removal in engineered wetlands. *Environmental Pollution* 157: 2830–6.

Babatunde, A.O., Y.Q. Zhao, Y. Yang, and P. Kearney. 2008. Reuse of dewatered aluminium-coagulated water treatment residual to immobilize phosphorus: Batch and column trials using a condensed phosphate. *Chemical Engineering Journal* 136: 108–15.

Bailey, S.E., T.J. Olin, R.M. Bricka, and D.D. Adrian. 1999. A review of potentially low-cost sorbents for heavy metals. *Water Research* 33: 2469–79.

Banat, I.M., P. Nigam, D. Singh, and R. Marchant. 1996. Microbial decolorization of textile-dyecontaining effluents: A review. *Bioresource Technology* 58: 217–27.

Bellier, N., F. Chazarenc, and Y. Comeau. 2006. Phosphorus removal from wastewater by mineral apatite. *Water Research* 40: 2965–71.

Bennett, E.M., S.R. Carpenter, and N.F. Caraco. 2001. Human impact on erodable phosphorus and eutrophication: A global perspective: Increasing accumulation of phosphorus in soil threatens rivers, lakes, and coastal oceans with eutrophication. *BioScience* 51: 227–34.

Benyoucef, S. and M. Amrani. 2011. Removal of phosphorus from aqueous solutions using chemically modified sawdust of Aleppo pine (Pinus halepensis Miller): Kinetics and isotherm studies. *The Environmentalist* 31: 200–7.

Bhatnagar, A., E. Kumar, and M. Sillanpaa. 2011. Fluoride removal from water by adsorption—a review. *Chemical Engineering Journal* 171: 811–40.

Biswas, B.K., K. Inoue, K.N. Ghimire, H. Harada, K. Ohto, and H. Kawakita. 2008. Removal and recovery of phosphorus from water by means of adsorption onto orange waste gel loaded with zirconium. *Bioresource Technology* 99: 8685–90.

Blaney, L.M., S. Cinar, and A.K. SenGupta. 2007. Hybrid anion exchanger for trace phosphate removal from water and wastewater. *Water Research* 41: 1603–13.

Blocher, C., C. Niewerschand, and T. Melin. 2012. Phosphorus recovery from sewage sludge with a hybrid process of low pressure wet oxidation and nanofiltration. *Water Research* 46: 2009–19.

Bohdziewicz, J. and E. Sroka. 2005. Integrated system of activated sludge–reverse osmosis in the treatment of the wastewater from the meat industry. *Process Biochemistry* 40: 1517–23.

Boujelben, N., J. Bouzid, Z. Elouear, M. Feki, F. Jamoussiand, and A. Montiel. 2008. Phosphorus removal from aqueous solution using iron coated natural and engineered sorbents. *Journal of Hazardous Materials* 151: 103–10.

Bowman, M.F., P.A. Chambers, and D.W. Schindler. 2007. Constraints on benthic algal response to nutrient addition in oligotrophic mountain rivers. *River Research and Applications* 23: 858–76.

Brix, H., C.A. Arias, and M. Del Bubba. 2001. Media selection for sustainable phosphorus removal in subsurface flow constructed wetlands. *Water Science and Technology* 44: 47–54.

Cai, R., X. Wang, X. Ji, B. Peng, C. Tan, and X. Huang. 2017. Phosphate reclaim from simulated and real eutrophic water by magnetic biochar derived from water hyacinth. *Journal of Environmental Management* 187: 212–19.

Carpenter, S.R., N.F. Caraco, D.L. Correll, R.W. Howarth, A.N. Sharpley, and V.H. Smith. 1998. Nonpoint pollution of surface waters with phosphorus and nitrogen. *Ecological Applications* 8: 559–68.

Chen, D., X. Xiao, and K. Yang. 2016. Removal of phosphate and hexavalent chromium from aqueous solutions by engineered waste eggshell. *RSC Advances* 6: 35332–9.

Chong, M.W.K., K.D. Gu, P.K.S. Lam, M. Yang, and W.F. Fong. 2000. Study on the cytotoxicity of microcystin-LR on cultured cells. *Chemosphere* 41: 143–7.

Clark, T., T. Stephenson, and P.A. Pearce. 1997. Phosphorus removal by chemical precipitation in a biological aerated filter. *Water Research* 31: 2557–63.

Corbridge, D.E. 2013. *Phosphorus: Chemistry, Biochemistry and Technology.* CRC Press, Boca Raton, FL..

Cordell, D., J.O. Drangert, and S. White. 2009. The story of phosphorus: Global food security and food for thought. *Global Environmental Change* 19: 292–305.

Cordell, D., A. Rosemarin, J.J. Schroderand, and A.L. Smit. 2011. Towards global phosphorus security: A systems framework for phosphorus recovery and reuse options. *Chemosphere* 84: 747–58.

Correll, D.L. 1998. The role of phosphorus in the eutrophication of receiving waters: A review. *Journal of Environmental Quality* 27: 261–6.

Craggs, R.J., W.H. Adey, K.R. Jenson, M.S.S. John, F.B. Green, and W.J. Oswald. 1996. Phosphorus removal from wastewater using an algal turf scrubber. *Water Science and Technology* 33: 191–8.

Daneshvar, E., A. Vazirzadeh, A. Niazi, M. Kousha, M. Naushad and A. Bhatnagar. 2017. Desorption of methylene blue dye from brown macroalga: Effects of operating parameters, isotherm study and kinetic modeling. *Journal of Cleaner Production* 152: 443–53.

De-Bashan, L.E. and Y. Bashan. 2004. Recent advances in removing phosphorus from wastewater and its future use as fertilizer (1997–2003). *Water Research* 38: 4222–46.

DeBusk, T.A., J.E. Peterson, and K.R. Reddy. 1995. Use of aquatic and terrestrial plants for removing phosphorus from dairy wastewaters. *Ecological Engineering* 5: 371–90.

De Lima, A.C.A., R.F. Nascimento, F.F. de Sousa, M. Josue Filho, and A.C. Oliveira. 2012. Modified coconut shell fibers: A green and economical sorbent for the removal of anions from aqueous solutions. *Chemical Engineering Journal* 185: 274–84.

Devey, D.G. and N. Harkness. 1973. The significance of man-made sources of phosphorus: Detergents and sewage. *Water Research* 7: 35–54.

Drizo, A., C. Forget, R.P. Chapuis, and Y. Comeau. 2006. Phosphorus removal by electric arc furnace steel slag and serpentinite. *Water Research* 40: 1547–54.

Drizo, A., C.A. Frost, J. Grace, and K.A. Smith. 1999. Physico-chemical screening of phosphate-removing substrates for use in constructed wetland systems. *Water Research* 33: 3595–602.

Dunets, C.S., Y. Zheng, and M. Dixon. 2015. Use of phosphorus-sorbing materials to remove phosphate from greenhouse wastewater. *Environmental Technology* 36: 1759–70.

Dunier, M. and A.K. Siwicki. 1993. Effects of pesticides and other organic pollutants in the aquatic environment on immunity of fish: A review. *Fish and Shellfish Immunology* 3: 423–38.

Farahbakhshazad, N. and G.M. Morrison. 2003. Phosphorus removal in a vertical upflow constructed wetland system. *Water Science and Technology* 48: 43–50.

Fisher, J. and M.C. Acreman. 2004. Wetland nutrient removal: A review of the evidence. *Hydrology and Earth System Sciences* 8: 673–85.

Forgacs, E., T. Cserhati, and G. Oros. 2004. Removal of synthetic dyes from wastewaters: A review. *Environment International* 30: 953–71.

Forni, C., J. Chen, L. Tancioni, and M.G. Caiola. 2001. Evaluation of the fern Azolla for growth, nitrogen and phosphorus removal from wastewater. *Water Research* 35: 1592–8.

Frossard, E., L.M. Condron, A. Oberson, S. Sinaj, and J.C. Fardeau. 2000. Processes governing phosphorus availability in temperate soils. *Journal of Environmental Quality* 29: 15–23.

Gray, S., J. Kinross, P. Read, and A. Marland. 2000. The nutrient assimilative capacity of maerl as a substrate in constructed wetland systems for waste treatment. *Water Research* 34: 2183–90.

Gupta, V.K., S. Khamparia, I. Tyagi, D. Jaspal, and A. Malviya. 2015. Decolorization of mixture of dyes: A critical review. *Global Journal of Environmental Science and Management* 1: 71–94.

Gustafsson, J.P., A. Renman, G. Renman, and K. Poll. 2008. Phosphate removal by mineral-based sorbents used in filters for small-scale wastewater treatment. *Water Research* 42: 189–97.

Haghseresht, F., S. Wang, and D.D. Do. 2009. A novel lanthanum-modified bentonite, Phoslock, for phosphate removal from wastewaters. *Applied Clay Science* 46: 369–75.

Hammer, M.J. 1986. *Water and Wastewater Technology*. 2nd Ed. Wiley, New York, NY.

Hano, T., H. Takanashi, M. Hirata, K. Urano, and S. Eto. 1997. Removal of phosphorus from wastewater by activated alumina adsorbent. *Water Science and Technology* 35: 39–46.

Havukainen, J., M.T. Nguyen, L. Hermann, M. Horttanainen, M. Mikkila, I. Deviatkin, and L. Linnanen. 2016. Potential of phosphorus recovery from sewage sludge and manure ash by thermochemical treatment. *Waste Management* 49: 221–9.

Heberer, T. 2002. Occurrence, fate, and removal of pharmaceutical residues in the aquatic environment: A review of recent research data. *Toxicology Letters* 131: 5–17.

Hedstrom, A. and L. Rastas. 2006. Methodological aspects of using blast furnace slag for wastewater phosphorus removal. *Journal of Environmental Engineering* 132: 1431–8.

Huett, D.O., S.G. Morris, G. Smith, and N. Hunt. 2005. Nitrogen and phosphorus removal from plant nursery runoff in vegetated and unvegetated subsurface flow wetlands. *Water Research* 39: 3259–72.

Hukari, S., L. Hermann, and A. Nattorp. 2016. From wastewater to fertilisers—technical overview and critical review of European legislation governing phosphorus recycling. *Science of the Total Environment* 542: 1127–35.

Ismail, Z.Z. 2012. Kinetic study for phosphate removal from water by recycled date-palm wastes as agricultural by-products. *International Journal of Environmental Studies* 69: 135–49.

Jayaweera, M.W. and J.C. Kasturiarachchi. 2004. Removal of nitrogen and phosphorus from industrial wastewaters by phytoremediation using water hyacinth (Eichhornia crassipes (Mart.) Solms). *Water Science and Technology* 50: 217–25.

Jöbgen, A., A. Palm, and M. Melkonian. 2004. Phosphorus removal from eutrophic lakes using periphyton on submerged artificial substrata. *Hydrobiologia* 528: 123–42.

Johansson, L. 1999. Industrial by-products and natural substrata as phosphorus sorbents. *Environmental Technology* 20: 309–16.

Johansson, L. and J.P. Gustafsson. 2000. Phosphate removal using blast furnace slags and opoka-mechanisms. *Water Research* 34: 259–65.

Jung, K.W., M.J. Hwang, K.H. Ahn, and Y.S. Ok. 2015. Kinetic study on phosphate removal from aqueous solution by biochar derived from peanut shell as renewable adsorptive media. *International Journal of Environmental Science and Technology* 12: 3363–72.

Jung, K.W., T.U. Jeong, H.J. Kang, and K.H. Ahn. 2016. Characteristics of biochar derived from marine macroalgae and fabrication of granular biochar by entrapment in calcium-alginate beads for phosphate removal from aqueous solution. *Bioresource Technology* 211: 108–16.

Kaasik, A., C. Vohla, R. Motlep, U. Mander, and K. Kirsimae. 2008. Hydrated calcareous oil-shale ash as potential filter media for phosphorus removal in constructed wetlands. *Water Research* 42: 1315–23.

Kapoor, A. and T. Viraraghavan. 1997. Nitrate removal from drinking water—review. *Journal of Environmental Engineering* 123: 371–80.

Karaca, S., A. Gurses, M. Ejder, and M. Acikyildiz. 2004. Kinetic modeling of liquid-phase adsorption of phosphate on dolomite. *Journal of Colloid and Interface Science* 277: 257–63.

Karageorgiou, K., M. Paschalis, and G.N. Anastassakis. 2007. Removal of phosphate species from solution by adsorption onto calcite used as natural adsorbent. *Journal of Hazardous Materials* 139: 447–52.

Khamparia, S. and D. Jaspal. 2016a. Investigation of adsorption of Rhodamine B onto a natural adsorbent Argemone mexicana. *Journal of Environmental Management* 183: 786–93.

Khamparia, S. and D. Jaspal. 2016b. Adsorptive removal of Direct Red 81 dye from aqueous solution onto Argemone mexicana. *Sustainable Environment Research* 26: 117–23.

Khamparia, S. and D.K. Jaspal. 2017a. Adsorption in combination with ozonation for the treatment of textile waste water: A critical review. *Frontiers of Environmental Science and Engineering* 11: 8.

Khamparia, S. and D.K. Jaspal. 2017b. Xanthium strumarium L. seed hull as a zero cost alternative for Rhodamine B dye removal. *Journal of Environmental Management* 197: 498–506.

Kim, Y.H., B. Bae, and Y.K. Choung. 2005. Optimization of biological phosphorus removal from contaminated sediments with phosphate-solubilizing microorganisms. *Journal of Bioscience and Bioengineering* 99: 23–9.

Kirkham, M.B. 1982. Agricultural use of phosphorus in sewage sludge. *Advances in Agronomy* 35: 129–63.

Kobayashi, H. and B.E. Rittman. 1982. Microbial removal of hazardous organic compounds. *Environment Science and Technology* 16: 170A–183A.

Koiv, M., M. Liira, U. Mander, R. Motlep, C. Vohla, and K. Kirsimae. 2010. Phosphorus removal using Ca-rich hydrated oil shale ash as filter material—the effect of different phosphorus loadings and wastewater compositions. *Water Research* 44: 5232–9.

Kumar, M. and A. Puri. 2012. A review of permissible limits of drinking water. *Indian Journal of Occupational and Environmental Medicine* 16: 40.

Kumar, P., S. Sudha, S. Chand, and V.C. Srivastava. 2010. Phosphate removal from aqueous solution using coir-pith activated carbon. *Separation Science and Technology* 45: 1463–70.

Kuroki, V., G.E. Bosco, P.S. Fadini, A.A. Mozeto, A.R. Cestari, and W.A. Carvalho. 2014. Use of a La (III)-modified bentonite for effective phosphate removal from aqueous media. *Journal of Hazardous Materials* 274: 124–31.

La Farre, M., S. Pérez, L. Kantiani, and D. Barceló. 2008. Fate and toxicity of emerging pollutants, their metabolites and transformation products in the aquatic environment. *Trends in Analytical Chemistry* 27: 991–1007.

Li, J.H., G.H. Lv, W.B. Bai, Q. Liu, Y.C. Zhang, and J.Q. Song. 2016. Modification and use of biochar from wheat straw (Triticum aestivum L.) for nitrate and phosphate removal from water. *Desalination and Water Treatment* 57: 4681–93.

Liao, P.H., W.T. Wong, and K.V. Lo. 2005. Release of phosphorus from sewage sludge using microwave technology. *Journal of Environmental Engineering and Science* 4: 77–81.

Mackenzie, F.T., L.M. Ver, and A. Lerman. 2002. Century-scale nitrogen and phosphorus controls of the carbon cycle. *Chemical Geology* 190: 13–32.

Mallampati, R. and S. Valiyaveettil. 2013. Apple peels—A versatile biomass for water purification. *ACS Applied Materials and Interfaces* 5: 4443–9.

Malviya, A., D. Jaspal, P. Sharma, and A. Dubey. 2015. Isothermal mathematical modelling for decolorizing water—A comparative approach. *Sustainable Environment Research* 25: 1.

Manivasakam, N. 2005. *Physico-Chemical Examination of Water Sewage and Industrial Effluents.* 5th Ed., Pragati Prakashan, Meerut, India.

Mann, R.A. and H.J. Bavor. 1993. Phosphorus removal in constructed wetlands using gravel and industrial waste substrata. *Water Science and Technology* 27: 107–13.

Masters, G.M. and W. Ela. 1991. *Introduction to Environmental Engineering and Science* (Vol. No. 3). Prentice Hall, Englewood Cliffs, NJ.

Mezenner, N.Y. and A. Bensmaili. 2009. Kinetics and thermodynamic study of phosphate adsorption on iron hydroxide-eggshell waste. *Chemical Engineering Journal* 147: 87–96.

Mittal, A., D. Kaur, and J. Mittal. 2008. Applicability of waste materials—bottom ash and deoiled soya—as adsorbents for the removal and recovery of a hazardous dye, brilliant green. *Journal of Colloid and Interface Science* 326: 8–17.

Mittal, A., D. Kaur, and J. Mittal. 2009a. Batch and bulk removal of a triarylmethane dye, Fast Green FCF, from wastewater by adsorption over waste materials. *Journal of Hazardous Materials* 163: 568–77.

Mittal, A., D. Kaur, A. Malviya, J. Mittal, and V.K. Gupta. 2009b. Adsorption studies on the removal of coloring agent phenol red from wastewater using waste materials as adsorbents. *Journal of Colloid and Interface Science* 337: 345–54.

Molle, P., A. Lienard, A. Grasmick, A. Iwema, and A. Kabbabi. 2005. Apatite as an interesting seed to remove phosphorus from wastewater in constructed wetlands. *Water Science and Technology* 51: 193–203.

Mor, S., K. Chhoden, and K. Ravindra. 2016. Application of agro-waste rice husk ash for the removal of phosphate from the wastewater. *Journal of Cleaner Production* 129: 673–80.

Mudd, S., A. Yoshida, and M. Koike. 1958. Polyphosphate as accumulator of phosphorus and energy. *Journal of Bacteriology* 75: 224.

Nash, W.P. 1984. Phosphate minerals in terrestrial igneous and metamorphic rocks. In *Phosphate Minerals*. J.O. Nriagu and P.H. Moore (ed.) Springer-Verlag, Heidelberg, Germany.

Nassef, E. 2012. Removal of phosphates from industrial waste water by chemical precipitation. *Engineering Science and Technology: An International Journal* 2: 409–13.

Naushad, M., M.A. Khan, Z.A. ALOthman, and M.R. Khan. 2014. Adsorptive removal of nitrate from synthetic and commercially available bottled water samples using De-Acidite FF-IP resin. *Journal of Industrial and Engineering Chemistry* 20: 3400–7.

Naushad, M., Sharma G., Kumar A., Sharma, S., Ghfar, A.A., Bhatnagar, A., Stadler, et al. 2018. Efficient removal of toxic phosphate anions from aqueous environment using pectin based quaternary amino anion exchanger. *International Journal of Biological Macromolecules* 106: 1–10.

Nguyen, T.A.H., H.H. Ngo, W.S. Guo, J. Zhang, S. Liang, and K.L. Tung. 2013. Feasibility of iron loaded 'okara' for biosorption of phosphorous in aqueous solutions. *Bioresource Technology* 150: 42–9.

Nygaard, K. and A. Tobiesen. 1993. Bacterivory in algae: A survival strategy during nutrient limitation. *Limnology and Oceanography* 38: 273–9.

Oladoja, N.A., R.O.A. Adelagun, A.L. Ahmad, and I.A. Ololade. 2015. Phosphorus recovery from aquaculture wastewater using thermally treated gastropod shell. *Process Safety and Environmental Protection* 98: 296–308.

Oladoja, N.A., R.O.A. Adelagun, A.L. Ahmad, and I.A. Ololade. 2017. Green reactive material for phosphorus capture and remediation of aquaculture wastewater. *Process Safety and Environmental Protection* 105: 21–31.

Oladoja, N.A., A.L. Ahmad, O.A. Adesina, and R.O.A. Adelagun. 2012. Low-cost biogenic waste for phosphate capture from aqueous system. *Chemical Engineering Journal* 209: 170–9.

Olguí, E.J. 2003. Phycoremediation: Key issues for cost-effective nutrient removal processes. *Biotechnology Advances* 22: 81–91.

Özacar, M. 2006. Contact time optimization of two-stage batch adsorber design using second-order kinetic model for the adsorption of phosphate onto alunite. *Journal of Hazardous Materials* 137: 218–25.

PainterOmoike, A.I. 1999. Removal of phosphorus and organic matter removal by alum during wastewater treatment. *Water Research* 33: 3617–27.

Pant, H.K. and K.R. Reddy. 2003. Potential internal loading of phosphorus in a wetland constructed in agricultural land. *Water Research* 37: 965–72.

Pant, H.K., K.R. Reddy, and F.E. Dierberg. 2002. Bioavailability of organic phosphorus in a submerged aquatic vegetation–dominated treatment wetland. *Journal of Environmental Quality* 31: 1748–56.

Pant, H.K., K.R. Reddy, and E. Lemon. 2001. Phosphorus retention capacity of root bed media of sub-surface flow constructed wetlands. *Ecological Engineering* 17: 345–55.

Pastorok, R.A. and G.R. Bilyard. 1985. Effects of sewage pollution on coral-reef communities. *Marine Ecology Progress Series* 21: 175–89.

Postgate, J. 1959. Sulphate reduction by bacteria. *Annual Reviews in Microbiology* 13: 505–20.

Powell, N., A.N. Shilton, S. Pratt, and Y. Chisti. 2008. Factors influencing luxury uptake of phosphorus by microalgae in waste stabilization ponds. *Environmental Science and Technology* 42: 5958–62.

Pratt, C., A. Shilton, S. Pratt, R.G. Haverkamp, and N.S. Bolan. 2007. Phosphorus removal mechanisms in active slag filters treating waste stabilization pond effluent. *Environmental Science and Technology* 41: 3296–301.

Prochaska, C.A. and A.I. Zouboulis. 2006. Removal of phosphates by pilot vertical-flow constructed wetlands using a mixture of sand and dolomite as substrate. *Ecological Engineering* 26: 293–303.

Ren, J., N. Li, L. Li, J.K. An, L. Zhao, and N.Q. Ren. 2015. Granulation and ferric oxides loading enable biochar derived from cotton stalk to remove phosphate from water. *Bioresource Technology* 178: 119–25.

Riahi, K., B.B. Thayer, A.B. Mammou, A.B. Ammar, and M.H. Jaafoura. 2009. Biosorption characteristics of phosphates from aqueous solution onto Phoenix dactylifera L. date palm fibers. *Journal of Hazardous Materials* 170: 511–19.

Roseth, R. 2000. Shell sand: A new filter medium for constructed wetlands and wastewater treatment. *Journal of Environmental Science and Health Part A* 35: 1335–55.

Rout, P.R., P. Bhunia, and R.R. Dash. 2015. A mechanistic approach to evaluate the effectiveness of red soil as a natural adsorbent for phosphate removal from wastewater. *Desalination and Water Treatment* 54: 358–73.

Sakadevan, K. and H.J. Bavor. 1998. Phosphate adsorption characteristics of soils, slags and zeolite to be used as substrates in constructed wetland systems. *Water Research* 32: 393–9.

Schröder, J.J., A.L. Smit, D. Cordelland, and A. Rosemarin. 2011. Improved phosphorus use efficiency in agriculture: A key requirement for its sustainable use. *Chemosphere* 84: 822–31.

Scott, J.P. and D.F. Ollis. 1995. Integration of chemical and biological oxidation processes for water treatment: Review and recommendations. *Environmental Progress* 14: 88–103.

Seliem, M.K., Komarneni, S., and Khadra, M.R.A. 2016. Phosphate removal from solution by composite of MCM-41 silica with rice husk: Kinetic and equilibrium studies. *Microporous and Mesoporous Materials* 224: 51–7.

Seo, D.C., J.S. Cho, H.J. Lee, and J.S. Heo. 2005. Phosphorus retention capacity of filter media for estimating the longevity of constructed wetland. *Water Research* 39: 2445–57.

Sharma, V.K. 2010. Oxidation of nitrogen-containing pollutants by novel ferrate (VI) technology: A review. *Journal of Environmental Science and Health Part A* 45: 645–67.

Simesen, M.G. 1970. Calcium, inorganic phosphorus, and magnesium metabolism in health and disease. In: *Clinical Biochemistry of Domestic Animals.* J.J. Kaneko and C.E. Cornelius (ed.), Academic Press, Inc., New York, NY. 1: 313–65.

Simons, R. 1979. The origin and elimination of water splitting in ion exchange membranes during water demineralisation by electrodialysis. *Desalination* 28: 41–2.

Smolders, G.J.F., J. Van der Meij, M.C.M. Van Loosdrecht, and J.J. Heijnen. 1994. Model of the anaerobic metabolism of the biological phosphorus removal process: Stoichiometry and pH influence. *Biotechnology and Bioengineering* 43: 461–70.

Sovik, A.K. and B. Klove. 2005. Phosphorus retention processes in shell sand filter systems treating municipal wastewater. *Ecological Engineering* 25: 168–82.

Stratful, I., S. Brett, M.B. Scrimshaw, and J.N. Lester. 1999. Biological phosphorus removal, its role in phosphorus recycling. *Environmental Technology* 20: 681–95.

Szogi, A.A. and M.B. Vanotti. 2009. Removal of phosphorus from livestock effluents. *Journal of Environmental Quality* 38: 576–86.

Takeda, E., Y. Taketani, N. Sawada, T. Sato, and H. Yamamoto. 2004. The regulation and function of phosphate in the human body. *Biofactors* 21: 345–55.

Talbot, P., G. Bélanger, M. Pelletier, G. Laliberté, and Y. Arcand. 1996. Development of a biofilter using an organic medium for on-site wastewater treatment. *Water Science and Technology* 34: 435–41.

Tanner, C.C., J.P.S. Sukias, and M.P. Upsdell. 1999. Substratum phosphorus accumulation during maturation of gravel-bed constructed wetlands. *Water Science and Technology* 40: 147–54.

Thakur, M., G. Sharma T. Ahamad A.A. Ghfar, D. Pathania, and M. Naushad. 2017. Efficient photocatalytic degradation of toxic dyes from aqueous environment using gelatin-Zr(IV) phosphate nanocomposite and its antimicrobial activity. *Colloids and Surfaces B: Biointerfaces* 157: 456–63.

Tiessen, H. 1995. *Phosphorus in the Global Environment: Transfers, Cycles and Management.* John Wiley and Sons, Chichester, UK.

Todd, D.K. and L.W. Mays. 2005. *Groundwater Hydrology*, Vol. No. 1625. John Wiley and Sons, New Jersey, NJ.

Urano, K. and H. Tachikawa, 1991. Process development for removal and recovery of phosphorus from wastewater by a new adsorbent. II. Adsorption rates and breakthrough curves. *Industrial and Engineering Chemistry Research* 30: 1897–9.

Valsami-Jones, E., ed., 2004. *Phosphorus in Environmental Technology.* IWA Publishing, London, UK.

Vanwijk, D.J. and T.H. Hutchinson. 1995. The ecotoxicity of chlorate to aquatic organisms: A critical review. *Ecotoxicology and Environmental Safety* 32: 244–53.

Vézie, C., J. Rapala, J. Vaitomaa, J. Seitsonen, and K. Sivonen. 2002. Effect of nitrogen and phosphorus on growth of toxic and nontoxic Microcystis strains and on intracellular microcystin concentrations. *Microbial Ecology* 43: 443–54.

Vohla, C., E. Poldvere, A. Noorvee, V. Kuusemets, and U. Mander. 2005. Alternative filter media for phosphorous removal in a horizontal subsurface flow constructed wetland. *Journal of Environmental Science and Health* 40: 1251–64.

Vymazal, J. 2007. Removal of nutrients in various types of constructed wetlands. *Science of the Total Environment* 380: 8–65.

Wahab, M.A., R.B. Hassine, and S. Jellali. 2011a. Posidonia oceanica (L.) fibers as a potential low-cost adsorbent for the removal and recovery of orthophosphate. *Journal of Hazardous Materials* 191: 333–41.

Wahab, M.A., R.B. Hassine, and S. Jellali. 2011b. Removal of phosphorus from aqueous solution by Posidonia oceanica fibers using continuous stirring tank reactor. *Journal of Hazardous Materials* 189: 577–85.

Wang, B. and C.Q. Lan. 2011. Biomass production and nitrogen and phosphorus removal by the green alga in simulated wastewater and secondary municipal wastewater effluent. *Bioresource Technology* 102: 5639–44.

Wang, S. and Y. Peng. 2010. Natural zeolites as effective adsorbents in water and wastewater treatment. *Chemical Engineering Journal* 156: 11–24.

Wang, Y.P., B.Z. Houlton, and C.B. Field. 2007. A model of biogeochemical cycles of carbon, nitrogen, and phosphorus including symbiotic nitrogen fixation and phosphatase production. *Global Biogeochemical Cycles* 21(1). http://dx.doi.org/10.1029/2006GB002797

Wang, C., X. Yu, H. Lv, and J. Yang. 2013. Nitrogen and phosphorus removal from municipal wastewater by the green alga Chlorella sp. *Journal of Environmental Biology* 34: 421.

Wang, W.Y., Q.Y. Yue, X. Xu, B.Y. Gao, J. Zhang, Q. Li, and J.T. Xu. 2010. Optimized conditions in preparation of giant reed quaternary amino anion exchanger for phosphate removal. *Chemical Engineering Journal* 157: 161–7.

Wauchope, R.D. 1978. The pesticide content of surface water draining from agricultural fields—a review. *Journal of Environmental Quality* 7: 459–72.

Weikard, H.P. and D. Seyhan. 2009. Distribution of phosphorus resources between rich and poor countries: The effect of recycling. *Ecological Economics* 68: 1749–55.

Wood, R.B. and C.F. McAtamney. 1996. Constructed wetlands for waste water treatment: The use of laterite in the bed medium in phosphorus and heavy metal removal. *Hydrobiologia* 340: 323–31.

Xie, R., Y. Chen, T. Cheng, Y. Lai, W. Jiang, and Z. Yang. 2016. Study on an effective industrial waste-based adsorbent for the adsorptive removal of phosphorus from wastewater: Equilibrium and kinetics studies. *Water Science and Technology* 73: 1891–900.

Xu, X., B. Gao, Q. Yue, and Q. Zhong. 2011. Sorption of phosphate onto giant reed based adsorbent: FTIR, Raman spectrum analysis and dynamic sorption/desorption properties in filter bed. *Bioresource Technology* 102: 5278–82.

Xu, H., P. He, W. Gu, G. Wang, and L. Shao. 2012. Recovery of phosphorus as struvite from sewage sludge ash. *Journal of Environmental Sciences* 24: 1533–8.

Yadav, D., M. Kapur, P. Kumar, and M.K. Mondal. 2015. Adsorptive removal of phosphate from aqueous solution using rice husk and fruit juice residue. *Process Safety and Environmental Protection* 94: 402–9.

Ye, H., F. Chen, Y. Sheng, G. Sheng, and J. Fu. 2006. Adsorption of phosphate from aqueous solution onto modified palygorskites. *Separation and Purification Technology* 50: 283–90.

Yeom, S.H. and K.Y. Jung. 2009. Recycling wasted scallop shell as an adsorbent for the removal of phosphate. *Journal of Industrial and Engineering Chemistry* 15: 40–4.

Yu, Y., R. Wu, and M. Clark. 2010. Phosphate removal by hydrothermally modified fumed silica and pulverized oyster shell. *Journal of Colloid and Interface Science* 350: 538–43.

Zhang, J., W. Shan, J. Ge, Z. Shen, Y. Lei, and W. Wang. 2011. Kinetic and equilibrium studies of liquid-phase adsorption of phosphate on modified sugarcane bagasse. *Journal of Environmental Engineering* 138: 252–8.

11 Technologies for the Treatment of Heavy Metal–Contaminated Groundwater

Ponnusamy Senthil Kumar and
Anbalagan Saravanan

CONTENTS

11.1 INTRODUCTION

Heavy metal contamination is an issue related to regions of intensified industry. Heavy metal–polluted groundwater is a genuine worry in most nations (Naushad et al., 2017; Ahamad et al., 2017). Environmental restoration of contaminated groundwater in the modern agricultural and urban domains is an extraordinary challenge in recent decades because of anthropogenic activities (Mahar et al., 2015). Anthropogenic activities, for example, mining, purifying operations, and farming, have unobtrusively increased the levels of heavy metals, for example, Cd, Co, Cr, Pb, Cu, Zn, As, and Ni, in soil up to hazardous levels (Sharma et al., 2014).

Heavy metals constitute an exceptionally heterogeneous group of components, widely varying in their compound properties and organic capacities. Heavy metals are classified as natural poisons because of their toxic effects on plants, animals, and people (Sharma et al., 2017). Heavy metal contamination of soil results from anthropogenic and also natural activities. Heavy metals are persistent in nature and hence, become aggregated in soils and plants. Dietary intake of numerous metals at high levels through the consumption of plants has long-term deleterious effects on human health. The effect of heavy metals on aquatic organisms is due to the development of contamination from different diffuse or point sources, which frequently combine in unforeseen ways in the ecosystem. In this way, they represent a hazard to maritime fauna, especially to fish, which constitute one of the genuine sources of protein-rich food for mankind.

Groundwater contamination, frequently because of contaminant leakage from transfer locales, is a major issue. Nonetheless, street routes, refuse disposal sites, and cars are now thought to be among the biggest sources of heavy metals. Numerous industrial activities from which chemicals or wastes might be discharged to the earth, either purposefully or accidentally, can possibly contaminate ground water. Industries such as plating, earthenware production, glass, mining, and battery assembly are viewed as the primary sources of excess metals in neighborhood water streams, which will result in the contamination of groundwater with heavy metals. Furthermore, heavy metals that are normally found in high concentration in landfill leachate are likewise a potential source of contamination for groundwater (Aziz et al., 2004).

The raised level of heavy metals in the groundwater represents a considerable hazard to users of nearby resources and possibly to the common habitat. At the point when groundwater winds up clearly polluted by contaminants, it causes illness, and enormous trouble is required to clean it. The World Health Organization (WHO) has already understood this issue and begun issuing standards for safe groundwater.

11.2 HEAVY METALS: SOURCES AND EFFECT IN THE ENVIRONMENT

Toxic metals, to a huge degree, are widespread in the earth due to industrial effluents, natural wastes, dumping of consumables, and the transport and power era. The sources of heavy metals in the environment are shown in Figure 11.1. They can be dispersed to places many miles from their sources by the wind, dependent on whether they are in airborne or particulate form. Metallic contamination is eventually washed

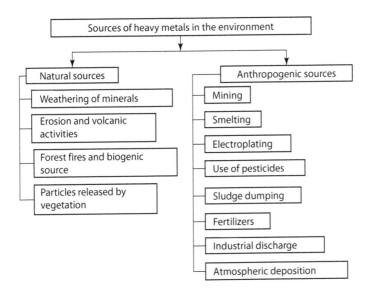

FIGURE 11.1   Sources of heavy metals in the environment.

out of the air onto land or the surface of roads. In this way, the air is an additional route for the contamination of the environment. Metal-containing mechanical effluents constitute a noteworthy source of metallic contamination of the hydrosphere. Another method for dispersal is the development of drainage water from catchment zones that have been polluted by waste from mining and purifying units.

Heavy metals enter plant, animal, and human tissues through breathing, diet, and manual handling. Motor vehicle exhausts are a noteworthy source of airborne contaminants, including arsenic, cadmium, copper, nickel, lead, zinc, mercury, manganese, and chromium. Water sources (groundwater, lakes, streams, and waterways) can be polluted by heavy metals draining from industrial and consumer waste; acid rain can fuel this procedure by releasing heavy metals trapped in soils. Plants are exposed to heavy metals through the uptake of water; animals eat these plants; the ingestion of plant- and animal-based foods is the biggest source of heavy metals in humans. The sources, toxic effects, and maximum concentration levels of various heavy metals are listed in Table 11.1.

11.3 HEAVY METAL TOXICITY

The essential heavy metals (Zn, Ca, Cu, Fe, Mn, Mg, and Mo) play biochemical and physiological roles in plants and animals. However, some heavy metals are also potent carcinogens and have a tendency to accumulate in biological systems. On account of their inherently persistent nature, when heavy metal particles (Cu, Cd, and Pb) are released or transported into the earth, they may undergo changes and can have an expanded natural, general health, and financial effect (Sang et al., 2008). If the final product contains a high concentration of heavy metals, it might be toxic to soil, plants, and human health (Alothman et al., 2013; Mittal et al., 2016). Metal toxicity is the

TABLE 11.1
Sources, Toxic Effects, and Maximum Concentration Limit of Various Heavy Metals

Metal	Sources	Toxic Effects	Maximum Contaminant Limit (mg/L)
Arsenic	Industrial dusts, medicinal uses of polluted water, production of pesticides, herbicides, and insecticides, as well as in semiconductor manufacturing	Perforation of nasal septum, respiratory cancer, peripheral neuropathy: dermatomes, muscle tenderness or weakness and changes in the skin pigmentation, intestinal pain, burning eyes and throat, diarrhea, dizziness	0.050
Chromium	Tannery, paints and pigments, electronics, mining, fertilizers, sewage sludge, landfill leachate, leather, photography	Ulcer, lung cancer, bone cancer, diarrhea, perforation of nasal septum, damage to kidney, liver, nervous system, skin irritation, dermatitis, respiratory tract, mutagenicity, and carcinogenicity	0.05
Zinc	Batteries, electroplating, paints and pigments, mining, fertilizers, alloy and steels, sewage sludge, paper and pulp, landfill leachate	Vomiting, diarrhea, bloody urine, icterus (yellow mucus membrane), liver failure, kidney failure, and anemia	0.80
Lead	Fumes and polluted food, old lead-pigment paints, batteries, industrial smelting and alloying, some types of solders, ayurvedic herbs, some toys and products from China, glazes on (foreign) ceramics, leaded (antiknock compound) fuels, bullets	Dysfunction in the kidney, reproductive system, liver, and brain, resulting in sickness and death. Inhibition of the synthesis of hemoglobin. Cardiovascular system. Acute and chronic damage to the central nervous system (CNS) and peripheral nervous system (PNS). Encephalopathy, peripheral neuropathy, central nervous disorders, anemia.	0.006
Cadmium	Fertilizers, batteries, electronics, mining, alloys and steels, sewage sludge, paper and pulp, landfill leachate	Pulmonary effects (emphysema, bronchiolitis, and alveolitis) and renal effects, nausea, vomiting, abdominal cramps, dyspnea and muscular weakness, anosmia, cardiac failure, cancers, cerebrovascular infarction, emphysema, osteoporosis, proteinuria, cataract formation in the eyes	0.01

(Continued)

TABLE 11.1 (CONTINUED)
Sources, Toxic Effects, and Maximum Concentration Limit
of Various Heavy Metals

Metal	Sources	Toxic Effects	Maximum Contaminant Limit (mg/L)
Nickel	Aerosols, alloys and steels, batteries, electroplating, mining, electronics, sewage sludge, paper and pulp, landfill leachate	Cancer, dermatitis, skin allergies, lung fibrosis, hyperallergenic responses, conjunctivitis, eosinophilic pneumonitis, asthma	0.20
Copper	Mining, electroplating, alloys and steels, sewage sludge, paper and pulp, landfill leachate, tannery	Gastrointestinal irritation and possible necrotic changes in the liver and kidney	0.25
Manganese	Fumes, mining and mineral processing (particularly nickel), emissions from alloy, steel, and iron production, combustion of fossil fuels	Central and peripheral neuropathies, systolic blood pressure, disturbed excretion of 17-ketosteroids, change in erythropoiesis and granulocyte formation	0.05
Mercury	Coal-fired power generation, medical waste incinerators and municipal waste combustors, manufacture of metals, alkalis, and cement	Spontaneous abortion, congenital malformation and gastrointestinal disorders (such as corrosive esophagitis and hematochezia), proteinuria, anorexia, numbness and paresthesias, headaches, hypertension, irritability and excitability, and immune suppression, possibly immune deregulation	0.00003

toxic effect on life of specific metals in specific compounds and concentrations. A few metals are dangerous when they form toxic solvent mixes. Certain metals have no natural role; that is, they are not basic minerals, or they are dangerous.

11.3.1 Heavy Metal Toxicity in Soil

Heavy metals are viewed as one of the significant sources of soil contamination. Excessive metal contamination of the soil is created by different metals, particularly Cu, Ni, Cd, Zn, Cr, and Pb (Karaca et al., 2010). Heavy metals have harmful consequences for soil microorganisms, consequently resulting in change to various qualities, population size, and general action of the soil microbial groups (Ashraf and Ali, 2007). The antagonistic impacts of heavy metals on the natural and bio-chemical properties of soil have been widely recorded. Soil properties, that is, natural substances, soil composition, and pH, have a real effect on the degree of the impact of metals on organic and biochemical properties. Heavy metals show a lethal impact on soil biota by influencing key microbial processes and reduce the number and movement of soil microorganisms. Heavy metals, by implication, influence soil enzymatic activities by changing the microbial group, which orchestrates enzymes (Shun-hong et al., 2009). An increase of metal concentration antagonistically influences soil microbial properties, for example, respiration rate and biosynthesis, which seem, by all accounts, to be exceptionally valuable indicators of soil contamination.

11.3.2　Heavy Metal Toxicity in Plants

Heavy metal toxicity in plants varies with plant species, the particular metal, its concentration, its compound structure, and soil composition and pH. The same heavy metals are thought to be fundamental to the development of all plants. A few metals, such as Cu and Zn, substitute as coenzymes and activators for protein responses. Heavy metal toxicity has been shown to increment the action of enzymes, for example, glucose-6-phosphate dehydrogenases and peroxidases, in the leaves of plants grown in contaminated soil. Some heavy metals, for example, Cd, Hg, and As, are unequivocally toxic to metal-sensitive compounds, causing developmental delay and death of animals. Some of these heavy metals, that is, As, Cd, Hg, Pb, and Se, are not fundamental to plant development, since they play no known physiological role in plants. Others, that is, Co, Cu, Fe, Mn, Mo, Ni, and Zn, are fundamental components required for the normal development and metabolism of plants; however, these components can easily cause harm when their concentrations are higher than ideal.

Ingestion by plant roots is one of the principal routes of heavy metals in the course of development. The adsorption and accumulation of heavy metals in plant tissue rely on many variables, which include temperature, moisture, natural materials, pH, and supplement accessibility (Jordao et al., 2006). Heavy metal accumulation by plants depends on the plant species, and the proficiency of various plants in absorbing metals is assessed by either plant uptake or soil-to-plant exchange of metallic elements (Khan et al., 2008). The uptake of heavy metals by plants and the ensuing accumulation along the food chain is a potential danger to animal and human health (Sprynskyy et al., 2007). These heavy metals aggregate in the environmentally evolved pathway through uptake at the essential producer level and after that, through use at the consumer levels.

Heavy metals have undesirable effects on the physiological and biochemical capacity of plants. The generally self-evident impacts are the inhibition of growth rate; chlorosis; decay; leaf rolling; changed stomatal activity; diminished water potential; efflux of cations; changes in photosynthesis; and inhibition of photosynthesis, respiration, metabolism, and the activities of a few key proteins (Ashfaque et al., 2016). Figure 11.2 shows the toxicity of heavy metals in plants.

Heavy metal contamination of vegetables cannot be ignored, as these foodstuffs are critical parts of the human diet. Vegetables are rich sources of vitamins, minerals, and fiber, and furthermore, have advantageous antioxidative effects. Notwithstanding, the ingestion of excessively contaminated vegetables may represent a hazard to human health. Heavy metal contamination of food stands out among

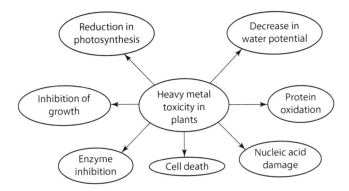

FIGURE 11.2 The toxicity of heavy metals in plants.

the most critical elements of food quality confirmation. Heavy metals are non-biodegradable and persistent environmental contaminants, which may collect on the soil surface and after that, be retained in the tissues of vegetables. Excess metal concentrations in vegetables from the marketplace have been checked and evaluated in some developed and developing nations.

It has long been perceived that heavy metal accumulation in soil may pose a potential health hazard to plants, carnivores, and people through an indirect or direct pathway, or by means of the food chain (Blakbern, 2003). An increased uptake of heavy metals by plants at concentrations below phytotoxic levels may pose potential dangers to developed ways of life when domestic animals are raised on contaminated soils.

11.3.3 Heavy Metal Toxicity in Aquatic Environment

Natural waters, especially estuaries and fresh water systems, are in general not currently being excessively polluted, but at the same time, they are subject to genuinely long-term contamination because of metals stored in silt from past human activities. Concerning the level of metal contamination in the aquatic system, it is low in open seas and increases steeply as it approaches waterfront waters and estuaries.

Heavy metals are profoundly persistent, are dangerous in trace amounts, and can conceivably cause extreme oxidative stress in aquatic life forms. Consequently, these contaminants are highly noteworthy regarding ecotoxicology. Heavy metals discharged into aquatic systems are for the most part bound to particulate matter, which in the long run, settles and ends up observably consolidated into silt. Surface silt, accordingly, is the most crucial repository or sink of metals and different contaminants in aquatic situations. As a large proportion of metals entering the oceanic environment in the end progress toward incorporation in the base silt, ecological contamination by metals can happen in territories where water quality criteria are not surpassed, yet living beings in or close to the sediments are unfavorably influenced.

Toxic metal pollution of water streams and groundwater represents a noteworthy ecological and medical issue that needs a powerful and controlled innovative approach. Trace heavy metals assume a fundamental role as micronutrients in organisms. In strongly polluted aquatic locations such as these, flora and fauna are by and large devastated or totally exposed; consequently, the pollution unfavorably influences aquatic biodiversity. In addition, metals are not subject to bacterial degradation and consequently, remain forever in the marine environment (Woo et al., 2009).

Heavy metal contaminants in aquatic systems stimulate the creation of reactive oxygen species (ROS), which can harm fish and other oceanic life. When heavy metals are amassed by an oceanic life form, they can be exchanged through the hierarchy of the natural food chain. Carnivores at the highest point of the food chain, including people, acquire the greater part of their heavy metal burden from the aquatic system as food, particularly where fish are available, so there exists the potential for significant biomagnification. There are five potential ways for a toxin to enter a fish: through food, non-food particles, the gills, oral intake of water, and the skin. Once the toxins are consumed, they are transported by the blood either to a storage point or to the liver for metabolism and disposal. In the event that the contaminants are metabolized by the liver, they might be stored there, discharged in the bile, returned to the blood for possible discharge by the gills or kidneys, or stored in fat, which is an additional hepatic tissue (Ayandiran et al., 2009).

11.3.4 Heavy Metal Toxicity in Humans

The fundamental dangers to human health from heavy metals are related to exposure to lead, cadmium, mercury, and arsenic. These metals have been broadly considered and their consequences for human health consistently surveyed by global bodies, for example, the WHO. Heavy metals have been used by people for a large number of years. Although a few antagonistic health impacts of heavy metals have been known for a long time, exposure to heavy metals persists and is even expanding in a few parts of the world, specifically in less developed nations; however emissions have declined in most developed nations during recent years. The use of food harvests polluted with heavy metals for human consumption is a noteworthy result of the natural hierarchy.

Living beings require different amounts of heavy metals. Co, Cu, Mg, Mo, and Zn are required by people. All metals are lethal at higher concentrations. Excessive levels can harm animals. Other heavy metals, for example, mercury, plutonium, and lead, are harmful metals that have no known indispensable or useful function in animals, and their accumulation for some time in the bodies of animals can cause genuine disease (Singh et al., 2011). The effect of these poisonous heavy metals on human health is at present an area of serious interest because of the omnipresence of their introduction. Heavy metals move toward becoming harmful when they are not used by the body and gather in sensitive tissues. The constant ingestion of dangerous metals has detrimental effects on people, and the related destructive effects progress toward becoming noticeable some time after their introduction (Khan et al., 2008).

Heavy metals upset metabolic functions in two ways:

- They accumulate and thus affect the functioning of indispensable organs, for example, the heart, cerebrum, kidneys, bone, liver, and so forth.

- They displace essential nutritional minerals from their specific sites, thereby impeding their functions within the body. It is, in any case, difficult to live in a domain free of heavy metals. There are numerous routes by which these toxins can be introduced into the body: for example, in food, in drink, through the skin, and through breathing the air.

By and large, the harmfulness of metal particles to humans is because of the reactivity of these particles with the cell's basic proteins and compounds and tissue systems. The organs subject to particular metal toxicities are typically those organs that accumulate the most noteworthy levels of the metal *in vivo*. This is generally dependent on the course of introduction and the properties of the metal compound, that is, its valency state, instability, lipid solubility, and so forth.

11.4 STEPS FOR ANALYZING HEAVY METALS IN WASTEWATER SAMPLES

Wastewater for testing may contain particulates or natural materials, which may require pre-treatment before spectrometric investigation. To break down the aggregate metal substance of a specimen, it is necessary to categorize metals as inorganically and naturally bound, dispersed, or particulate. Generally, different steps must be followed for metal analysis; each step is critical and a potential source of error if not performed in the correct sequence. Figure 11.3 shows the steps needed to analyze the heavy metals in wastewater.

After choosing an applicable sample preparation step, the most helpful methods are clarified underneath, such as atomic absorption spectrometry (AAS), inductively coupled plasma optical emission spectrometry (ICP-OES), inductively coupled plasma mass spectrometry (ICP-MS), laser initiated breakdown spectroscopy (LIBS), and anodic stripping.

11.5 TREATMENT METHODOLOGIES

Cadmium, zinc, copper, nickel, lead, mercury, and chromium are regularly detected in industrial wastewaters. They originate from metal plating, mining activities, purifying, battery manufacturing, tanneries, oil refining, paint fabrication, pesticides, pigment manufacture, printing and photographic enterprises, and so on. The lethal metals, presumably existing in high concentrations (even up to 500 mg/L), must be adequately treated/expelled from the wastewater. In the unlikely event that the wastewater is released straight into the receiving waters, it will constitute an awesome hazard for the aquatic environment, while immediate release into the sewerage system may adversely influence the ensuing natural wastewater treatment (Wan Ngah and Hanafiah, 2008).

Some advances have been made in the remediation of heavy metal–contaminated groundwater, and the results are indisputable. Figure 11.4 illustrates the clear outcome from the remediation process (treatment of heavy metal–contaminated groundwater).

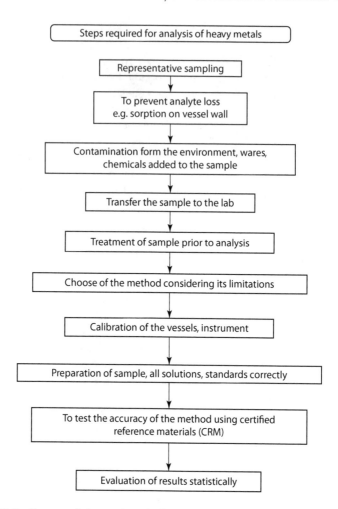

FIGURE 11.3 Steps needed to analyze the heavy metals in wastewater.

Heavy metal removal from inorganic profluent can be accomplished by ordinary treatment forms. The removal of heavy metals from industrial wastewaters can be accomplished through different treatment choices, including such unit operations as compound chemical precipitation, coagulation, complexation, activated carbon adsorption, ion exchange, solvent extraction, electro-winning, cementation, and membrane operations. Figure 11.5 demonstrates treatment methodologies for the removal of heavy metals. The advantages and disadvantages of different treatment methodologies for heavy metal removal from groundwater are listed in Table 11.2.

11.5.1 CHEMICAL PRECIPITATION

Chemical precipitation stands out among the most widely used methods for heavy metal removal from inorganic emissions in industry because of its straightforward operation. The conventional forms of chemical precipitation deliver insoluble forms

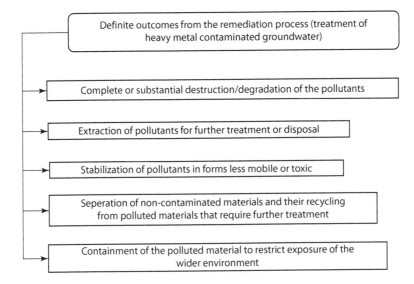

FIGURE 11.4 Definite outcomes from the remediation process (treatment of heavy metal–contaminated groundwater).

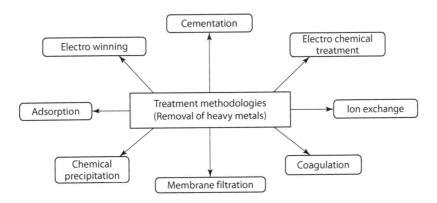

FIGURE 11.5 Treatment methodologies for the removal of heavy metals.

of heavy metals, such as hydroxide, sulfide, carbonate, and phosphate. This procedure relies on the creation of insoluble metal precipitation by reacting dispersed metals in the system with a precipitant (Veeken et al., 2003). To change the dispersed metals into a strong molecular form, a precipitation reagent is added to the blend. A coalescence response, activated by the reagent, causes the dispersed metals to form strong particles. Filtration can then be used to remove the particles from the blend (Ku and Jung, 2001; Fu and Wang, 2011). In hydroxide precipitation, a commonly used compound precipitation process, calcium or sodium hydroxide is used as the reagent to produce strong metal hydroxides. Nonetheless, it can be hard to make hydroxides from dispersed metal particles in wastewater on the grounds that numerous wastewater arrangements contain blended metals.

TABLE 11.2

Advantages and Disadvantages of Different Treatment Methodologies for Heavy Metal Removal from Groundwater

Treatment Methodology	Advantages	Disadvantages
Chemical precipitation	Some treatment chemicals are relatively inexpensive. Minimal maintenance and skilled operator requirement are needed for completely enclosed systems	Extensive amount of sludge containing metals; financially expensive; high synthetic requirements; parameters must be observed amid this strategy, which is hazardous; transport of large amounts of chemicals; addition of treatment chemicals, especially lime, may increase the volume of waste sludge up to 50%
Ion exchange	Specific removal of metal particles	Expensive (synthetic resins), large-scale application, smaller number of metal particles removed, high upkeep and operational cost, regeneration can cause serious secondary pollution, cannot be used on a large scale
Adsorption	High effectiveness of metal particle removal, low energy, broad scope of target toxins, easy operation, minimization of biological sludge, no additional nutrient requirements, prospect of metal recovery, and moreover, the adsorbent material is cheap	High cost of activated carbon, execution relies on the adsorbent
Membrane filtration	Extremely high selectivities, little space required	Membrane fouling, capital cost, maintenance, limited flow rates, lower efficiency at lower metal particle concentration, process complexity, membrane fouling, and low permeate flux
Coagulation	Dewatering qualities	More sludge settling, costly reagents, chemical use
Electrochemical treatment	No substance use	Necessity of filtration process for flocs, practical and running expenses, high initial capital investment, and expensive electricity supply
Electro-winning	Environmental compatibility	Waste generation, recycling
Cementation	Recovery of metals in relatively pure metallic form, simple control requirements, low energy consumption, and is in general a low-cost process	Excess sacrificial metal consumption

Some metal salts are insoluble in water, which is promoted when a counter-anion is included. Despite the fact that the procedure is expensive, its productivity is influenced by low pH and the proximity of different salts (particles). The procedure requires large amounts of different chemicals, which at long last prompts the period of a high–water content sludge, which is costly to transfer. Precipitation with lime or bisulfide, or particle exchange, is non-specific and is insufficient for the removal of metal particles at low concentration. The most commonly used precipitation method is hydroxide treatment because of its relative straightforwardness, ease of precipitant (lime), and simplicity of programmed pH control. The solubilities of the different metal hydroxides are limited at pH in the range of 8.0 to 11.0 (Alvarez-Ayuso et al., 2003).

Mixed precipitation procedures are suggested when fluids with a rather high salt content must be dealt with. As a result of wide experience over many years in the treatment of radioactive fluid waste by substance precipitation, the provision of equipment should not represent any irregular issues. Expenditure is, however, an indicator that should be painstakingly considered due to the broad use of an extensive variety of concentrated synthetic reagents over a wide range of pH values.

Concerning process costs, there is, right now, an inclination to replace dissipation with different systems (compound or new processes) beyond what many would consider possible, and for the most part, when huge volumes of fluid radioactive wastes must be dealt with. Synthetic precipitation is viewed as the least expensive process (20–50 times more affordable than dissipation). The costs of particle exchange treatment are situated between those for dissipation and substance precipitation.

11.5.2 Ion Exchange

Ion exchange is another technique used effectively in the business for the removal of heavy metals from effluents. Despite the fact that it is moderately costly when contrasted with alternate strategies, it can accomplish parts per billion levels of purification while taking care of a generally expansive volume. It is a reversible compound reaction in which the removal of heavy metals is accomplished by the exchange of particles on a bed for those in wastewater. At the point when the resin is saturated, it must be recovered with an acid or alkaline medium to remove the metal particles from the resin bed. The regenerated salt water is lower in volume and higher in concentration than the initial wastewater; however, these metals should then be satisfactorily treated or recovered.

Ion exchange, in the form of a particle exchanger, is highly suitable for exchanging either cations or anions from the surrounding materials. The systems normally used for ion exchange are synthetic organic ion exchange resins (Alyuz and Veli, 2009). Generally, resins are fabricated in a spherical, stress-free shape to counter physical degradation. They are stable at high temperatures and over a wide pH range. Ion exchange resins, which are totally insoluble in many fluids and natural systems, comprise a cross-connected polymer lattice to which charged residues are attached by covalent bonding. There is an assortment of resins for particular applications with different metals. Synthetic organic resins are most regularly used because of their capacity to be made for particular applications.

The ion exchange technique can remove trace particle contamination from water and process streams and give a result of the desired quality. Ion exchangers are widely used in investigative science, hydrometallurgy, anti-infection agents, cleaning, and the separation of radioisotopes and find extensive application in water treatment and contamination control (Cavaco et al., 2007; Inglezakis et al., 2002). The list of metals that are recovered and filtered on a mechanical scale by particle exchange methods includes uranium, thorium, rare earth elements (REEs), gold, silver, chromium, copper, zinc, nickel, cobalt, and tungsten (Luca et al., 2009).

In some of these cases, the size of operations is moderately small, for example, in the rare earth elements or noble metals, yet estimates of recovered metals are high. The ion exchange process is especially suitable for cleaning metal particles with high value and low charge. It can also be used as a procedure for the large-scale recovery of metals from waste streams, for example, cadmium and mercury, chromium, or copper and zinc. The use of ion exchange methods in hydrometallurgy is high and constantly evolving. It is related basically to the advances in new specific chelating ion exchangers containing complexing ligands.

The drawback of this strategy is that it cannot deal with concentrated metal systems, as the lattice is easily fouled by organics and different solids in the wastewater. Also, particle exchange is non-selective and is exceptionally sensitive to the pH of the system.

Issues commonly encountered with ion exchange treatment include:

- Metallic fouling (from Fe, Mn, Cu, etc.) of the particle exchange medium.
- Fouling because of oil, residues, dirt, colloidal silica, natural materials, and organisms. The decision to undertake an authorized cleaning procedure can re-establish a great part of the lost efficiency.
- The proximity of free acid diminishes the efficiency of the operation.
- Fairly high operational costs exist.

11.5.3 Adsorption

Adsorption has developed as a compelling, conservative, and ecofriendly treatment strategy (Barakat, 2011; Zamboulis et al., 2011). It is a procedure sufficiently powerful to satisfy water reuse commitment and high effluent standards in industry. Adsorption is fundamentally a mass transfer process by which a substance is exchanged from the fluid stage to the surface of a strong adsorbent and winds up noticeably bound by physical as well as chemical interactions (Babel and Kurniawan, 2003). It is a segment procedure in which couple of parts of the fluid stage are migrated to the surface of the strong adsorbents. All adsorption techniques are dependent on strong fluid balance and on mass exchange rates. The adsorption system can be batch, semi-batch, or continuous. The adsorption process offers adaptability in outline and operation and much of the time, will create good-quality treated effluent. Moreover, in light of the fact that adsorption is now and again reversible, adsorbents can be recovered by an appropriate desorption process. On the molecular level, adsorption takes place for the most part because of attractive interfaces between a surface and the compound being retained.

In this context, adsorption has emerged as a promising method, with advantages such as

- High efficiency even with low metal concentrations
- Low cost
- No additional nutrient requirements
- Easy operation
- Potential metal recovery
- No detrimental effects on the environment

Dependent on the type of intermolecular attractive forces, adsorption can take place due to

- Physical adsorption
- Chemical adsorption

11.5.3.1   Physical Adsorption

This is a general occurrence and happens in any solid/fluid or solid/gas system. Physical adsorption is a procedure in which the attachment of the adsorbate on the adsorbent surface is brought about by van der Waals powers of attraction. The electronic structure of the atom or molecule is not really involved in physical adsorption. Van der Waals forces start from the associations between incited, perpetual, or transient electric dipoles. Physical adsorption must involve low-temperature solids, and under suitable conditions, gas-phase atoms can undergo multilayer adsorption. Industrial adsorbents use physical adsorption for their surface attachment.

11.5.3.2   Chemical Adsorption

This is an after-effect of synthetic collaboration between the solid and the adsorbed substance. It is additionally called *actuated adsorption*. It is irreversible. It is particularly essential in catalysis. In this way, the energy of chemisorption is considered as a chemical reaction. It may be exothermic or endothermic, with procedures ranging from small to extensive in size. The basic steps in chemisorption regularly include high activation energy.

11.5.3.3   Low-Cost Adsorbents

Many minimal-effort adsorbents have been created and tried for heavy metal removal, which reflect changing adsorption efficiencies relying on the sort of adsorbents used. Activated carbon has been the most used adsorbent, although it is generally costly. Scanning for minimal-effort and easily accessible adsorbents to remove heavy metal particles has become a principal research focus. To date, many reviews on the use of minimal-effort adsorbents have been distributed. Agricultural wastes, industrial by-products and wastes, and natural substances have been examined as adsorbents for heavy metal wastewater treatment.

A wide assortment of agricultural waste and residues has been investigated for the removal of heavy metals. This chapter assesses the practicality of using agricultural waste biomass for heavy metal removal in light of

- Heavy metal adsorption limits of agricultural waste biomass
- The impacts of working parameters for process advancement
- Adsorption instruments
- Change strategies for creating better adsorbents

11.5.4 MEMBRANE FILTRATION

Membrane filtration innovations with various sorts of films indicate awesome guarantee for heavy metal removal due to their high efficiency, simple operation, and space-saving features (Kurniawan et al., 2006; Barakat and Schmidt, 2010). The membrane forms used to remove metals from wastewater are ultra-filtration (UF), reverse osmosis (RO), nanofiltration, and electro-dialysis.

11.5.4.1 Ultra-Filtration

UF is a membrane procedure working at low transmembrane weights for the removal of dispersed and colloidal material. UF uses films that permit the passage of water and low–atomic weight solutes while retaining macromolecules and hydrated metal particles, which are bigger than the pore size of the membrane (Samper et al., 2009). Since the pore sizes of UF films are bigger than dispersed metal particles as hydrated particles or as low–atomic weight constructions, these particles would easily pass through UF films. To achieve high removal efficiency for metal particles, micellar enhanced ultra-filtration (MEUF) and polymer improved ultra-filtration (PEUF) have been proposed.

MEUF has turned out to be a successful separation strategy to remove metal particles from wastewater. This separation strategy uses the addition of surfactants to wastewater. Whenever the concentration of surfactants in fluid systems is above the critical micelle concentration (CMC), the surfactant particles will assemble into micelles that can bind metal particles to form extensive metal-surfactant structures (Landaburu-Aguirre et al., 2009). The micelles containing metal particles can be retained by a UF film with pore sizes smaller than micelle sizes, while the untrapped species promptly go through the UF film. To achieve the best results, surfactants of electric charge opposite to that of the particles to be removed must be used. Sodium dodecyl sulfate (SDS), an anionic surfactant, is regularly chosen for the comprehensive removal of heavy metal particles in MEUF. For MEUF, a surfactant is used to make micelles, while for complexation–ultrafiltration, complexation agents, for example, poly(ethylenimine) (PEI) and poly(acrylic acid), are used to specifically bind cations.

PEUF has likewise been proposed as an achievable strategy to separate an awesome assortment of metal particles from fluid streams. PEUF uses water-solvent polymer to complex metallic particles and form a macromolecule having a higher sub-atomic weight than the sub-atomic weight cut-off of the film. The macromolecules will be retained when they are pumped through UF film. From that point onward, the retentate can be dealt with, keeping in mind the end goal to recover metallic particles and also to reuse the polymeric agent.

11.5.4.2   Nanofiltration

Nanofiltration (NF) has remarkable properties. The detachment system includes both steric (pore measure) and electrical mechanisms (oppositely charged surface groups). The layers used for NF can be portrayed as "low-weight turn around osmosis layers." NF is an intermediate procedure between UF and RO. NF is a promising innovation for the removal of excess metal particles, for example, nickel, chromium, copper, and arsenic, from wastewater (Muthukrishnan and Guha, 2008; Murthy and Chaudhari, 2008; Ahmad and Ooi, 2010). NF treatment benefits from simplicity of operation, dependability, and low energy use as well as high effectiveness of contamination removal.

11.5.4.3   Reverse Osmosis

RO treatment uses a semi-permeable membrane, permitting the liquid that is being cleaned to go through it while rejecting the contaminants. RO is a prevalent wastewater treatment alternative in synthetic and ecological design. RO is a weight-driven film treatment whereby water can unreservedly go through the membrane while cationic compounds are retained (Greenlee et al., 2009). RO film pore sizes can be as little as 0.1 nm. RO can be used for systems with low levels of dispersed metal (micromolar to millimolar range).

11.5.5   Coagulation

Coagulation stands out among the most essential techniques for wastewater treatment, yet the primary objects of coagulation are just hydrophobic colloids and suspended particles. With the specific end goal to remove both dissolved heavy metals and insoluble substances effectively by coagulation, a sodium xanthogenate group was joined to polyethyleneimine. This new kind of coagulant was an amphoteric polyelectrolyte (Chang and Wang, 2007).

11.5.6   Electrochemical Treatment

Electrochemical treatment is an electrically supplemented system such as ultra-filtration. It is widely used to remove toxic heavy metal particles from wastewater (Wang et al., 2007). It generally requires a vast capital venture to start the process, supplemented by long-term operational and support costs, and the cost of the electrical supply limits its applicability. The electricity can be used in various procedures, for example, electro-coagulation, electro-floatation, and electro-dialysis.

11.5.6.1   Electro-Coagulation

Electro-coagulation is an electrochemical approach with a reactive anode and cathode (iron or aluminum electrode), which uses an electrical current to remove metals from a solution. The electro-coagulation method is additionally powerful in removing suspended solids, dispersed metals, tannins, and colors. The contaminants found in wastewater are kept in solution by electrical charges (Heidmann and Calmano, 2008). At the point when these and other target particles are neutralized by particles of opposite electrical charge originating from the electro-coagulation system, they move toward becoming destabilized and forming a stable shape. After this, metallic particles combine into greater flocs and can be removed easily (Rajkumar and Palanivelu, 2004; Borba et al., 2006). Water particles are hydrolyzed at the cathode at the same time. Two diverse components proposed for iron anodes can be found in the following mechanisms:

Mechanism I
In the anode:

$$4\,Fe_{(s)} \rightarrow 4\,Fe^{2+}_{(aq)} + 8e^- \tag{11.1}$$

In the cathode:

$$8\,H^+_{(aq)} + 8\,e^- \rightarrow 4\,H_{2(g)} \tag{11.2}$$

In solution:

$$4\,Fe^{2+}_{(aq)} + 10\,H_2O_{(l)} + O_{2(g)} \rightarrow 4\,Fe(OH)_{3(s)} + 8\,H^+_{(aq)} \tag{11.3}$$

Mechanism II
In the anode:

$$Fe_{(s)} \rightarrow Fe^{2+}_{(aq)} + 2e^- \tag{11.4}$$

In the cathode:

$$2\,H_2O_{(l)} + 2e^- \rightarrow H_{2(g)} + 2\,OH^-_{(aq)} \tag{11.5}$$

In solution:

$$Fe^{2+}_{(aq)} + 2\,OH^-_{(aq)} \rightarrow Fe(OH)_{2(s)} \tag{11.6}$$

Metal ions created at the anode and hydroxide ions created at the cathode react in the aqueous medium to create different hydroxide species depending on the pH; for example, $Fe(OH)_2$, $Fe(OH)_3$, $Fe(OH)^{2+}$, $Fe(OH)^{2+}$, and $Fe(OH)_4^-$. The iron hydroxides coagulate and settle to the bottom of the solution (Mollah et al., 2004; Vasudevan et al., 2011).

11.5.6.2   Electro-Flotation

Electro-flotation uses modest increases of hydrogen and oxygen gases produced by water electrolysis, keeping in mind the end goal to buoy toxins up to the surface of the effluent. Electro-flotation can be used as part of a blend with aluminum electro-coagulation.

11.5.6.3 Electro-Dialysis

Electro-dialysis is a sort of film detachment treatment in which the ionized species in the aqueous solution pass following the use of an electric potential. The anions present in the solution move toward the anode, while the cations move toward the cathode, crossing the diversely planned membranes. The process efficiency is advanced by increasing voltage and temperature; nonetheless, it diminishes at higher flow rates. Most commonly, electro-dialysis innovation uses ion exchange membranes to isolate metal particles dispersed in wastewater. Cation and anion exchange membranes can be used to accomplish copper removal efficiency of 94%–97%, and the performance can be upgraded using high particle exchange for the removal of the selected metal.

11.5.7 Electro-Winning

Electro-winning, also called *electroextraction*, is the electrode deposition of metals from their minerals that have been dissolved in solution or melted. Electro-winning is one of the numerous innovations used to remove metals from process water streams. This procedure uses power to pass a current through a fluid metal-bearing solution containing a cathode plate and an insoluble anode. Charged metallic particles stick to the oppositely charged cathode, leaving a metal deposit that can be stripped and recovered. A recognizable disadvantage was that consumption could become a serious restriction, whereby cathodes would often have to be replaced.

Electro-winning is generally used as part of the mining and metallurgical mechanical operations for pile filtering and corrosive mine seepage. It is likewise used as part of the metal reclamation, gadget, and electrical businesses for the removal and recovery of metals. Metals such as Ag, Au, Cd, Co, Cr, Ni, Pb, Sn, and Zn occurring in effluents can be recovered by electro-winning using insoluble anodes.

11.5.8 Cementation

Cementation is used as a general term to depict the procedure whereby a metal is precipitated from a solution of its salts by another electropositive metal with unconstrained electrochemical reduction to its essential metallic state, accompanied by the oxidation of a sacrificial metal, for the recovery of more costly and more valuable dispersed metal species in fluid solutions. Cementation is a metal-substitution treatment in which a solution containing the dispersed metallic ion(s) interacts with a more dynamic metal, for example, iron. Cementation is, in a way, another precipitation technique, suggesting an electrochemical system in which a metal having a higher oxidation potential goes into solution; for example, the oxidation of metallic iron, $Fe(0)$, to ferrous $Fe(II)$, to replace a metal having a lower oxidation potential. Copper is mainly isolated by cementation alongside other valuable metals, for example, Ag, Au, and Pb, and also As, Cd, Ga, Pb, Sb, and Sn can be recovered in this way.

The reaction for copper and iron is

$$Cu^{2+} + Fe^{\circ} \rightarrow Cu^{\circ} + Fe^{2+} \tag{11.7}$$

The advantages of the cementation process are as follows:

- Basic control necessities. The interest in treatment of the substance is dependent on the rate at which the target toxin enters the system. In the iron cementation of copper, the rate of iron use shifts according to the rate at which copper particles are released into the system. This dispenses with the requirement for close checking of the waste stream development and outside control of the supply rate of the treatment reagent.
- Low energy use.
- Recovery of important high-purity metals, for example, copper.

11.6 CONCLUSION

Heavy metal contamination of wastewater is one of the most vital ecological issues all around the world. To meet the expanded, increasingly stringent ecological directives, an extensive variety of treatment innovations, for example, chemical precipitation, coagulation flocculation, electrochemical treatment, ion exchange, and membrane filtration, have been created for heavy metal removal from wastewater. Albeit all the heavy metal wastewater treatment procedures can be used to remove heavy metals, they each have their own favorable circumstances and constraints. Albeit every strategy described can be used for the treatment of heavy metal wastewater, it is essential to specify that the choice of the most appropriate treatment procedures relies on the underlying metal concentration, the type of wastewater, capital speculation and operational cost, plant adaptability, consistent quality, ecological effects, and so on.

REFERENCES

Ahamad, T., Naushad, M., AlMaswari B. M., et al. 2017. Synthesis of a recyclable mesoporous nanocomposite for efficient removal of toxic Hg^{2+} from aqueous medium. *Journal of Industrial and Engineering Chemistry* 53: 268–75.

Ahmad, A. L. and B. S. Ooi. 2010. A study on acid reclamation and copper recovery using low pressure nanofiltration membrane. *Chemical Engineering Journal* 156: 257–63.

Al Othman, Z. A., Alam, M. M., and Naushad, M., 2013. Heavy toxic metal ion exchange kinetics: Validation of ion exchange process on composite cation exchanger nylon 6,6 Zr(IV) phosphate. *Journal of Industrial and Engineering Chemistry* 19: 956–960.

Alvarez-Ayuso, E., Garcia-Sanchez, A., and X. Querol. 2003. Purification of metal electroplating waste waters using zeolites. *Water Research* 37: 4855–62.

Alyuz, B. and S. Veli. 2009. Kinetics and equilibrium studies for the removal of nickel and zinc from aqueous solutions by ion exchange resins. *Journal of Hazardous Materials* 167: 482–8.

Ashfaque, F., Inam, A., S. Sahay, et al. 2016. Influence of heavy metal toxicity on plant growth, metabolism and its alleviation by phytoremediation—a promising technology. *Journal of Agriculture and Ecology Research International* 6(2): 1–19.

Ashraf, R. and T. A. Ali. 2007. Effect of heavy metals on soil microbial community and mung beans seed germination. *Pakistan Journals of Botany* 39(2): 629–36.

Ayandiran, T. A., Fawole, O. O., S. O. Adewoye, et al. 2009. Bioconcentration of metals in the body muscle and gut of Clarias gariepinus exposed to sublethal concentrations of soap and detergent effluent. *Journal of Cell and Animal Biology* 3(8): 113–18.

Aziz, H. A., Yusoff, M. S., M. N. Adlan, et al. 2004. Physico-chemical removal of iron from semi-aerobic landfill leachate by limestone filter. *Waste Management* 24(4): 353–8.

Babel, S. and T. A. Kurniawan. 2003. Various treatment technologies to remove arsenic and mercury from contaminated groundwater: An overview. In: *Proceedings of the First International Symposium on Southeast Asian Water Environment*, Bangkok, Thailand, 24–25 October: 433–40.

Barakat, M. A. 2011. New trends in removing heavy metals from industrial wastewater. *Arabian Journal of Chemistry* 4: 361–77.

Barakat, M. A. and E. Schmidt. 2010. Polymer-enhanced ultrafiltration process for heavy metals removal from industrial wastewater. *Desalination* 256: 90–3.

Blakbern, A. A. 2003. Accumulation and migration of trace elements along trophic chains in ecosystems of the Chaktal Biosphere Reserve (the Western Tien Shan, Uzbekistan). *Russian Journal of Ecology* 34(1): 68–71.

Borba, C. E., Guirardello, R., E. A. Silva, et al. 2006. Removal of nickel (II) ions from aqueous solution by biosorption in a fixed bed column: Experimental and theoretical breakthrough curves. *Biochemical Engineering Journal* 30: 184–91.

Cavaco, S. A., Fernandes, S., M. M. Quina, et al. 2007. Removal of chromium from electroplating industry effluents by ion exchange resins. *Journal of Hazardous Materials* 144: 634–8.

Chang, Q. and G. Wang. 2007. Study on the macromolecular coagulant PEX which traps heavy metals. *Chemical Engineering Science* 62: 4636–43.

Fu, F. and Q. Wang. 2011. Removal of heavy metal ions from wastewaters: A review. *Journal of Environmental Management* 92(3): 407–18.

Greenlee, L. F., Lawler, D. F., B. D. Freeman, et al. 2009. Reverse osmosis desalination: Water sources, technology, and today's challenges. *Water Research* 43: 2317–48.

Heidmann, I. and W. Calmano. 2008. Removal of Zn(II), Cu(II), Ni(II), Ag(I) and Cr(VI) present in aqueous solutions by aluminium electro coagulation. *Journal of Hazardous Materials* 152: 934–41.

Inglezakis, V. J., Loizidou, M. D., and H.P. Grigoropoulou. 2002. Equilibrium and kinetic ion exchange studies of Pb^{2+}, Cr^{3+}, Fe^{3+} and Cu^{2+} on natural clinoptilolite. *Water Research* 36: 2784–92.

Jordao, C. P., Nascentes, C. C., P. R. Cecon, et al. 2006. Heavy metal availability in soil amended with composted urban solid wastes. *Environmental Monitoring and Assessment* 112: 309–26.

Karaca, A., Cetin, S. C., O. C. Turgay, et al. 2010. Effects of heavy metals on soil enzyme activities. In: I. Sherameti and A. Varma (Ed), *Soil Heavy Metals, Soil Biology*, Springer-Verlag, Berlin, Germany 19: 237–65.

Khan, S., Cao, Q., Y. M. Zheng, et al. 2008. Health risks of heavy metals in contaminated soils and food crops irrigated with wastewater in Beijing, China. *Environmental Pollution* 152: 686–92.

Ku, Y. and I.-L. Jung. 2001. Photocatalytic reduction of Cr (VI) in aqueous solutions by UV irradiation with the presence of titanium dioxide. *Water Research* 35(1): 135–42.

Kurniawan, T. A., Chan, G. Y. S., and W.-H. Lo. 2006. Physico-chemical treatment techniques for wastewater laden with heavy metals. *Chemical Engineering Journal* 118: 83–98.

Landaburu-Aguirre, J., Garcia, V., E. Pongracz, et al. 2009. The removal of zinc from synthetic wastewaters by micellar-enhanced ultrafiltration: Statistical design of experiments. *Desalination* 240: 262–9.

Luca, C., Vlad, C. D., and I. Bunia. 2009. Trends in weak base anion exchangers resins. *Revue Roumaine de Chimie*, 54(2): 107–17.

Mahar, A., Wang, P., R. Li, et al. 2015. Immobilization of lead and cadmium in contaminated soil using amendments: A review. *Pedosphere* 25: 555–68.

Mittal, A., Naushad, M., Sharma, G., Alothman, Z. A., Wabaidur, S. M., and Alam, M. 2016. Fabrication of MWCNTs/ThO2 nanocomposite and its adsorption behavior for the removal of Pb(II) metal from aqueous medium. *Desalin Water Treat* 57: 21863–21869.

Mollah, Y. A., Morkovsky, P., J. A. G. Gomes., et al. 2004. Fundamentals, present and future perspectives of electrocoagulation. *Journal of Hazardous Materials* 114(1–3): 199–210.

Murthy, Z. V. P. and L. B. Chaudhari. 2008. Application of nanofiltration for the rejection of nickel ions from aqueous solutions and estimation of membrane transport parameters. *Journal of Hazardous Materials* 160: 70–7.

Muthukrishnan, M. and B. K. Guha. 2008. Effect of pH on rejection of hexavalent chromium by nanofiltration. *Desalination* 219: 171–8.

Naushad, M., Ahamad, T., B. M. Al-Maswari, et al. 2017. Nickel ferrite bearing nitrogen-doped mesoporous carbon as efficient adsorbent for the removal of highly toxic metal ion from aqueous medium. *Chemical Engineering Journal* 330: 1351–60.

Rajkumar, D. and K. Palanivelu. 2004. Electrochemical treatment of industrial wastewater. *Journal of Hazardous Materials*, 113(1–3): 123–9.

Samper, E., Rodriguez, M., and M. A. De la Rubia. 2009. Removal of metal ions at low concentration by micellar-enhanced ultrafiltration (MEUF) using sodium dodecyl sulfate (SDS) and linear alkylbenzene sulfonate (LAS). *Separation and Purification Technology* 65: 337–42.

Sang, Y., Li, F., Q. Gu, et al. 2008. Heavy metal-contaminated groundwater treatment by a novel nanofiber membrane. *Desalination* 223: 349–60.

Sharma, G., Pathania, D., M. Naushad, et al. 2014. Fabrication, characterization and antimicrobial activity of polyaniline Th(IV) tungstomolybdophosphate nanocomposite material: Efficient removal of toxic metal ions from water. *Chemical Engineering Journal (Amsterdam, Netherlands)* 251: 413–21.

Sharma, G., Naushad, M., A. H. Al-Muhtaseb, et al. 2017. Fabrication and characterization of chitosan-crosslinked-poly(alginic acid) nanohydrogel for adsorptive removal of Cr(VI) metal ion from aqueous medium. *International Journal of Biological Macromolecules* 95: 484–93.

Shun-hong H., Bing P., Y. Zhi-hui, et al. 2009. Chromium accumulation, microorganism population and enzyme activities in soils around chromium-containing slag heap of steel alloy factory. *Transactions of Nonferrous Metals Society of China* 19: 241–8.

Singh, R., Gautam, N., A. Mishra, et al. 2011. Heavy metals and living systems: An overview. *Indian Journal of Pharmacology* 43(3): 246–53.

Sprynskyy, M., Kosobucki, P., T. Kowalkowski, et al. 2007. Influence of clinoptilolite rock on chemical speciation of selected heavy metals in sewage sludge. *Journal of Hazardous Materials* 149: 310–16.

Vasudevan, S., Lakshmi, J., and G. Sozhan. 2011. Effects of alternating and direct current in electrocoagulation process on the removal of cadmium from water. *Journal of Hazardous Materials* 192: 26–34.

Veeken, A. H. M., de Vries, S., A. van der Mark, et al. 2003. Selective precipitation of heavy metals as controlled by a sulfide-selective electrode. *Separation Science and Technology* 38: 1–19.

Wan Ngah, W. S. and M. A. K. M. Hanafiah. 2008. Removal of heavy metal ions from wastewater by chemically modified plant wastes as adsorbents: A review. *Bioresource Technology* 99: 3935–48.

Wang, L. K., Hung, Y. T., and N. K. Shammas. 2007. Advanced physicochemical treatment technologies. In: *Handbook of Environmental Engineering*, vol. 5. Humana, Totowa, NJ.

Woo, S., Yum, S., H. S. Park, et al. 2009. Effects of heavy metals on antioxidants and stress-responsive gene expression in Javanese medaka (Oryzias javanicus). *Comparative Biochemistry and Physiology Part C* 149: 289–99.

Zamboulis, D., Peleka, E. N., N. K. Lazaridis, et al. 2011. Metal ion separation and recovery from environmental sources using various flotation and sorption techniques. *Journal of Chemical Technology and Biotechnology* 86: 335–44.

12 Life Cycle Assessment Applied to Support Sustainable Wastewater Treatment in Developing Countries

Lineker Max Goulart Coelho,
Hosmanny Mauro Goulart Coelho, and
Liséte Celina Lange

CONTENTS

12.1   INTRODUCTION: SUSTAINABILITY CHALLENGES IN DEVELOPING COUNTRIES

Sustainable development means a balance between economic efficiency, social equity, and environmental protection. According to the first definition of this term, provided by the World Commission on Environment and Development (WCED) in 1991, sustainable development consists of a "development that meets the needs of the present without compromising the ability of future generations to meet their own needs" (WCED, 1991).

Nowadays, the world is facing challenges related to the three pillars of sustainability: economy, society, and environment. In this context, to move from the current scenario toward a sustainable development, a new way of thinking focused on social inclusion needs to be implemented. In addition, changes in consumption and production patterns, incentives for the preservation of natural resources, and inequality reduction are crucial to the sustainability of humanity (United Nations (UN), 2013).

For developing countries particularly, sustainable development is even more challenging, as they face significant urban environmental issues due to rapid urbanization, population growth, inability to manage climate and environmental risks, inefficient governance, corruption, and investment restrictions (Ameen and Mourshed, 2017).

So, unfortunately, sustainability concepts in developing countries remain only theoretical rather than finding practical application. Indeed, due to the aforementioned factors, currently, sustainable development concepts are in reality scarcely observed in most developing countries.

12.2   WASTEWATER AND SUSTAINABILITY

Nowadays, sustainability has become a core issue of wastewater management (Massoud et al., 2009). For Andersson et al. (2016), few areas of investment today have as much to offer to the global shift toward sustainable development as sanitation and wastewater management. The role of wastewater management in a sustainable society is to act as an instrument to minimize environmental burdens, preserve human health, and create business opportunities.

Actually, the importance of wastewater for sustainability involves not only the traditional function of pollutant emission reduction but also the development of resource recovery practices. According to Mateo-Sagasta et al. (2015), the world's daily generation of municipal wastewater is estimated to be 900 million m^3. The same author also estimates that the wastewater from a city with a population of 10 million inhabitants presents enough nutrients to fertilize about 500,000 hectares of agricultural land. The correct estimation of pollutant emission and the determination of resource recovery potential are crucial figures that directly affect the definition of strategies that aim for sustainability of the wastewater management system.

Indeed, for the United Nations Environment Programme (UNEP, 2010), a crucial point in better understanding water and sanitation challenges is the quality of data, as it allows the identification of problems and opportunities contributing to the development of sustainable solutions. However, a consolidated database grouping water and wastewater monitoring information has not yet been implemented at regional or global level.

So, the adoption of decision-making tools is required for the implementation of a sustainable way of thinking in wastewater planning, as they can support the investigation and analysis of several scenarios by means of a quantitative and scientific methodology. In addition, the consideration of regional and local specificities is very important for the selection of the most sustainable solution.

12.3   WASTEWATER TREATMENT IN DEVELOPING COUNTRIES

The implementation of sustainable wastewater management systems is a major challenge in developing countries, as sanitation data from these countries indicates critical conditions of population in terms of incomes, health, hygiene, and social issues. For Pradel et al. (2016), the release of raw wastewater into the environment causes sanitary problems due to the increase of fecal pathogens in water resources. Indeed, Prüss-Ustün et al. (2014) estimate that annually more than 800,000 people die due to diarrheal diseases, many of them as a result of inadequate sanitation infrastructure. Moreover, according to Andersson et al. (2016), one of the main causes of death in children under the age of five in developing countries is diarrhea, caused mainly by poor sanitation and hygiene conditions. Waddington et al. (2009) carried out research to estimate the influence of sanitation and hygiene improvements on diarrheal diseases, and concluded that such interventions result in reductions greater than 30%.

For Massoud et al. (2009), the adoption of developed countries' strategies for wastewater management is neither appropriate nor viable for developing countries, as there are huge differences between them in terms of political structures, national priorities, socioeconomic conditions, cultural traits, and financial resources. According to Flores et al. (2008), aside from the cost, conventional wastewater treatment technologies are normally not appropriate in developing countries, because they are focused on the removal of organic compounds to protect water resources, whereas the main concern in developing countries is the protection of human health by pathogen removal. So, in aiming for sustainability of sanitation systems, the specificities of developing countries must to be considered in the definition of wastewater management strategies.

It is important to note that in developing countries, most wastewater generated is normally released into water bodies after a low-efficiency treatment or none at all. In addition, the coverage of the wastewater collection system is very limited. Often, the reason for the lack of wastewater treatment is financial (Mara, 2003). However, there are already several low-cost wastewater treatment processes available that could support treatment rate improvements in developing countries.

Another important aspect is simplicity of operation and maintenance. Treatment technologies that demand complex operation systems and advanced maintenance procedures are not a sustainable alternative due to the difficulty of finding a qualified workforce in developing countries that can operate the sanitary infrastructure demanded. So, low-cost alternatives and simple operation systems, as well as investment in professional qualification and training, are key points to improve sanitation conditions in such countries.

Lohri et al. (2014) indicate the following aspects as the main reasons for problems when wastewater treatment plants (WWTPs) are implemented in developing countries: unsuitable technologies, lack of ownership and responsibility of operators, absence of professional and academic networks, obstructive legislation, lack of institutional support, design and scale that do not match the availability of feedstock, lack of local skills for operation, and absence of maintenance and service support.

Furthermore, the scarcity of studies dedicated to developing countries leads to the selection of inappropriate treatment technologies in terms of climatic conditions, financial and human resource capabilities, and social or cultural acceptability (Massoud et al., 2009).

These findings reveal the urgent need to develop less expensive, more efficient, and easier-to-maintain technologies for wastewater management, which could contribute to the sustainability of the systems (Lutterbeck et al., 2017).

Developing countries from tropical and subtropical regions, for example, present a climate that makes the use of anaerobic technology applicable and less expensive (Foresti et al., 2001).

According to Mara (2003), upflow anaerobic sludge blanket reactors (UASBs) are high-rate anaerobic wastewater treatment units and have been extensively used in developing countries such as Brazil, Colombia, and India. However, for Torres et al. (2012), further research is needed to improve and upgrade the existing anaerobic treatment methods, adapting them to each particular context.

Constructed wetlands also meet the main characteristics required for an adequate treatment process for developing countries. Zhang et al. (2014) provided a detailed review of previous research involving constructed wetlands in developing countries. Lutterbeck et al. (2017), in turn, presented results related to the evaluation of a wetland in the context of rural areas.

Hawkins et al. (2013) analyzed the pros and cons of traditional centralized systems and decentralized solutions for wastewater management in developing countries and concluded that the former are normally preferred for political reasons, despite the latter often being the most suitable option. Simplified and decentralized systems, such as individual septic tanks, for example, are promising solutions for developing countries and have been widely used in Brazil.

12.4 WASTEWATER LIFE CYCLE ASSESSMENT STUDIES IN DEVELOPING COUNTRIES

Life cycle assessment (LCA) is a tool already widely used around the world in the analysis of wastewater management systems, because it allows the negative and positive impacts of a project to be quantified by a scientific methodology. Indeed, LCA has been proved to be a suitable instrument to assess the environmental effects of WWTPs in both design and operation phases (Zang et al., 2015). Providing a better understanding of the environmental impacts generated by wastewater management, LCA enables the minimization of negative effects contributing to environmental protection (Limphitakphong et al., 2016).

However, most studies related to this theme have been carried out considering the context of developed countries, mainly Europe and North America (Wang et al., 2012). Consequently, the results of these studies could not be simply extended to developing countries such as Brazil, China, India, and South Africa, as these present different characteristics in terms of region, policies, and the economy. As previously discussed, developing countries present some issues in the context of sanitation that are normally not present in developed economies, such as budget limitations, intense growth of urbanization, health and sanitation problems, and scarcity of specialized and qualified professionals in this field. So, LCA studies dedicated to developing countries need to take these differences into account to really support wastewater management by considering regional particularities. For instance, studies focused on identifying treatment technologies that combine low costs, high efficiencies, and lower environmental impacts are encouraged in the context of developing countries (Lutterbeck et al., 2017). However, it is important to note that even among developing countries, there are also several differences. For example, concerning the BRICS (Brazil, Russia, India, China, and South Africa), the energy mix, an aspect that could directly affect the LCA results, is very distinct among these developing countries. In Brazil, for example, hydropower is the predominant source of electricity, whereas in China, India, Russia, and South Africa, fossil fuel dominates the energy mix. Furthermore, even among countries in which fossil fuel predominates, differences in energy mix are noted; for example, for China, the main fuel used is coal, whereas for Russia, natural gas is the main source of energy. Other differences related to regional aspects are also relevant. For instance, the climate in Russia and Brazil is completely different, which directly affects the wastewater management solution suitable for each one.

Therefore, it is necessary to be careful to avoid generalizations related to the applications of LCA results not only from developed countries to developing ones, but also among developing countries.

Concerning the existing studies dedicated to LCA application in wastewater in developing countries, 39 papers on this theme were found in the literature.

12.4.1 PREVIOUS STUDIES

It is important to note that previous works have already presented reviews related to LCA applications in wastewater management. Table 12.1 shows a summary of the major aspects of each review evaluated in the present study.

As presented in Table 12.1, Larsen et al. (2007) performed a review including 22 studies reporting scope, impact categories, and LCA methodologies, focusing on discussions involving toxicity impacts and energy-related emissions. Corominas et al. (2013), in turn, investigated 45 studies involving LCA applications in wastewater treatment and compiled general information about LCA, such as scope of the study, functional unit, impact categories, and LCA methodology. The review provided in Zang et al. (2015) analyzed 53 studies focused on LCA of activated sludge systems and evaluated the papers considering scope, impact categories, and LCA methodologies. Concerning only LCA involving sludge treatment, Yoshida et al (2013) and Pradel et al. (2016) carried out reviews focused on studies involving this theme.

TABLE 12.1

Description of Existing Review Studies on LCA Applications to Wastewater Management

Review Paper	Number of Papers Analyzed	Scope of the Studies Recorded	Main Contribution
Larsen et al. (2007)	22	Wastewater treatment	Discussion about LCA categories, toxicity impact, and energy emissions
Corominas et al. (2013)	45	Wastewater treatment	Overview of LCA structure
Yoshida et al. (2013)	35	Sludge treatment	Overview of LCA structure
Zang et al. (2015)	53	Activated sludge treatment plants	Overview of LCA structure
Heimerson et al. (2016)	62	Wastewater and sludge treatment and disposal	Overview of C, N, and P WWTP-related emissions
Pradel et al. (2016)	44	Sludge treatment	Scope, objectives, and functional unit
Hernandez-Padilla et al. (2017)	46	Wastewater treatment	Overview of LCA methods and type of treatment assessed

Heimersson et al. (2016) also presented a review in which they evaluated 62 papers related to LCA applied to wastewater and sludge management. These last authors carried out a review in a different way from the other aforementioned papers; they performed an analysis dedicated to the flows of C, N, and P in wastewater and sludge treatment systems. Hernandez-Padilla et al. (2017) presented a short review based on 46 studies. This paper provides an interesting discussion related to LCA regionalization, including a case study for developing countries from Latin America.

It is important to note that these review papers considered existing studies from all over the world. This chapter, in turn, presents a more specific review than the previous ones, focused on studies involving LCA applications in wastewater in the context of developing countries.

12.4.2 REVIEW PROCESS

The review process consisted of four steps: selection of articles, data classification, frequency, and critical analysis.

A total of 39 papers concerning LCA applied to developing countries were compiled for the present study. First, the papers used in this research were obtained from the following electronic databases: American Society of Civil Engineers, Sage Journals, Science Direct, Scopus, Springer Link, Taylor Francis, and Wiley Online Library. In addition, to expand the coverage of the research, environmental reports from governmental organizations and international events were also consulted. The papers were selected based on a keyword search using the following keywords: "wastewater," "LCA," and "developing countries."

For each article selected, the year of publication and data related to the LCA model were recorded. The information obtained from each study was classified considering some characteristics that are presented in Table 12.2. These characteristics were defined to enable the evaluation of the articles in terms of the main steps of an LCA study. It is important to note that Table 12.2 presents several categories used to classify the articles according to the characteristics considered. Furthermore, comments related to special aspects of each article were recorded. To avoid errors in the information acquisition process, the definition of the categories used to classify the studies was performed by a double independent analysis involving two researchers. The results of both were compared, and in the case of disagreement, the results were discussed until consensus was obtained.

After the classification process, to assist the overall evaluation of the data and to provide a quantitative measure of the information acquired, a frequency analysis was carried out to determine the number of papers in each class of a characteristic considered. Thus, a critical analysis was performed based on the information acquired in the classification process and in the results of the frequency analysis. This critical assessment aims to indicate the main insights and limitations of the approaches presented in the papers analyzed, which includes global trends noted as well as the results of specific studies. Lastly, from the strengths and shortcomings observed, recommendations for upcoming works were provided.

12.4.3 TEMPORAL EVOLUTION

Referring to temporal evolution, the earliest recorded paper on using LCA for wastewater applications in developing countries was Zhang and Wilson (2000), while the earliest study addressing the same topic in developed countries was published in 1995

TABLE 12.2
Characteristics and Categories Evaluated in Each Paper Analyzed

Characteristic	Categories
Functional unit	Cubic meters of wastewater, Person Equivalent, kilograms of chemical oxygen demand (COD), kilograms of sludge
LCA approach	Midpoint, endpoint, both
LCA model	BSM2, CML, Eco-Indicator, Ecological Scarcity, EDP, LIME (Life-cycle Impact assessment Method based on Endpoint modeling), Impact, ReCiPe, TRACI, others
Scope of the study	Collection, treatment, both
Impact categories considered	Abiotic depletion, water resources depletion, fossil fuel depletion, ecotoxicity, human toxicity, land use, acidification, eutrophication, photochemical oxidation, ozone depletion, global warming, land use, ionizing radiation, particulate material formation, respiratory effects, embodied energy
Adoption of ISO standards	Yes or no
Normalization process	Yes or no
Sensitivity analysis	Impact categories, input data, LCA model

(Emmerson et al., 1995). In fact, other studies on this theme involving developed countries have been present in literature since the 1990s, as detailed in Corominas et al. (2013) and Heimerson et al. (2016).

From 2000 to the present, it was noted that LCA applications to wastewater in the context of developing countries have attracted an increasing amount of interest from research groups. Indeed, a growth in the number of publications was recently observed, mainly after 2010, as presented in Figure 12.1. However, as already mentioned, only 39 studies on this topic were recorded. This shows that studies on this subject are still very scarce; it is an important field to be further explored in future research.

12.4.4 REGION OF RESEARCH

Concerning regional distribution, as presented in Figures 12.2 and 12.3, most papers recorded originated from Asian countries, mainly from China, the top-ranked country in terms of number of publications. To a lower extent, studies were also found from other Asian countries, such as India (Kalbar et al., 2012, 2013), Singapore (Zhang and Wilson, 2000), and Thailand (Limphitakphong et al., 2016). America also presented several publications, principally from Brazil (Gutierrez et al., 2014; Lopes et al., 2014; Lutterbeck et al., 2017) and Mexico (Garcia et al., 2011; Musharrafie et al., 2011; Romero-Hernandez, 2005).

Similarly to the observations for Asia, papers were recorded from several other countries in South America, such as Argentina (Ontiveros and Campanella, 2013), Bolivia (Cornejo, 2015), Chile (Ledon et al., 2017), and Colombia (Tatiana et al., 2016). Particularly, Hernandez-Padilla et al. (2017) carried out research involving developing countries from Latin America.

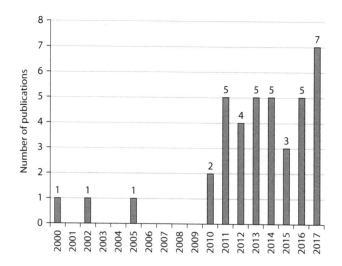

FIGURE 12.1 Temporal evolution of publications related to LCA applied to wastewater management in developing countries.

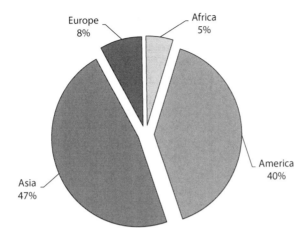

FIGURE 12.2 Regional distribution of publications related to LCA applied to wastewater management in developing countries.

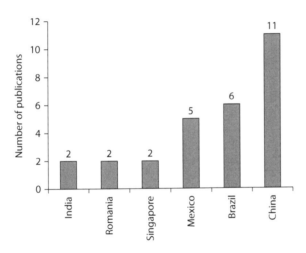

FIGURE 12.3 LCA applied to wastewater management in developing countries classified by country.

This interest in LCA applications to wastewater in Asia and America demonstrates that researchers in these countries already have some know-how regarding this approach. However, the dissemination of LCA still depends on the development of inventories, characterization factors, and methodologies adapted to regional specificities.

Concerning Africa, only two studies were recorded, one related to Egypt (El-Sayed et al., 2010) and another involving the South African context (Nikadimeng, 2015). The low level of research in Africa is probably related to the absence of regional models and databases and due to the low number of regional research groups dedicated to LCA studies.

Very few papers from European developing countries were found, which was unexpected, as in general, review papers, such as Corominas et al. (2013), indicate that most studies on LCA in wastewater were carried out in Europe. In addition, Europe is a traditional region in terms of LCA research, being the home continent of several LCA models widely used in the world and presenting regional inventories, databases, and regional characterization factors. So, the expertise in LCA from European developed countries needs to be shared with developing ones. Indeed, the contributions of LCA studies could be much more useful for the latter than for the former, as developing countries normally present a wastewater management system under development. So, studies such as Barjoveanu et al. (2014) and Teodosiu et al. (2016) are strongly encouraged.

Concerning only a particular group of developing countries, the BRICS, formed by Brazil, Russia, India, China, and South Africa, all of them presented papers published on this subject, but it is noted that China and Brazil were ahead in the number of publications in comparison with the others, indicating that research groups in these countries are currently active and interested in this theme. The importance and relevance of the development of studies in these countries are related to their influence. Indeed, BRICS is formed by developing countries with high regional and international influence and with high economic potential. So, the adoption of scientific methodologies such as LCA and the development of wastewater management in BRICS could be very influential for the adherence of other developing countries to these practices. Tables 12.3 and 12.4 present overall information about the studies reviewed.

12.4.5 RESEARCH SCOPE

The objective and scope for most studies recorded was LCA applied to the operation phase of wastewater treatment. Indeed, 36 studies compiled were focused on resources consumption and emissions from the operation of a WWTP. Only three studies evaluated present a larger scope and comprise not only the treatment but also the collection system.

This preference for treatment is probably related to the fact that the operation phase of treatment plants could directly impact the consumption of energy and the emission of pollutants such as greenhouse gases, nitrogen, and phosphate. Indeed, according to Wang et al. (2012), impacts from the operation phase of a WWTP are more relevant than those from the construction and final disposal phases. Limphitakphong et al. (2016) carried out a study in the context of Thailand and found that the environmental burden associated with the operating stage of a WWTP represents 80% of the total impact generated by it, whereas those associated with construction and final disposal correspond to 20% and 0.1%, respectively.

Moreover, one of the main applications of LCA in wastewater in developing countries is to support the selection of alternatives by the comparison of different treatment scenarios. In fact, LCA applied to define the wastewater treatment method is a convenient approach in terms of environmental protection. In developing countries, the treatment phase is still a major issue in wastewater management, as already discussed. So, in this context, LCA could be very helpful during the decision-making

TABLE 12.3

General Information about the LCA Studies Considered in the Review Analysis

Study	Country	Research Scope	Functional Unit	ISO Standards Adoption
Zhang and Wilson (2000)	Singapore	Treatment	Cubic meters of wastewater	No
Stromberg and Paulsen (2002)	Russia	Treatment	Person equivalent	Yes
Romero-Hernandez (2005)	Mexico	Treatment	—	No
El-Sayed et al. (2010)	Egypt	Collection and treatment	Cubic meters of wastewater	Yes
Zhang et al. (2010)	China	Treatment	Cubic meters of wastewater	Yes
Musharrafie et al. (2011)	Mexico	Treatment	Cubic meters of wastewater	No
Garcia et al. (2011)	Mexico	Treatment	Cubic meters of wastewater	Yes
Patricia et al. (2011)	Mexico	Treatment	Cubic meters of wastewater	No
Alarcon et al. (2011)	Mexico	Treatment	Cubic meters of wastewater	Yes
Pan et al. (2011)	China	Treatment	Cubic meters of wastewater	Yes
Yıldırım and Topkaya (2012)	Turkey	Treatment	Person equivalent	No
Kalbar et al. (2012)	India	Treatment	Person equivalent	Yes
Wang et al. (2012)	China	Treatment	Cubic meters of wastewater	No
Pirani et al. (2012)	United Arab Emirates	Treatment	Cubic meters of wastewater	No
Ontiveros and Campanella (2013)	Argentina	Treatment	Cubic meters of wastewater	No
Kalbar et al. (2013)	India	Treatment	Person equivalent	Yes
Li et al. (2013)	China	Treatment	Cubic meters of wastewater	Yes
Chen et al. (2013)	China	Treatment	Cubic meters of wastewater	Yes
Tong et al. (2013)	China	Treatment	Cubic meters of wastewater	No
Barjoveanu et al. (2014)	Romania	Collection and treatment	Cubic meters of wastewater	No

(Continued)

TABLE 12.3 (CONTINUED)
General Information about the LCA Studies Considered in the Review Analysis

Study	Country	Research Scope	Functional Unit	ISO Standards Adoption
Tourinho (2014)	Brazil	Treatment	Cubic meters of wastewater	Yes
Lopes (2014)	Brazil	Treatment	Cubic meters of wastewater	Yes
Gutierrez (2014)	Brazil	Treatment	Cubic meters of wastewater	Yes
Ng et al. (2014)	Singapore	Treatment	Cubic meters of wastewater	No
Nkadimeng (2015)	South Africa	Treatment	Kilograms COD	Yes
Cornejo (2015)	Bolivia	Treatment	Cubic meters of wastewater	Yes
Resende et al. (2015)	Brazil	Treatment	Cubic meters of wastewater	Yes
Teodosiu et al. (2016)	Romania	Treatment	Cubic meters of wastewater	Yes
Tatiana et al. (2016)	Colombia	Treatment	Cubic meters of wastewater	Yes
Limphitakphong et al. (2016)	Thailand	Collection and treatment	Cubic meters of wastewater	Yes
Chiu et al. (2016)	China	Treatment	Tonnes of sludge	Yes
Buonocore et al. (2016)	China	Treatment	Cubic meters of wastewater	Yes
Hernandez-Padilla et al. (2017)	Latin America	Treatment	Cubic meters of wastewater	Yes
Ledon et al. (2017)	Chile	Treatment	Person equivalent	Yes
Schwaickhardt et al. (2017)	Brazil	Treatment	Cubic meters of wastewater	Yes
Lutterbeck et al. (2017)	Brazil	Treatment	—	No
Lu et al. (2017)	China	Treatment	Cubic meters of wastewater	Yes
Bai et al. (2017)	China	Treatment	Cubic meters of wastewater	No
Li et al. (2017)	China	Treatment	Cubic meters of wastewater	Yes

TABLE 12.4
LCA Characteristics of the LCA Studies Reviewed

Study	LCA approach	LCA model	Normalization	Sensitivity Analysis
Zhang and Wilson (2000)	Midpoint	—	No	No
Stromberg and Paulsen (2002)	Midpoint	CML 2001	Yes	No
Romero-Hernandez (2005)	Midpoint	SETAC	No	Yes
El-Sayed et al. (2010)	Endpoint	EcoIndicator 99	No	No
Zhang et al. (2010)	Endpoint	EcoIndicator 99	No	No
Musharrafie et al. (2011)	Midpoint	CML 2001	No	No
Garcia et al. (2011)	Midpoint	TRACI	No	No
Patricia et al. (2011)	Midpoint	CML 2001	No	No
Alarcon et al. (2011)	Midpoint	TRACI	No	No
Pan et al. (2011)	Midpoint	—	No	No
Yıldırım and Topkaya (2012)	Midpoint	CML 2001	Yes	No
Kalbar et al. (2012)	Midpoint	CML 2001	No	No
Wang et al. (2012)	Midpoint	—	No	No
Pirani et al. (2012)	Midpoint	EcoIndicator 99	Yes	Yes
Ontiveros and Campanella (2013)	Midpoint	CML 2001	No	No
Kalbar et al. (2013)	Midpoint	CML 2001	No	No
Li et al. (2013)	Midpoint	CML 2001	Yes	No
Chen et al. (2013)	Midpoint	CML 2001 and Ecological Scarcity	Yes	Yes
Tong et al. (2013)	Midpoint	CML 2001	No	No
Barjoveanu et al. (2014)	Midpoint	CML 2001	Yes	Yes
Tourinho (2014)	Midpoint and endpoint	ReCiPe	Yes	Yes
Lopes (2014)	Midpoint	CML 2001	Yes	Yes
Gutierrez (2014)	Midpoint	ReCiPe	No	No
Ng et al. (2014)	Midpoint	CML 2001	Yes	No
Nkadimeng (2015)	Midpoint	CML 2001	No	Yes
Cornejo (2015)	Midpoint	EcoIndicator 95	No	Yes
Rezende et al. (2015)	Midpoint	CML 2001	No	Yes
Teodosiu et al. (2016)	Midpoint and endpoint	ReCiPe and Ecological Scarcity	Yes	Yes
Tatiana et al. (2016)	Midpoint	EDP 2013	No	No
Limphitakphong et al. (2016)	Endpoint	LIME	No	No
Chiu et al. (2016)	Midpoint	ReCiPe	No	No
Buonocore et al. (2016)	Midpoint	ReCiPe	Yes	No
Hernandez-Padilla et al. (2017)	Midpoint	ReCiPe and Impact	No	Yes
Ledon et al. (2017)	Midpoint	IPCC	No	No

(*Continued*)

TABLE 12.4 (CONTINUED)
LCA Characteristics of the LCA Studies Reviewed

Study	LCA approach	LCA model	Normalization	Sensitivity Analysis
Schwaickhardt et al. (2017)	Midpoint and endpoint	ReCiPe	Yes	No
Lutterbeck et al. (2017)	Midpoint and endpoint	ReCiPe	Yes	No
Lu et al. (2017)	Endpoint	EcoIndicator 99	No	No
Bai et al. (2017)	Midpoint	CML 2001	Yes	No
Li et al. (2017)	Midpoint	CML 2001	Yes	No

process, providing information about environmentally friendly alternatives for future projects. In addition, the expansion of the scope to other components of the system could provide relevant information, as is noted in Barjoveanu et al. (2014), El-Sayed et al. (2010), and Limphitakphong et al. (2016).

12.4.6 Functional Unit

The selection of the functional unit is acknowledged as a crucial point in an LCA. So, from the review, it is noted that in the huge majority of papers analyzed in which LCA was carried out, the volume of wastewater managed was expressed in cubic meters as functional unit. Indeed, from Figure 12.4, it is noted that 29 of 39 studies adopted cubic meters of wastewater as functional unit. In second position, but with a much lower occurrence, it is observed that six studies used person equivalent (PE) as functional unit. This preference for relating resource consumption and emissions to wastewater volume was expected, as it is a measure normally available from the operational monitoring database.

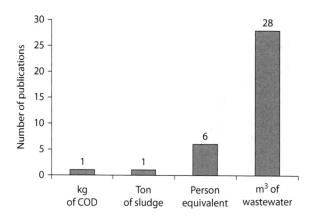

FIGURE 12.4   Functional unit adopted in the LCA studies reviewed.

12.4.7   ISO STANDARDS ADOPTION

International Organization for Standardization (ISO) standards related to LCAs, mainly ISO 14040 and 14044, are among the main references developed to offer an organized methodology to execute an LCA study independently of the LCA model adopted, as they present global recommendations that aim to provide reliability to the study. So, from the results obtained, it is noted that 65% of the recorded studies followed ISO LCA standards, which represents the majority of the analyzed papers. The adoption of ISO LCA standards is not necessarily obligatory to guarantee the quality of the LCA results, but is a good practice, as it provides directions about crucial steps. In the context of developing countries, in which LCA studies are still scarce and this approach is not completely mastered, the adoption of ISO procedures or other standards that provide guidance to LCA studies is recommended and encouraged.

12.4.8   LCA APPROACH AND METHODOLOGY

In terms of LCA approach, 80% of the studies analyzed, that is, the huge majority, adopted a midpoint approach, 10% used endpoint analysis, and 10% considered both approaches. The preference for midpoint analysis is likely due to the fact that it explicitly presents the potential impact caused, which facilitates interpretation, whereas the endpoint approach presents the results aggregated in terms of protection areas affected. However, if the objective of the study is to obtain overall information about the impact of the wastewater system proposed, endpoint is likely more suitable.

Concerning LCA methodology, as presented in Figure 12.5, CML was the most widely used LCA model in the recorded studies. CML was developed by Guinée et al. (2002) and follows a midpoint approach. The predominance of this method was expected, as CML was one of the first LCA models to be disseminated worldwide, and it is still one of the preferred approaches. However, if the study aims to adopt

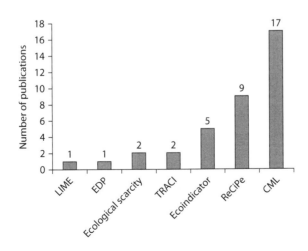

FIGURE 12.5   LCA methodologies adopted in the studies reviewed.

an endpoint approach, this method is not suitable. From Figure 12.5, it is also noted that in second and third position are ReCiPe and Eco-Indicator, respectively. The former was developed based on CML and the Eco-Indicator model, allowing midpoint and endpoint approaches to be integrated and harmonized, whereas the latter follows an endpoint approach. Actually, Eco-Indicator was one of the early methods to adopt endpoint thinking.

So, observing the three main methodologies adopted, it is noted that most papers preferred to adopt traditional LCA methodologies. Indeed, 78% of papers adopted one of the following LCA models: CML, Eco-Indicator, and ReCiPe. This probably occurs due to the fact that LCA applications in most developing countries are a recent area, which is still in development. So, the adoption of acknowledged methodologies is justified, as it is likely to guarantee the robustness of the models as well as to provide comparability of the studies. It seems that the choice of one of these three methods was based on the intended approach: endpoint, midpoint, or both.

It is important to note that other LCA methods, developed for specific regions of the world, were also adopted, such as TRACI and LIME. For instance, Limphitakphong et al. (2016) carried out a study in the context of Thailand adopting LIME, which is an LCA method developed for Japan. Alarcon et al. (2011) and Garcia et al. (2011), in turn, selected TRACI, an LCA model focused on the U.S. context, to investigate wastewater systems. In general, the adoption of regional models is recommended, as they can consider specific impact categories and provide regional characterization factors, improving the accuracy of results. However, country-specific LCA models for other countries, even those close to the home country of the LCA model, need to be used with caution, as differences in some aspects, such as country energy mix and normalization factors, could be strongly distinct. This is why the development of regional LCA characterization factors on a regional level for Africa, Latin America, and Asia is urgently recommended, as they will better represent the conditions of these regions. Indeed, improvements provided by the regionalization of LCA models will be very useful to support sustainable solutions in wastewater management in developing countries.

12.4.9 IMPACT CATEGORIES

The ten impact categories most often adopted in the studies analyzed are presented in Figure 12.6.

It is noted that the most often used impact categories are the ones originating from CML, which is expected, as it is the most widely used LCA model in the papers analyzed.

The three most often used impact categories were global warming, acidification, and eutrophication. This result is in agreement with the findings of Corominas et al. (2013), who carried out a review involving LCA in wastewater without limiting the papers to developing countries. For Heimerson et al. (2016), the interest in research on these impact categories is probably linked to the emissions of N, P, and C from wastewater treatment, which directly impact them. Indeed, the global warming category

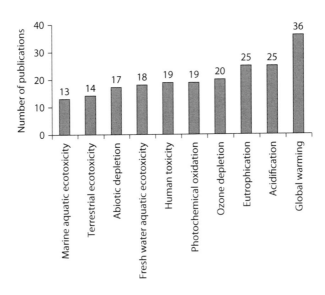

FIGURE 12.6 Impact categories most often adopted in the studies reviewed.

is strongly impacted by emissions of CO_2, CH_4, and N_2O from wastewater treatment plants (Heimersson et al., 2016). Furthermore, N and P releases are important for eutrophication. Acidification, in turn, is impacted, for example, by hydrogen sulfide flow from anaerobic processes, as well as by ammonia emissions from land application.

It was observed that 36 works, almost all studies analyzed, included global warming as an impact category. Indeed, some studies are dedicated to studying only this impact category, such as Pan et al. (2011), which studied global warming–related emissions from a wastewater treatment system in China. The interest of research on global warming in developing countries is probably related to the fact that some countries of this group are at the top of the list of the countries that present high CO_2 emission rates, such as Brazil, China, India, and Russia. Moreover, according to the Intergovernmental Panel on Climate Change (IPCC, 2014), more than half of the total amount of greenhouse gas emissions from waste comes from wastewater management.

Acidification and eutrophication are in second position, as both presented 25 occurrences. Acidification is an important impact category in WWTPs in developing countries because of the hydrogen sulfide emissions related to anaerobic treatment, which is the preferred approach for treatment, as it does not demand energy related to aeration and simplifies sludge treatment.

Furthermore, the high presence of the eutrophication impact category in several studies is justified by the fact that it is considered a crucial impact category in LCA applied to WWTP analysis (Corominas et al., 2013; Limphitakphong et al., 2016). The importance of eutrophication in WWTP is explained by the fact that wastewater treatment could prevent the release of nutrients to the environment, avoiding the intensification of eutrophication processes (Wang et al., 2012).

It is important to note that several studies have adopted a large number of impact categories, such as El-Sayed et al. (2010), Musharrafie et al. (2011), Barjoveanu et al. (2014), and Resende et al. (2015).

The main advantage of considering several impact categories in a study is to guarantee that all possible impacts of the system will be investigated. On the other hand, the workload will be proportional to the number of impact categories evaluated, which could justify the selection of those with the greatest impact.

So, some studies preferred to focus only on one or two impact categories, such as Limphitakphong et al. (2016), which studied wastewater systems in Thailand addressing global warming and eutrophication, and Ledon et al. (2017), which focused on evaluating global warming and resource consumption in wastewater systems in Chile.

According to the International Reference Life Cycle Data System (ILCD) guide (ILCD, 2010), previous knowledge based on experience acquired from detailed studies for similar systems may indicate that one or more of the default impact categories are not relevant for the system being analyzed. However, the elimination of impact categories needs to be explicitly justified, as it could directly affect the conclusions of the study. So, in the absence of a previous study that could support the selection of some impact categories, the adoption of several categories is encouraged. In the case of developing countries, as previous studies are scarce, there is a low chance of finding similar studies that could support the elimination of impact categories. So, an important aspect for future research in these countries is to study the influence of each impact category on the overall impact of the system.

12.4.10 NORMALIZATION PROCEDURES

According to Stranddorf et al. (2005), the goal of normalization is to relate the impact scores to a common reference to enable the comparison of different environmental impacts. From ISO guidelines, normalization is an optional step of life cycle impact assessment (LCIA), which allows results to be presented after the characterization step using a common reference impact (Benini et al., 2014). The review results show that 39% of papers included the normalization step in the LCA study.

The predominance of papers that do not adopt normalization is probably related to the uncertainty involved in the normalization factors, which express the total impact occurring in a reference region for a certain impact category within a reference year (Benini et al., 2014). It is important to note that the determination of normalization factors is a very challenging task, which demands significant effort and time as well as an intimate knowledge of data availability and quality (Kim et al., 2013). As normalization factors vary from region to region, and for most developing countries there is no specific data for their regions, normalization is not recommended without the development of regional factors.

12.4.11 SENSITIVITY ANALYSIS

According to Figure 12.7, only 28% of studies presented a step involving sensitivity analysis. So, although it is an important step in LCA, it was observed that the huge

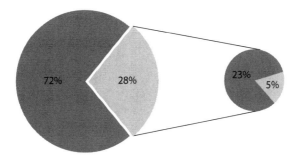

FIGURE 12.7 Number of studies involving LCA applied to wastewater management in developing countries that adopts sensitivity analysis and type of sensitivity assessed.

majority of the existing studies have completely neglected this analysis. This low frequency of sensitivity is probably related to the fact that it greatly increases the time and work involved in the study, as it normally involves several scenarios and changes in input data or model premises. Indeed, for Yoshida et al. (2013), the analysis of uncertainty involved in a LCA study is an important step to provide integrity and robustness to the evaluation.

Therefore, as sensitivity analysis is a crucial step to confirm the robustness of results (Qureshi et al., 1999; Syamsuddin, 2013), to guarantee the consistency of future studies, this step must be considered an essential procedure.

Concerning the studies that performed a sensitivity analysis, from Figure 12.7, it is noted that eight of them evaluated the effect of criteria by changing input data such as energy mix and operational conditions of the wastewater system. In addition, only two studies (Barjoveanu et al., 2014; Teodosiu et al., 2016) examined the sensitivity related to the use of different LCA models (CML, ReCiPe, EcoIndicator). Hernandez-Padilla et al. (2017) performed a sensitivity analysis for both input data and LCA methodology. However, the sensitivity of the LCA results is not only related to these two aspects. Actually, subjective steps such as definition of boundary conditions and normalization process can be as impacting as, or even more impacting than, input data and the LCA model. So, to really assess the sensitivity of a model and to guarantee its robustness, it is suggested to carry out such analysis for future works considering the aforementioned aspects.

12.5 RECOMMENDATIONS AND PERSPECTIVES

12.5.1 Consideration of Adapted Treatment Technologies

With a focus on supporting sustainable wastewater management systems, an important recommendation for future studies is to consider technologies adapted to the local conditions in terms of environmental, economic, and social aspects. Most studies reported in the review take into account traditional wastewater methods, which are not necessarily the best alternatives in the context of developing countries. LCA investigations dedicated to treatment methods that require low investment and present simplified operation and maintenance procedures could be useful to identify opportunities for promising techniques that meet the reality of most developing countries. High-performance tertiary treatment in LCA results should be considered with caution, as normally, wastewater treatment in developing countries is mainly focused on primary and secondary steps. So, to model a WWTP without a tertiary stage or with a simplified system could better reflect emissions from actual systems, increasing fidelity to reality. In addition, as in several developing countries the rural population is very representative and sometimes predominant, LCA studies involving wastewater systems focused on rural areas are encouraged. Lutterbeck et al. (2017), for example, assessed constructed wetlands applied to treat wastewater in rural areas in Brazil.

12.5.2 CONSIDERATION OF REGIONAL CHARACTERISTICS

One of the most critical concerns in LCA studies is establishing consistent inventory data (Yoshida et al., 2014). From the existing studies involving LCA applications to wastewater in developing countries, a common issue was observed: the lack of regional models and databases. With the exception of developing countries connected with Europe, which could adopt LCA from the European database, for most developing countries, characterization factors and life cycle inventories (LCIs) are scarce or nonexistent. As already discussed, previous papers assessed wastewater management scenarios in developing countries using LCA methodologies, inventories, and characterization factors developed mainly for North American and European countries, which could increase uncertainty and limits the consistency of interpretation of results. So, the development of LCI adapted to developing countries, as well as the determination of regional characterization factors, could strongly improve the reliability of LCA results and the conclusions obtained in future research. For Teodosiu et al. (2016), who performed research on LCA in a wastewater system in Romania, regionalized impacts could improve LCA effectiveness in supporting wastewater decision-making, as a better representation of local conditions enables better estimations of impacts generated by the alternatives evaluated.

Regionalization of LCI is required mainly due to differences in technologies from one country to another, which includes the determination of LCI for regionally specific technologies as well as the revision of LCI for traditional technologies to consider country adaptations that could impact resource consumption and/or emissions. Energy mix is also an important regional and even local characteristic that needs to be incorporated into all LCA studies. As previously remarked for BRICs, developing countries present a wide range of electricity mix arrangements that could greatly impact the final results and conclusions, as presented in Hernandez-Padilla et al. (2017).

For Corominas et al. (2013), the development of LCA methodologies that consider specific local factors is especially critical for the determination of eutrophication impact, one of the main impact categories in the wastewater management context. Hernandez-Padilla et al. (2017), assessing Latin American countries, showed how the eutrophication impacts of an adopted wastewater solution can vary from one country to another if country-specific factors are considered.

By the way, Hernandez-Padilla et al. (2017) is one of the few works that tried to consider regional conditions by using a country-specific energy mix and specific characterization factors for eutrophication in Latin American countries. The same study shows the differences in results if regional and local data are considered, which reinforces the need for regionalization of data.

However, in practice, the establishment of regional and even local specific inventories for all substances that cause adverse health and environmental impacts is prohibitively expensive, and in many cases, it is not even feasible (Yoshida et al., 2014).

So, these findings suggest the necessity of developing methodologies that indicate how to determine what data must be analyzed at regional and local level. Initiatives as presented in Hernandez-Padilla et al. (2017) are encouraged, as they provide a methodology based on a contribution analysis of impacts, to determine the regional data that needs to be collected.

12.5.3 STANDARDIZATION OF STUDIES

As already discussed, the high variability and diversity of procedures used in previous studies is one of the major challenges to the development of research in this area and consequently, to its use as a decision-supporting tool. For Teodosiu et al. (2016), who carried out an LCA study involving a wastewater system in Romania, these differences among studies impose serious difficulties on the comparison and extrapolation of results. It is important to note that this lack of standardization is not restricted to the context of developing countries. Corominas et al. (2013) provided similar conclusions considering studies from developed countries and argued that the creation of standardized procedures in wastewater LCA studies is required to ensure the quality of the application of the LCA methodology. It is important to note that a standardization procedure does not mean carrying out studies using a default process without considering the specificities of region and location. Actually, standardization aims to provide comparability and guidance to LCA studies to increase their reliability. So, standardizing functional units, boundaries definition, impact categories, and LCA methodology could be an interesting initiative. On the other hand, characterization, factors, LCIs, energy mix, and other characteristics are highly dependent on local conditions and must be analyzed case by case.

12.5.4 CONSIDERATION OF COSTS AND SOCIAL ISSUES

Considering that the main objective of LCA studies is to contribute to the sustainable development of sanitation systems in developing countries, it is recommended that the three pillars of sustainability, economy, and environment and society have to be considered in future works. Nowadays, LCA studies carried out in these countries are only focused on environmental aspects, although LCA presents variations that allow the analysis to be extended to social and economic points of view. The inclusion of both aspects is crucial in the context of developing countries, as they present several socially related issues and financial limitations. Indeed, life cycle costing (LCC) and social life cycle assessments (S-LCA) are robust methodologies already widely used in other areas to provide useful information based on the life cycle thinking approach. Rebtzer et al. (2003) was one of the first studies to incorporate LCC in LCA studies aiming to support the sustainable development of wastewater systems. So, the incorporation of LCC and S-LCA in future LCA works on wastewater systems in developing countries is crucial to assess the sustainability of the alternatives considered. For instance, Limphitakphong et al. (2016), who carried out an LCA study involving a WWTP in Thailand, considered operational costs in the investigation performed, acknowledging that life cycle cost assessment can support decision-makers in investment and policy planning.

12.5.5 GENERAL RECOMMENDATIONS

From the data and discussions presented in this chapter, to support sustainable development in developing countries, the following recommendations are provided for future LCA studies involving wastewater management:

- Perform sensitivity analysis of inputs, LCA methodologies' premises, normalization, and other sources of uncertainty, which is crucial to the robustness of results.
- Develop regionalized LCI and characterization factors for developing countries.
- Carry out research focused on local reality in terms of technologies, investments, and social conditions.
- Investigate treatment alternatives that are low cost, simple to operate, and designed for rural areas.
- Carefully select the impact categories to be considered to avoid information losses.
- Incorporate economic and social aspects in the LCA.
- Develop guidelines and standard procedures to assess wastewater management by LCA.

12.6 CONCLUSIONS

This chapter presented an overview of the challenges and perspectives of LCA in terms of its application to sustainable wastewater treatment in developing countries. A review of the previous works addressing this theme was provided, and discussions and recommendations for future research were presented.

REFERENCES

Alarcon, J., Rodriguez Martinez, A., and Herrera Orozco, I. 2011. Life cycle assessment to municipal wastewater treatment plant. *Chemical Engineering Transactions* 24, 1345–1350.
Ameen, R. F. M. and Mourshed, M. 2017. Urban environmental challenges in developing countries—A stakeholder perspective. *Habitat International* 64: 1–10.
Andersson, K., Rosemarin, A., Lamizana, B., Kvarnström, E., McConville, J., Seidu, R., Dickin, S., and Trimmer, C. 2016. *Sanitation, Wastewater Management and Sustainability: From Waste Disposal to Resource Recovery.* Nairobi and Stockholm: United Nations Environment Programme and Stockholm Environment Institute. 156.
Bai, S., Wang, X., Huppes, G., Zhao, X., and Ren, N. 2017. Using site-specific life cycle assessment methodology to evaluate Chinese wastewater treatment scenarios: A comparative study of site-generic and site-specific methods. *Journal of Cleaner Production* 144: 1–7.

Barjoveanu, G., Comandaru, I. M., Rodriguez-Garcia, G., Hospido, A., and Teodosiu, C. 2014. Evaluation of water services system through LCA. A case study for Iasi City, Romania. *International Journal of Life Cycle Assessment* 19: 449–462.

Benini, L., Mancini, L., Sala, S., Manfredi, S., Schau, E. M., and Pant, R. 2014. *Normalisation Method and Data for Environmental Footprints.* Luxembourg: European Union. EUR, 26842.

Buonocore, E., Mellino, S., De Angelis, G., Liu, G., and Ulgiati, S. 2016. Life cycle assessment indicators of urban wastewater and sewage sludge treatment. *Ecological Indicators.* http://dx.doi.org/10.1016/j.ecolind.2016.04.047

Chen, S. and Chen, B. 2013. Net energy production and emissions mitigation of domestic wastewater treatment system: A comparison of different biogas-sludge use alternatives. *Bioresource Technology* 144: 296–303.

Chiu, S. L. H., Lo, I. M. C., Woon, K. S., and Yan, D. Y. S. 2016. Life cycle assessment of waste treatment strategy for sewage sludge and food waste in Macau: Perspectives on environmental and energy production performance. *International Journal of Life Cycle Assessment* 21: 176–189.

Cornejo, P. K. 2015. Environmental Sustainability of Wastewater Treatment Plants Integrated with Resource Recovery: The Impact of Context and Scaleniversity of South Florida, Tampa, USA, 197.

Corominas, L., Foley, J., Guest, J. S., Hospido, A., Larsen, H. F., Morera, S., and Shaw, A. 2013. Life cycle assessment applied to wastewater treatment: State of the art. *Water Research* 47: 5480–5492.

El Sayed., M. M. M., van der Steen, N. P., Abu-Zeid, K., and Vairavamoorthy, K. 2010. Towards sustainability in urban water: a life cycle analysis of the urban water system of Alexandria City, Egypt. *Journal of Cleaner Production*, 18(10): 1100–1106.

Emmerson, R. H. C., Morse, G. K., Lester, J. N., and Edge, D. R. 1995. The life-cycle analysis of small-scale sewage-treatment processes. *Water and Environment Journal* 9: 317–325.

Flores, A., Buckley, C., and Fenner, R. 2008. Selecting wastewater systems for sustainability in developing countries. In: *The 11th International Conference on Urban Drainage*, Edinburgh, Scotland.

Foresti, E. Perspectives on anaerobic treatment in developing countries. 2001. *Water Science and Technology* 44 (8): 141–148.

Garcrn, J. S., Herrera, I., and Rodrdrali, A. 2011. Análisis de Ciclo de Vida de uma Planta de Tratamiento de Aguas Residuales Municipales. Caso: PTARM de Yautepec (Morelos, México). *Informes Técnicos Ciemat* 1226.

Guinée, J., Gorrée, M., Heijungs, R., Huppes, G., Kleijn, R., De Koning, A., Van Oers, L., Sleeswijk, A., Suh, S., de Haes, H. A. U., de Briujn, H., Van Duin, R., and Huigbregts, M. A. J. 2002. *Life cycle assessment: an operational guide to ISO standards. I: LCA in perspective. IIa: Guide. IIb: operational annex. III: scientific background.* Kluwer Academic Publishers, Dordrecht.

Gutierrez, K. G. 2014. An14.rrezademic Publishersoperational guide to ISOTratamento de Esgoto Domrspective. IIa: Guide. IIb: operational an. *Thesis.* School of Engineering. Federal University of Minas Gerais. Brazil. 129.

Hawkins, P., Blackett, I. , and Heymans, C. 2013. *Poor-Inclusive Urban Sanitation: An Overview. Water and Sanitation Program.* World Bank, Washington DC.

Heimersson, S., Svanström, M., Laera, G., and Peters, G. 2016. Life cycle inventory practices for major nitrogen, phosphorus and carbon flows in wastewater and sludge management systems. *The International Journal of Life Cycle Assessment* 21(8): 1197–1212.

Hernández-Padilla, F., Margni, M., Noyola, A., Guereca-Hernandez, L., and Bulle, C. 2017. Assessing wastewater treatment in Latin America and the Caribbean: Enhancing life cycle assessment interpretation by regionalization and impact assessment sensibility. *Journal of Cleaner Production* 142: 2140–2153.

ILCD. 2010. *General Guide for Life Cycle Assessment-Detailed Guidance*. Luxembourg. JRC Publishing, 417.

IPCC (Intergovernmental Panel on Climate Change). 2014. *Fifth Assessment Report. Contribution of Working Groups I, II, and III to the Fifth Assessment Report of the Intergovernmental Panel on Climate Change*. IPCC publications, Geneva, Switzerland, 151.

Kalbar, P. P., Karmakar, S., and Asolekar, S. R. 2012. Estimation of environmental footprint of municipal wastewater treatment in India: Life cycle approach. *International Conference on Environmental Science and Technology*, Singapore 30–34.

Kalbar, P. P., Karmakar, S., and Asolekar, S. R. 2013. Assessment of wastewater treatment technologies: Life cycle approach. *Water and Environment Journal* 27(2): 261–268.

Kim, J., Yang, Y., Bae, J., and Suh, S. 2013. The importance of normalization references in interpreting life cycle assessment results. *Journal of Industrial Ecology* 17(3): 385–395.

Larsen, H. F., Hauschild, M., Wenzel, H., and Almemark, M. 2007. NEPTUNE—new sustainable concepts and processes for optimization and upgrading municipal wastewater and sludge treatment, Work Package 4—assessment of environmental sustainability and best practice. Deliverable4.1-homogeneousLCAmethodologyagreed by NEPTUNE and INNOWATECH. Contract No. 036845.

Ledrn, Y. C., Rivas, A., Lvas, D., and Vidal, G. 2017. Life-cycle greenhouse gas emissions assessment and extended exergy accounting of a horizontal-flow constructed wetland for municipal wastewater treatment: A case study in Chile. *Ecological Indicators* 74: 130–139.

Li, H., Jin, C., Zhang, Z., O'Hara, I., and Mundree, S. 2017. Environmental and economic life cycle assessment of energy recovery from sewage sludge through different anaerobic digestion pathways. *Energy* 126: 649–657.

Limphitakphong, N., Pharino, C., and Kanchanapiya, P. 2016. Environmental impact assessment of centralized municipal wastewater management in Thailand. *International Journal of Life Cycle Assessment* 21: 1789–1798.

Lohri, C. R., Rodic, L., and Zurbrügg, C. 2013. Feasibility assessment tool for urban anaerobic digestion in developing countries. *Journal of Environmental Management* 126: 122–131.

Lopes, T. A. S. 2014. AvaliaS.of Environmental Managementan anaerobic digestion in developing countries Thailandough. *Thesis*. Polytechnic School. Federal University of Bahia. Brazil. 136p.

Lu, B., Du, X., & Huang, S. 2017. The economic and environmental implications of wastewater management policy in China: From the LCA perspective. *Journal of Cleaner Production*, 142: 3544–3557.

Lutterbeck, C. A., Kist, L. T., Lopez, D. R., Zerwes, F. V., and Machado, E. L. 2017. Life cycle assessment of integrated wastewater treatment systems with constructed wetlands in rural areas. *Journal of Cleaner Production* 148: 527–536.

Mara, D. 2003. *Domestic Wastewater Treatment in Developing Countries*. EarthScan, London. 310.

Massoud, M. A., Tarhini, A., and Nasr, J. A. 2009. Decentralized approaches to wastewater treatment and management: Applicability in developing countries. *Journal of Environmental Management* 90: 652–659.

Mateo-Sagasta, J., Raschid-Sally, L., and Thebo, A. 2015. Global wastewater and sludge production, treatment and use. In *Wastewater: Economic Asset in an Urbanizing World*, P. Drechsel, M. Qadir and D. Wichelns, eds. London: Springer.

Musharrafie, A., Güereca, P. L., Padilla, A., Morgan, J. M., and Noyola, A. 2011. A comparison of two wastewater treatment plants: Stabilization ponds and activated sludge with a social perspective impacts. Instituto de Ingeniería, Universidad Nacional Autónoma de México, México.

Ng, B. J. H., Zhou, J., Giannis, A., Chang, V. W. C., and Wang, J. 2014. Environmental life cycle assessment of different domestic wastewater streams: Policy effectiveness in a tropical urban environment. *Journal of Environmental Management* 140: 60–68.

Nkadimeng, L. S. 2015. Maximising energy recovery from the brewery wastewater treatment system: A case study evaluating the anaerobic digestion wastewater treatment plant at SAB's Newlands Brewery. *Thesis*. University of Cape Town. South Africa. 163.

Ontiveros. G. A. and Campanella, E. A. 2013. Environmental performance of biological nutrient removal processes from a life cycle perspective. *Bioresource Technology* 150: 506–512.

Pan, T., Zhu, X., and Ye, Y. 2011. Estimate of life-cycle greenhouse gas emissions from a vertical subsurface flow constructed wetland and conventional wastewater treatment plants: A case study in China. *Ecological Engineering* 37: 248–254.

Patricia L. G., Musharrafie, A., Martínez, E., Padilla, A., Morgan, J. M., and Noyola Robles, A. 2011. Comparative life cycle assessment of a wastewater treatment technology considering two inflow scales. Instituto de Ingeniería, Universidad Nacional Autónoma de México. Mexico.

Pirani, S., Natarajan, L., Abbas, Z. , and Arafat, H. 2012. Life cycle assessment of membrane bioreactor versus CAS wastewater treatment: Masdar city and beyond. *JIChEC06 - The Sixth Jordan International Chemical Engineering Conference*, Amman, Jordan.

Pradel, M., Aissani, L., Villot, J., Baudez, J., and Laforest, V. 2016. From waste to added value product: Towards a paradigm shift in life cycle assessment applied to wastewater sludge—a review. *Journal of Cleaner Production* 131: 60–75.

Prüss-Üstün, A., Clasen, T., Clumming, O., Bonjour S., De France, J., Freeman, M. C., Hunter, P. R., Mathers, C., Medlicott, K., and Wolf, J. 2014. Burden of disease from inadequate water, sanitation and hygiene in low- and middle-income settings: A retrospective analysis of data from 145 countries. *Tropical Medicine and International Health* 19(8): 894–905.

Qureshi, M. E., Harrison, S.R., and Wegener, M. K. 1999. Validation of multicriteria analysis models. *Agricultural Systems* 62(2): 105–116.

Rebtzer, G., Hunkeler, D. , and Jolliet, O. 2003. LCC—the economic pillar of sustainability: Methodology and application to wastewater treatment. *Environmental Progress* 22: 241–249.

Resende, J. D., Rodrigues, P. F. M. A., Pacca, S. A., and Nolasco, M. A. 2015. Life Cycle Assessment of Wastewater Treatment Systems for Conventional Activated Sludge and UASB Reactor followed by Activated Sludge. *5th International Workshop Advances in Cleaner Production*. São Paulo, Brazil.

Rezende, J. D., Rodrigues, P. F. M. A., Pacca, S. A., and Nolasco, M. Life Cycle Assessment of Wastewater Treatment Systems for Conventional Activated Sludge and UASB Reactor followed by Activated Sludge. *International workshop advances on cleaner production*. São Paulo, Brazil, 2015.

Romero-Hernandez, O. 2005. Applying life cycle tools and process engineering to determine the most adequate treatment process conditions. A tool in environmental policy. *International Journal of Life Cycle Assessment* 10 (5): 355–363.

Schwaickhardt, R. O., Machado, E. L., and Lutterbeck, C. A. 2017. Combined use of VUV and UVC photoreactors for the treatment of hospital laundry wastewaters: Reduction of load parameters, detoxification and life cycle assessment of different configurations. *Science of the Total Environment* 590–591: 233–241.

Stranddorf, H. K., Hoffmann, L. , and Schmidt, A. 2005. LCA technical report: impact categories, normalization and weighting in LCA. Update on selected EDIP97-data. Danish Ministry of the Environment.

Strômberg, L. and Paulsen, J. 2002. LCA application to Russian conditions. *International Journal of Life Cycle Assessment* 7(6): 349–357.

Syamsuddin, I. 2013. Multicriteria evaluation and sensitivity analysis on information security. *International Journal of Computer Applications* 69(24): 22–25.

Tatiana, F. Pinilla P. Bojacá, V. Pinilla, R. Ortiz, and J. Acevedo P. 2016. Life cycle assessment to identify environmental improvements in an anaerobic waste water treatment plant, *Chemical Engineering Transactions*, 49: 493–498.

Teodosiu, C., Barjoveanu, G., Sluser, B. R., Popa, S. A. E., and Trofin, O. 2016. Environmental assessment of municipal wastewater discharges: A comparative study of evaluation methods. *International Journal of Life Cycle Assessment* 21: 395–411.

Tong, L., Liu, X., Liu, X., Yuan, Z., and Zhang, Q. 2013. Life cycle assessment of water reuse systems in an industrial park. *Journal of Environmental Management* 129: 471–478.

Torres, P. 2012. Perspectivas del Tratamiento Anaerobio de Aguas Residuales Domésticas en Países en Desarrollo. *Revista EIA* 18: 115–129.

Tourinho, T. C. O. 2014. Avaliação Comparativa do Ciclo de Vida de Processos de Tratamento de Efluentes Domésticos. *Thesis.* Polytechnic School and Chemical School. Federal University of Rio de Janeiro. Brazil. 182.

United Nations Environment Programme (UNEP). 2010. Clearing the Waters. A Focus on Water Quality Solutions. United Nations Environment Programme UNEP publishing, Nairobi.

United Nations (UN). 2013. *World Economic and Social Survey—Sustainable Development Challenges.* United Nations: New York. 216p.

Waddington, H., Snilstveit, B., White, H. , and Fewtrell, L. 2009. Water, sanitation and hygiene interventions to combat childhood diarrhoea in developing countries. International Initiative for Impact Evaluation. Synthetic Review 001. World Bank.

Wang, X., Liu, J., Ren, N. , and Duan, Z. 2012. Environmental profile of typical anaerobic/anoxic/oxic wastewater treatment systems meeting increasingly stringent treatment standards from a life cycle perspective. *Bioresource Technology* 126: 31–40.

World Commission on Environment and Development (WCED). 1991. *Our Common Future.* Oxford University Press, Oxford.

Yıldırım, M. and Topkaya, B. 2012. Assessing environmental impacts of wastewater treatment alternatives for small-scale communities. *Clean Soil, Air, Water* 40(2): 171–178.

Yoshida, H., Christensen, T. H. , and Scheutz, C. 2013. Life cycle assessment of sewage sludge management: A review. *Waste Management & Research* 31: 1083–1101.

Yoshida, H., Clavreul, J., Scheutz, C.H. , and Christensen, T. H. 2014. Influence of data collection schemes on the Life Cycle Assessment of a municipal wastewater treatment plant. *Water Research* 56: 292–303.

Zang, Y., Li, Y., Wang, C., Zhang, W. , and Xiong, W. 2015. Towards more accurate life cycle assessment of biological wastewater treatment plants: A review. *Journal of Cleaner Production* 107: 676–692.

Zhang, Z., and Wilson F. 2000. Life-cycle assessment of sewage-treatment plant in South-East Asia. *J. CIWEN* 14: 51–56.

Zhang, Q. H., Wang, X. C., Xiong, J. Q., Chen, R. , and Cao, B. 2010. Application of life cycle assessment for an evaluation of wastewater treatment and reuse project—Case study of Xi'an, China. *Bioresource Technology* 101: 1421–1425.

Zhang, D. Q., Jinadasa, K. B. S. N., Gersberg, R. M., Liu, Y., Ng, W. J. , and Tan, S. K. 2014. Application of constructed wetlands for wastewater treatment in developing countries—A review of recent developments (2000–2013). *Journal of Environmental Management* 141: 116–131.

13 Life Cycle Analysis of Anaerobic Digestion of Wastewater Treatment Plants

Rosalía Rodríguez, Juan José Espada,
Raúl Molina, and Daniel Puyol

CONTENTS

13.1 INTRODUCTION

13.1.1 AD AS A KEY IN CIRCULAR ECONOMY OF WASTEWATER MANAGEMENT

Anaerobic digestion (AD) is a highly mature technology. The conversion of particulates and soluble organic components from waste and wastewater into methane has been approached from the 19th century (Abbasi et al., 2012). Originally, AD was used for the stabilization of manure and human excreta solids, but was then largely applied to treat biomass solid waste from biological processes for domestic wastewater treatment until the 1970s. The energy recovery concept by AD was not included until the early 1970s, which was concurrent with the economic crisis caused by the rise of fossil fuel prices, and was the starting point to consider AD as a feasible technology to convert residual waste or energy crops into bioenergy (as biogas or biohydrogen). The energy balance is therefore a key step to address sustainable biogas production in novel applications of AD (such as, for example, lignocellulosic biomass conversion). However, new paradigms in the production system have opened up possibilities to expand the concept of AD into the field of resource recovery from waste sources.

The engagement of AD with the circular economy concept was probably introduced, but certainly supported with data, by Zeeman and Lettinga (1999). The circular economy concept for sustainable industrial production encompasses the life cycle of a product "cradle-to-cradle" by partially reusing components or completely recycling the basic components to build new raw materials (Gregson et al., 2015). Sustainable water management adds water recycling and resource recovery into the storyline. The introduction of "full" sustainability of water management necessitated setting prices for water components and, most importantly, to the effort for sustainable water management (Verstraete and Vlaeminck, 2011). But the energetic considerations are imperative for achieving sustainability in a short timeframe. According to the waste-to-energy supply chain (WTE), solving the dilemma of energy demand, waste management, and greenhouse gas (GHG) emission for communities globally is not only an energetic opportunity (especially for fossil fuel–importing countries) but ultimately, a human need (Pan et al., 2014). In this context, AD emerges as the core part.

Increasing energy demands has driven the change from fossil fuels to renewable energetic sources as primary energy. Biomass is the only one to be currently integrated into the production system at all levels. AD plays a key role in transforming the biomass from different origins (domestic and industrial organic waste and wastewater, sewage waste, cattery waste, and agroforestry waste) into a transportable and fully valuable energy source (in the form of biogas or ultimately, biomethane) (Weiland, 2010). Among the possible applications for biomass transformation into energy, AD is considered the most sustainable from a GHG emissions perspective, especially for closed reactors, with an average CO_2 emission potential of 54.7 kg CO_2-eq. GJ^{-1} (Fruergaard et al., 2009). This can save up to 196.2 kg of CO_2-eq per ton of biomass treated compared with current processes for waste management (Masullo, 2017). But more prominent is the energetic efficiency of AD processes, especially in novel developments.

The energy balance of the AD process has been recursively calculated to be positive. This essentially means that the transformation of low-value waste into bioenergy as biogas is always energetically and economically favorable. Methane has a calorific value around 50–55 MJ kg^{-1}, which is the second highest among the common fuels, lower only than hydrogen. This turns AD into a highly attractive process even from an economic perspective. By analogy, the application of AD for wastewater treatment with organic contamination should be energetically positive as well. As an example, a single-stage AD was used for treating beer factory wastewater, achieving an energetic potential of 90 KJ L^{-1} of wastewater treated with a methane production of 2.5 L CH_4 L^{-1} and a mass balance of 86% of chemical oxygen demand converted into methane (Nishio and Nakashimada, 2007). In this way, wastewater (which has essentially no value as per the existing perspective) is fully treated to discharge limits, and energy is produced during the process. The concept of positive energy in wastewater treatment, considered as non-credible just a while ago, has become fully possible by basing the treatment on AD (Batstone and Virdis, 2014).

But other emerging applications of AD have been proposed recently to completely envelope AD into the global circular economy, as is the case with the potential resource recovery from wastewater.

Another direct application of AD besides biogas production is to use the digestate (which is the remnant residue after the digestion) to extract the commodities that remain in this line (Verstraete et al., 2009). Traditional applications of digestate are the use of its solid fraction as a natural biofertilizer and more recently, the use of the liquid fraction containing the majority of the N and P released on hydrolysis to recover these nutrients as struvite ($NH_4MgPO_4x6H_2O$), a mineral which contains an equilibrated amount of N and P but requires, in most cases, the addition of external Mg^{2+} salts (Vaneeckhaute et al., 2017). Other options to reclaim N are ammonia stripping and electro-dialysis (Batstone and Virdis, 2014). These processes are not directly related to the structure of AD metabolic pathways and do not require control and optimization but just enhance hydrolysis. However, new research lines have emerged recently that are focused on the specialization of AD processes' metabolism to produce high value-added products based on the recovery of carbon (single-cell proteins, bioplastics, organic acids ...) or metals from wastewater (Puyol et al., 2017b).

13.1.2   CONVENTIONAL APPLICATION OF AD

Biological wastewater treatment entails the transformation of soluble contamination (essentially C, N, and P) into particulate biomass. The disposal of sewage waste sludge represents a major problem in wastewater treatment plants (WWTPs), costing around 50% of the total expenses of the plant. AD is the most used method for stabilizing the sludge, also considerably reducing its volume by between 20% and 70% (Appels et al., 2008). Typical operations of waste sludge include thickening before feeding into the digester. Then, the gas line (biogas) is stored before direct use or upgraded to obtain biomethane. The supernatant (liquid phase) is commonly submitted to nitrification-denitrification to remove excess ammonium and is recycled into the main line. The residual sludge (solid phase) is dewatered and further dried before combustion or land application as biofertilizer. Characteristic configurations of AD reactors include continuous stirred-tank reactor (CSTR), high-rate CSTR with active mixing, and two-stage hydrolysis+methanogenesis. Most applied temperature is in the mesophilic range (30–37°C), though thermophilic processes (50–65°C) are becoming an interesting option to enhance the hydrolysis step in low-biodegradable feedstocks, considerably reducing the reactor's volume (De la Rubia et al., 2012). Another option is two-stages thermophilic+mesophilic (TPAD), where the thermophilic stage improves the hydrolysis through imposing very short solid retention times (SRT, around 2–4 d), whereas the mesophilic reactor produces most of the biogas (Wang et al., 2017). However, new applications of AD for low-biodegradable feedstocks have promoted the need to pre-treat the biomass to enhance the biogas production.

Hydrolysis is the limiting step in the AD process treating solid feedstocks, as some microorganisms must release specialized enzymes into the medium (hydrolytic enzymes) to solubilize the solid organics, which are further used by other microbial communities in catabolic processes (as methanogenesis) and biomass growth. To enhance the process, it is quite common to promote "artificial" hydrolysis that can speed up the methane production rate and increase the biochemical methane potential by converting non- or low-biodegradable solid substrate into easily accessible soluble substrate (Appels et al., 2008). More typical pretreatments include thermal treatment, mostly at high temperature ($>100°C$) and constant or variable pressure ("steam explosion"); mechanical treatment, mainly by ultrasound; and chemical treatment, such as addition of acid or alkali and wet oxidation. Novel treatments include microwave irradiation, advanced oxidation processes at mild temperature, pulsed electric fields, and free nitrous acid addition (Carlsson et al., 2012; Wang et al., 2017).

AD has been also applied to treat domestic, industrial, and agroforestry wastewater with high organic composition, such as that coming from the food, beverage, alcohol distillery, and pulp and paper industries, which accounts for more than 2000 anaerobic reactors installed worldwide (van Lier et al., 2015). Domestic wastewater has been intensively treated by anaerobic technologies especially in various parts of the tropical world, notably in Latin America and India (Chernicharo et al., 2015), as will be further discussed. But most experience is in the application of high-rate anaerobic reactors to treat industrial wastewater.

Since its appearance in the early 1980s, the upflow anaerobic sludge blanket reactor (UASB) has been widely applied to treat wastewater worldwide (van Lier et al., 2015). The original concept is described as a compact reactor in which the biomass concentration is greatly increased due to a natural association of microorganisms, cations, and inert particulates to form highly compact and structured biogranules, which are formed as a result of the hydraulic stress imposed by a high upflow velocity (usually higher than 0.5 m h^{-1}) (Zeeman and Lettinga, 1999). As a result, hydraulic retention time (HRT) is decoupled from SRT, and this leads to a high increase in the volumetric activity inside the reactor (with specific activities typically around $0.3–1.0$ kgCOD-CH$_4$ kgVSS^{-1} d^{-1}), which ultimately leads to obtaining high chemical organic demand (COD) removal efficiencies in treating industrial wastewater with high organic content by imposing HRT similar or even lower than in typical activated sludge processes (van Lier et al., 2015). Some variants of the original UASB concept have appeared so far, including the expanded granular sludge bed (EGSB) reactor, which applies a much higher upflow rate (around $6–30$ m h^{-1}), leading to an almost perfect mix, and the internal circulation (IC) reactor, where the produced biogas is separated from the liquid halfway through the reactor by means of an in-built gas-liquid-solid separator device and conveyed upward through a pipe to a degasifier unit or expansion device. Another option is a hybrid UASB configuration upgraded with an external liquid-solid membrane separation (van Lier et al., 2015). Though UASB-based high-rate anaerobic reactors have quite high COD removal efficiency in compact configurations, there are still some constraints that have to be considered, as the partial dissolution of CH$_4$ in the effluent (highly inversely correlated with the influent COD concentration), which increases the GHG emission potential; the potential production of reduced gases, such as H$_2$S, which can cause odor problems; and their inherent limitations at low temperature and high sulfate concentrations, which reduce the biogas production potential considerably (Chernicharo et al., 2015).

To fulfill these limitations and expand the niche of anaerobic technologies, novel concepts of anaerobic wastewater treatment have emerged recently.

13.1.3 NOVEL CONCEPTS IN ANAEROBIC DIGESTION

The year 2014 was the 100th anniversary of the activated sludge (AS) process. During these years, biological wastewater treatment has been settling and consolidating, and nowadays, it is unusual to find a WWTP with no AS treatment. However, there is increasing concern about limitations of the AS process, mainly due to its high energy demand, high GHG emission potential, and limited recovery potential for C and other resources. Therefore, the fourth generation of WWTPs is becoming a reality, and four strong alternatives to the AS process have been recently proposed (Figure 13.1), all of them related to AD processes and the circular economy in some way (Puyol et al., 2017b):

- McCarty et al. (2011) proposed low-energy anaerobic wastewater treatment in the main line. The concept is based on substituting the AS process by an anaerobic secondary treatment, leading to a 100% higher biogas potential, a 50% reduction of sludge to be disposed, and a net energy production within the whole plant. This concept is considered as "low cost" and has been successfully applied in India and Latin America, especially by using UASB-type reactors (Chernicharo et al., 2015). However, the concept does not include efficient management of nutrient recovery, and the effluent is merely proposed to be used for irrigation. An interesting option is to combine this concept with the anaerobic ammonium oxidation (anammox) process to remove excess N, but still the P recovery is inefficient (Winkler et al., 2012).
- van Loosdrecht and Brdjanovic (2014) introduced the concept of transforming a WWTP into a biofactory to reclaim the resources contained in the wastewater, particularly the organic contamination, and produce high value-added resources that can be re-introduced into the market in a cradle-to-cradle fashion. As direct transformation of the organic components into human-usable products is hardly possible due to pathogens and heavy metal contamination, some interesting options have appeared so far, such as the indirect combination of the C coming from AD of sewage waste with hydrogen from solar-driven water electrolysis to grow chemolithotrophic biomass that can be further used to extract edible single-cell proteins or prebiotics (Matassa et al., 2015). This concept leads to full use of the residual N from wastewater.
- The first concept in considering all the wastewater components as valuable resources was proposed by Verstraete et al. (2009). The basis consists of a first step of physical separation of water from both soluble and particulate components by sequenced ultra- and nanofiltration followed by reverse osmosis (main line, 90% of flow) and a second step of up-concentration of a minor flow (10% of flow), whereby all the wastewater components are concentrated. In this process, AD plays a fundamental role, as it is the first step where all the biodegradable organics are converted into biogas. The

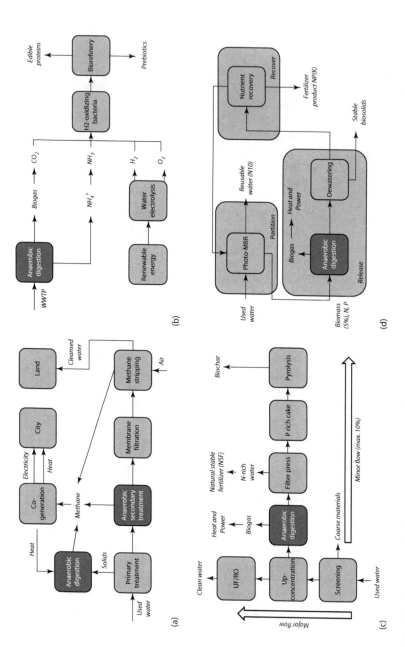

FIGURE 13.1 Novel concepts for wastewater treatment with energy recovery through anaerobic digestion and improved resource recovery. (a) "Low-energy" main line concept (From McCarty, P.L., et al., *Environ. Sci. Technol.*, 45, 7100–6, 2011); (b) wastewater factory concept (From Matassa, S., et al., *Water Res.*, 68, 467–78, 2015); (c) separation of streams concept (From Verstraete, W., et al., *Bioresour. Technol.*, 100, 5537–45, 2009); (d) partition-release-recover concept (From Batstone, D.J., et al., *Chemosphere*, 140, 2–11, 2015a).

resulting digestate is separated by press filtration, after which the N- and P-rich liquid phase is used for natural stable fertilizer production, and the solid phase is pyrolyzed to obtain biochar. Despite the novelty of this concept, there is a clear economic imbalance, as the physical separation processes are highly energy consuming, which could impose up to 92% of total plant costs.

- An improved version of the Verstraete concept was proposed by Batstone et al. (2015a). The first step of separation is performed by microbial organisms (partition step) in an enhanced membrane bioreactor. The microbial biomass is then submitted to anaerobic digestion, where biogas is produced, and the nutrients are released (release step). Finally, the digestate is separated into two phases, closely following the Verstraete concept (recovery step). Unlike the "mother" concept, the partition-release-recovery proposal has been proved to be energetically positive if the inlet wastewater concentration exceeds 600 mgCOD L^{-1}.

A recurring technological solution for novel wastewater concepts is the use of the anaerobic membrane bioreactor (AnMBR). Although the AnMBR concept is not new, with the first attempts made between the 1970s and the 1980s, novel development of efficient membrane operation procedures in aerobic MBRs has improved the operability and sustainability of their anaerobic counterparts (Liao et al., 2006). There are three AnMBR configurations: side-stream (external, cross-flow), submerged within the reactor, and submerged in a separate chamber. The membrane filtration is the economic limitation of the process; typical energy demand values are between 0.2 and 1 kW h m^{-3}. However, this has not been an impediment for process applications to a wide range of industrial wastewater, such as slaughterhouse, molasses, landfill leachate, dairy manure, pharmaceutical, and diverse types of food industry wastewater. The main advantage over CSTR configurations is the separation between SRT and HRT, leading to AD operation of short HRT (a few days) and long SRT (between 20 and 700 d) (Dvořák et al., 2015). In these reactors, biomass concentration can be increased up to 75 g L^{-1}, which is like the biomass concentration in granular UASB-type reactors.

New anaerobic processes have been recently proposed to be part of the framework of wastewater treatment by AD, although these processes are not related to biogas production through methanogenesis (Batstone et al., 2015b). Instead, these are more oriented toward direct resource recovery and the decrease of the energetic demand of WWTPs (Puyol et al., 2017b), and therefore are dedicated to including wastewater treatment into a circular economy concept in water management:

- *Fermentation targeting chain elongation.* Particulate and soluble organic matter contained in wastewater can be transformed anaerobically into CO_2, volatile fatty acids (VFAs), and light alcohols by fermentative bacteria. These compounds can be further chain-elongated to high value-added products following three different pathways: (1) homoacetogenesis of CO_2 to acetate, (2) succinate formation from glycerol, and (3) reverse β oxidation of VFAs and light alcohols to form n-butyrate and n-caproate

(Spirito et al., 2014). To maximize the selectivity of the elongated products, three technological possibilities are emerging: (1) artificial removal or addition of electrons by microbial electro-catalysis, (2) separation of the products by in-line extraction technology to enhance product yield, and (3) immediate conversion of the elongated products to increase the product value (biorefinery concept).

- *Use of anoxygenic purple phototrophic bacteria (PPB).* PPB are very versatile bacteria, able to perform a wide range of metabolic pathways, though their most interesting mechanism is anaerobic photoheterotrophy, in which simple organics such as VFAs, alcohols, and sugars are assimilated, with infrared (IR) light as the energy carrier of the process (Puyol et al., 2017a). It has been proposed to use these bacteria as the key part in the partition step of the partition-release-recovery concept due to their high redox and C recycle efficiencies, high biomass yield, and high C/N/P growth ratio, thus optimizing C and nutrient recovery via assimilation instead of dissipation (Batstone et al., 2015a). Compared with other phototrophs such as algae and cyanobacteria, PPB can use low-energy IR light, which greatly decreases the energy demand of the process (Hülsen et al., 2014). These bacteria have been used for both domestic (Hülsen et al., 2014) as well as industrial (Chitapornpan et al., 2013) wastewater treatment with resource recovery. Also, their metabolism can be altered to enhance biohydrogen production with an excess of organic electron sources, nutrient deficiency, and lack of ammonium in the medium (Ghosh et al., 2017).
- *Use of the anammox process to low-energy main line with complete nitrogen depletion in a single-stage process.* The anammox process entails the anaerobic oxidation of ammonium to dinitrogen gas with nitrite as the electron source. In domestic wastewater applications, complete removal of C and N is feasible with partial oxidation of ammonium to nitrite (nitritation) and anammox with concomitant C removal by fermentation and microaerophilic organic oxidation in a delicate equilibrium, where the pH and the temperature play a critical role. Anammox bacteria are very sensitive to temperature changes and concentrations of unionized nitrogen forms (free nitrous acid and free ammonia), so that pH must remain in a narrow range of 7–8.5 (Cao et al., 2017).
- *Engagement of sulfate reduction with methanogenesis.* The sulfur cycle in bacteria is key for very high-strength and sulfate-rich wastewater treatment by AD technologies, basically including the sulfate reduction and the sulfide oxidation processes (Batstone et al., 2015b). Many food industry wastewaters, such as those coming from distilleries and fermentation processes, contains high amounts of COD and sulfate, the latter coming from the use of sulfuric acid in chemical processes. Sulfate reduction implies the oxidation of hydrogen or VFA with concomitant sulfate reduction to hydrogen sulfide. This means that sulfate reduction bacteria (SRB) can compete directly with methanogens for substrate (hydrogen and acetate) and therefore, critically affects the methane potential in the treatment of these

wastewaters. Also, sulfide inhibits the biological activity of anaerobic bacteria and methanogens. Thus, it has been included as a key mechanism to control the AD process (Barrera et al., 2015). Considering sulfide as a key element in AD systems is essential to understanding the physicochemical interactions in the inorganic chemistry of wastewater, which is especially relevant to predicting the fate of inorganic nutrients such as S, P, and Fe to optimize nutrient recovery (Flores-Alsina et al., 2016).

Another emerging field of study in AD is the physicochemical framework, not only for single AD processes, but for plant-wide modeling approaches focused on the analysis of the fate of key inorganic components (Flores-Alsina et al., 2015). What happens in a WWTP has deep consequences for water bodies and the sewage system. For example, aluminum and iron sulfate have been intensively used as coagulants in wastewater treatment, and this has caused a tremendous impact on the economy of water management, as added sulfate is the main source of corrosion in the sewage system due to the SRB activity producing sulfide (Pikaar et al., 2014). From a circular economy perspective, the physicochemistry of the wastewater treatment processes affects mainly the inorganic resource recovery potential, especially for P, Mg, and Fe. A systematic study of precipitation processes for multiple minerals in wastewater treatment, including amorphous calcium phosphate, calcium carbonate monohydrate, dicalcium phosphate hydrate, octacalcium phosphate, and struvite, has been conducted by Kazadi Mbamba et al. (2015). The mineral precipitation can predict not only the potential P recovery by struvite formation but also the pH of the system. This approach has been highly useful to predict the real fate of P, Ca, P, Mg, S, and Fe in AD processes by including the Fe-S-P interaction that can be derived from SRB activity, since sulfide can compete with other anions such as phosphate and carbonate to entrap mono- and bivalent cations such as Fe^{2+}, Ca^{2+}, Mg^{2+}, and K^+ (Flores-Alsina et al., 2015). This chemistry module has been used for accurately predicting actual pH and P, NH_4^+, and Mg^{2+} fate in the digestate on AD, especially relevant to P recovery potential as struvite (Flores-Alsina et al., 2016).

13.2 LIFE CYCLE ASSESSMENT (LCA) OF ANAEROBIC DIGESTION OF URBAN AND INDUSTRIAL WASTEWATER TREATMENT

WWTPs are implemented to decrease the environmental impacts of municipal and industrial wastewater discharges. Quality requirements in both cases are completely satisfied with the conventional activated sludge (CAS) technology, traditionally used for urban wastewater (UWW) (Pintilie et al., 2016). CAS technology shows good sewage treatment efficiency with low operating cost. One recently emerging technology for both urban and industrial wastewater treatment is the use of membrane bioreactors (MBRs), in particular aerobic membrane bioreactors (AeMBRs). This technology allows the complete degradation of microcontaminants or pollutants compared with conventional biological systems, showing other advantages over the CAS process: excellent effluent quality, good disinfection capability, reduced footprint and sludge production, and so on (Ioannou-Ttofa et al., 2016).

However, energy requirements (including electricity consumption) should be taken into account regarding the sustainable concept of energy-sufficient WWTPs. Thus, some studies determine that aerobic processes (such as CAS or AeMBR) represent a large percentage of the total WWTP energy consumption (Gu et al., 2017; Pretel et al., 2016). Moreover, aerobic WWTP does not exploit the potential energy contained in the organic matter and the fertilizer value of nutrients (Pretel et al., 2016).

In this sense, anaerobic reactors have emerged as an alternative for the sustainability treatment of industrial as well as urban wastewaters, especially since they can be operated under high SRT and very low HRT, providing them with great potential for application (Martinez-Sosa et al., 2011). In this context, although anaerobic reactors for wastewater treatment have been known and used for over 100 years, a critical technological advance that expanded AD was the development of methods to concentrate methanogenic biomass in the reactor, especially with very low solids concentration in the wastewater (1%–2%). Other advantages are that only a small portion of the organic material is converted into microbial biomass as compared with aerobic systems, the excess sludge usually being more concentrated, with better dewatering characteristics. This is due to the transformation of the organic matter into biogas, which is used afterward for the production of electricity and energy (Weiland, 2010).

Anaerobic technologies for wastewater treatments combine different attractive features (high efficiency in chemical oxygen demand [COD] removal, low values of HRT, biogas production) with a major disadvantage; that is, the energy requirements. However, this problem could be solved with the self-produced biogas. Additionally, CH_4, N_2O, and CO_2 emissions (as escaped gases or dissolved in the permeate) can be derived from the WWTP operation, depending on the reaction system and the characteristics of the influent (Meneses-Jácome et al., 2016). For that reason, environmental evaluation of these processes must be carried out.

LCA methodology is applied to determine the environmental feasibility of processes and products from the "cradle to grave perspective." LCA performs the evaluation of all types of potential impacts to the environment, from depletion of natural resources and energy requirements to emissions to air, land, or water (eutrophication, acidification, etc.) as well as resources recovery from the process. All of these are assessed within a consistent framework according to the International Organization for Standardization (ISO) standards (Hospido et al., 2012; Smith et al., 2014; Massara et al., 2016). LCA analysis can also be used as a tool for the evaluation of different scenarios for improving processes under study, identifying process bottlenecks and possible opportunities toward which specific research efforts should be focused. However, although LCA in the field of wastewater treatment has been applied for the last 21 years (Corominas et al., 2013), its application to anaerobic wastewater treatment has several constraints. On the one hand, comparison between studies is difficult, since functional units or references for calculation differ between works focused on biogas production (energy units) and those devoted to water treatment (volume or treat capacity units). On the other hand, interpretation of results is difficult and is not always harmonized due to the lack of unified criteria in the selection of the evaluation method (Renou et al., 2008; Corominas et al., 2013; Meneses-Jácome et al., 2016).

Different life cycle impact assessment (LCIA) methodologies are reported in LCA studies on anaerobic wastewater treatments. Among them, CML and ReCiPe impact assessment methods are usually applied.

13.2.1   LCA of Anaerobic Digestion of Urban Wastewater Treatment

Urban WWTPs (UWWTPs) are conceived to reduce the environmental impacts of municipal wastewater, which is characterized by low organic strength (Pintilie et al., 2016; Martinez-Sosa et al., 2011). As previously described, anaerobic processes show several advantages over aerobic ones: they obtain biogas as an additional source of energy, the sludge production is lower than in aerobic systems, and it is possible to reuse the inorganic nutrients contained in the effluent stream (Martinez-Sosa et al., 2011).

The main objective of anaerobic processes is to operate with a long SRT due to the slow growth of microorganisms (related to the low organic strength of municipal wastewater). In this sense, AnMBRs are a promising alternative, as they can decouple HRT, reducing the reactor volume and maintaining sludge concentration (Martinez-Sosa et al., 2011). As reported elsewhere, these systems can also reduce the sludge production compared with an aerobic process, eliminate the aeration energy consumption, and generate methane (Pretel et al., 2016).

From 2008 until 2016, the number of patents related to AnMBR systems has grown constantly (Krzeminski et al., 2017), as they present some advantages over other systems. Thus, biogas of excellent quality can be obtained by converting a large amount of input COD (Skouteris et al., 2012). Using mesophilic AnMBR technology, the methane production is reported to be in the range of 110–320 ml/g COD (Pretel et al., 2013), which can be increased by using novel technologies (Gu et al., 2017; Wei et al., 2014).

In comparison with other anaerobic systems, AnMBRs' anaerobic membranes can also operate in mesophilic conditions at lower temperatures (15–30°C) or at around 55°C in the thermophilic range. They do not show the limitation of operating at high temperatures to increase the microorganisms' growth rate that is usually registered in anaerobic UWW systems.

On the other hand, AnMBR technologies show some disadvantages, such as membrane fouling and high cost of membranes. Regarding the former, energy is needed to prevent membrane fouling, which could reduce membrane permeate fluxes, causing an increase in the operating costs (Skouteris et al., 2012). To limit this problem, gas bubbling is used to increase the flux level. Additionally, AnMBRs require the recovery of nutrients and methane from the effluent to improve the environmental feasibility of this technology. Finally, AnMBRs show operating difficulties when fluctuations in wastewater composition occur, or when the influent presents toxic compounds (Skouteris et al., 2012; Wei et al., 2014; Smith et al., 2014).

As described previously, AnMBR systems can be installed in three configurations, two of which correspond to immersed configuration. These systems are called

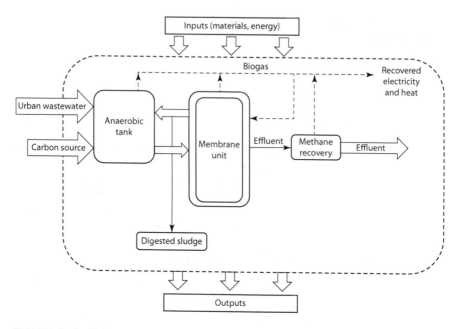

FIGURE 13.2   Scheme of SAnMBR process for urban WWTPs.

the *submerged anaerobic membrane bioreactor* (SAnMBR) technology, recently explored as a promising technology for urban wastewater treatment.

Pretel et al. (2013) report an environmental study of SAnMBR technology at different temperatures using LCA analysis. The process scheme is illustrated in Figure 13.2.

The reported SAnMBR system is constituted by an anaerobic reactor connected to two membrane tanks, which include commercial ultrafiltration membranes. The plant operates with an SRT of 70 days, and the studied temperatures are 20 and 30°C.

The following system boundaries are considered in that work in the context of LCA evaluation:

- The wastewater treatment process is included. The contribution of the stages involved in this phase to the environmental impact is decisive for LCA studies.
- The operating phase is also included, as it significantly contributes to environmental impact (the construction phase, including membrane investment cost, is not considered).
- Final effluent composition is determined to evaluate whether nutrient recovery is possible.
- Pre-treatment processes are not included in this LCA analysis, because they are well known for all the WWTP LCA studies.
- The demolition phase is not taken into account, as it has been demonstrated that it shows a negligible contribution.

- Global warming potential (GWP) is estimated as GWP100. In this sense, electricity consumption is considered as the main factor in the impact on GHG.
- The potential thermal impact of the effluent is not included in this work.

The functional unit selected in this work is 1 m³ of wastewater, as it is usual in LCAs on AD of municipal wastewater (Ioannou-Ttofa et al., 2016).

Life cycle inventory (LCI) analysis is generated by using energy, raw materials, and air, land and water emissions (Ioannou-Ttofa et al., 2016).

The methodology chosen to analyze environmental impacts is CML 2 baseline 2000. The studied factors relevant to doing an environmental evaluation of an urban wastewater treatment process are the energy consumption, the obtained energy associated with biogas production, the composition of the final effluent, and the sludge disposal.

After LCA analysis, SAnMBR for the treatment of medium-strength wastewater shows the highest GWP compared with other processes. Thus, LCA analysis shows that there is a problem with energy-related emissions. This energy consumption decreases when the system operates at ambient temperature, because the associated heat energy needed to operate with the SAnMBR system at 33°C is very high. However, temperature decrease alone is not the key to reducing energy emissions.

The solution consists in recovering the nutrient and the dissolved methane from the effluent. AnMBR effluent is a suitable stream to be used in potable applications, as it contains a low level of suspended solids and colloidal material, which can be used for reverse osmosis treatment. However, in that case, the potential advantages of nutrient removal could be lost (Smith et al., 2014).

On the other hand, methane losses (dissolved methane that escaped in the permeate) drastically increase the carbon footprint of the process (the loss of methane in municipal wastewater treatment can be around 30%–40% of the generated methane). For this reason, the main approach for anaerobic treatment processes is to recover this dissolved fraction (the effluent-dissolved methane). This methane can be used for energy generation by employing an additional technology (Smith et al., 2014; Krzeminski et al., 2017; Martinez-Sosa et al., 2011).

Thus, Krzeminski et al., (2017) report LCA results comparing anaerobic membranes with traditional membranes. As can be seen, fresh-water and marine eutrophication show high impact values, whereas MBR systems imply higher energy-related emissions. In submerged AnMBR systems, the procedures previously described to reduce energy-related emissions are included. For that reason, the emissions from energy consumption in MBR technologies may have higher values.

13.2.2   LCA of Anaerobic Digestion of Industrial Wastewater Treatment

Wastewater streams generated in the industrial sector are usually characterized by high and concentrated organic content and the presence of complex and slowly biodegradable organic compounds, which are not easy to treat, depending on the factory under study (Appels et al., 2008). The design of WWTPs, particularly for industrial effluents, needs to consider those factors related to the characteristics of the wastewater. However, the concept of the circular economy has moved WWTP concepts to other technologies with overall low environmental impacts, cost, and investment operations and high energy efficiency. Thus, WWTP should produce not only a clean effluent fulfilling the environmental legislation, but also a treated stream able to be reused, recovering energy and nutrients during the process (Massara et al., 2016).

In the case of industrial effluents, anaerobic treatment has been proposed in different configurations and different industries (Massara et al., 2016; Dvořák et al., 2015), such as the dairy industry (Georgiopoulou et al., 2008; Bialek et al., 2014), food-processing plants (Wu et al., 2010), the corn starch industry (Vera et al., 2015), or olive mill wastewater and agro-industrial wastewaters in general (Meneses-Jácome et al., 2016; Jaouad et al., 2016).

The LCA inventory of some anaerobic reactor configurations, such as AnMBR, lacks critical data, such as energy demand, since literature on full-scale application is scarce, and even data reported at pilot-plant scale are not completely accepted, as the limited size could compromise the reactor's energy attainment (Krzeminski et al., 2017). Nevertheless, Table 13.1 summarizes the most relevant studies regarding anaerobic reactors in the treatment of industrial wastewaters.

Foley et al. (2010) evaluated the potential environmental impacts of three industrial wastewater treatment alternatives: a high-rate anaerobic sludge reactor with biogas generation (with full-scale data); a microbial fuel cell treatment with direct electricity generation (pilot plant–scale data); and a microbial electrolysis cell with hydrogen peroxide production (data obtained from a laboratory-scale reactor). The results demonstrated a major negative impact associated with electricity consumption in all cases. The microbial electrolysis cell with hydrogen peroxide production was the best option of the three technologies, mainly due to the positive impact of the production of chemicals. In the anaerobic reactor, the positive impact derived from biogas and energy production outweighed the negative environmental impacts. Actually, the benefits coming from the displacement of fossil fuel–dependent resources outweigh the environmental cost of constructing and operating the treatment plant in all three cases, although the limited scale of the data available for one of the processes critically influenced this result (Foley et al., 2010). However, the environmental benefits obtained from anaerobic biological reactors do not always balance the negative potential impacts. Vera et al. (2015) performed the environmental evaluation of a corn starch WWTP with simultaneous microbial oil production as compared with a non-oil-producing treatment scheme. The implementation of the new system required substantial inputs of electrical energy, increasing the indirect GWP related to the electricity consumption by about 2330%. The production of microbial oil by a fermentation process increased the COD removal, reducing the negative effect. But finally, it was necessary to implement the system using corn stover biomass as a renewable source of energy to obtain environmental benefits by reducing the GWP impact and improving the local economy.

It is evident that the interpretation of LCA results is crucial in this methodology, but this is not easy to perform. For example, a comparison of the environmental performance of different configurations for pulp and paper effluent treatment based on six processes, including an upflow anaerobic sludge blanket (UASB) reactor, revealed

TABLE 13.1
LCA Studies in the Application of Anaerobic Reactors for Industrial Wastewater Treatment

Wastewater	Anaerobic System[a]	LCIA Method	LCIA Categories[b]	Reference
Dairy industry	UASB reactor	CML 2 baseline 2000, World 1995 normalization set	ADP, GWP, ODP, HTP, FAETP, MAETP, TETP, POCP, AP and EP.	Georgiopoulou et al. (2008)
Simulated industrial wastewater	HR-AS	IMPACT 2002+ (v.2.03)	Carcinogens, TETP, GWP, ODP, AP, NRE and end-points (human health, climate change, ecosystem quality, resources)	Foley et al. (2010)
Food-processing industry	Un-specified anaerobic reactor	Hybrid LCA (economy and energy flow)	Grey relational analysis	Wu et al. (2010)
Simulated medium-strength wastewater	AnMBR	Life cycle costing (LCC), net energy balance(NEB), and life cycle assessment (LCA) methods.	Tools for the Reduction and Assessment of Chemical and Other Environmental Impacts (TRACI)developed by the U.S. Environmental Protection Agency	Smith et al. (2014)
Pulp and paper industry	UASB	CML-IA baseline 4.1, with the addition of water extraction	EP, HTTP, FAETP, and the GWP100 method for GHG emissions	O'Connor et al. (2014)
Agro-industrial wastewaters	UASB / EGSB	–	Multi-criteria analysis (MCA) for sustainable development indicators (SDIs)	Meneses-Jácome et al. (2016)
Starch wastewater	Un-specified anaerobic reactor	CML 2001	GWP and EP	Vera et al. (2015)

[a] AnMBR, anaerobic membrane biological reactor; EGSB, expanded granular sludge bed; HR_AS, high-rate anaerobic system; UASB, upflow anaerobic sludge blanket.

[b] ADP, abiotic depletion potential; AP, acidification potential; EP, eutrophication potential; FAETP, fresh-water aquatic ecotoxicity potential; GWP, global warming potential; HTTP, human toxicity potential; MAETP, marine aquatic ecotoxicity potential; NRE, non-renewable energy; ODP, ozone layer depletion potential; POCP, photochemical oxidation potential; TETP, terrestrial ecotoxicity potential.

no single optimal configuration in terms of the evaluated impact categories: GHG emissions, water recovery, fresh-water aquatic ecotoxicity, and eutrophication discharge impact (O'Connor et al., 2014). Nevertheless, the authors demonstrated that wastewater pre-treatment in the UASB before the activated sludge process resulted in an overall reduction of GHG emissions and eutrophication potential impacts as compared with the non-pre-treated system.

LCA methodology has been also applied as a tool for the analysis of the best available technology for the treatment of wastewater generated in the dairy industry (Georgiopoulou et al., 2008). Among six different systems (activated sludge, high-rate and extended aerated sludge, activated sludge with an anoxic reactor before the aeration basin, an aerated lagoon, and a UASB reactor), AD was the most environmentally friendly and economic option, based on lower GHG emissions and energy requirements. In this work, the combination of economy and environment was already established, something that has been further developed in other works. Thus, it has been accepted that LCA should be combined with technical and economic aspects (Krzeminski et al., 2017; Hospido et al., 2007), taking also into account the social dimension, to obtain more complete indicators of sustainable development and a holistic evaluation of the process (Meneses-Jácome et al., 2016; Smith et al., 2014). In this sense, different approaches can be found linking all these aspects: LCA, net energy balance, and life cycle cost as tools to determine the most cost-effective option among different wastewater treatment alternatives. Thus, Smith et al. found that energy consumption in AnMBR systems was four times higher than energy recovery, and the GWP was higher than for other biological reactors for wastewater treatment. But at the same time, the life cycle costs of the energy recovery system were lower than those of conventional biological wastewater processes (Smith et al., 2014). Wu et al. applied a combination of LCA, a fuzzy recognition system, and a grey relational model to assess the potential environmental impacts related to the operational efficiency and economic benefits of a food-processing industry in China (Wu et al., 2010). The results obtained in this study revealed that an increase in benefits with a reduction in consumption could be obtained by looking in the first place for an improvement in the energy use ratio, which would secondarily ensure higher COD removal.

To summarize all these studies, LCA combined with economic evaluation is a critical tool for the evaluation of anaerobic biological reactors for the treatment of industrial wastewater. This technology exhibits promising characteristics regarding the sustainability of the process, such as GHG emissions mitigation and energy recovery from the methane produced during the operation. However, each particular technology and application should be evaluated, as the balance between benefits and negative impacts is greatly influenced by materials and energy flows.

13.3   ANAEROBIC DIGESTION OF WASTEWATER SLUDGE (ADWWS)

As mentioned previously, the increasing concern to protect the environment from harmful effects of WWTPs has led governments to adopt strict regulations regarding water quality specifications. Thus, as an example, in EU countries, more effective WWTPs have been implemented according to specific regulations, increasing the production of sewage sludge (Pradel et al., 2016; Gourdet et al., 2016). In this context, in EU countries, about 10 million tonnes of dry solids of sludge is produced yearly (Heimersson et al., 2017). According to these figures, it can be inferred that sewage sludge management is a key aspect that must be considered in WWTP, as the disposal of these wastes may represent up to 50% of the operating costs of a WWTP. Sludge is a source of energy, carbon, and nutrient content (mainly nitrogen and phosphorus); on the other hand, it may contain toxic substances, pathogens, heavy metals, and organic contaminants that can be harmful to the environment, and for that reason, it requires specific treatment to recover profitable materials (Buonocore et al., 2016). Traditionally, direct disposal in landfill has been the most frequently used sludge management method, but this approach is in decline, because the concept of sludge as waste has shifted toward the consideration of sludge as an end-product from which energy and/or materials can be recovered (Pradel et al., 2016).

Energy recovery is a key aspect to improve the energy and environmental performance of traditional WWTP (Li et al., 2017). For this purpose, there are different technologies, such as incineration, gasification, pyrolysis, and AD (Cao et al., 2017). ADWWS is extensively used in WWTPs, as it allows the volume of the sludge to be stabilized and reduced (Gourdet et al., 2016). Moreover, and this is the main feature of this treatment, ADWWS fulfills one of the key objectives of wastewater sludge management: the integrated use of the sludge. Thus, as is well known, AD allows the recovery of biogas (which can be used to produce electricity and/or heat) and a digestate with a high content of nutrients (nitrogen and phosphorus), which can be further applied as fertilizer (Mills et al., 2014; Heimersson et al., 2017).

13.3.1   LCA Modeling of Wastewater Sludge Management Systems

The modeling of wastewater sludge management requires a change of perspective regarding the main objective of LCA application. Waste management systems are focused on determining the environmental aspects related to the process of handling the waste. Hence, the functional unit (FU) in these systems is defined in terms of amount (mass or volume) of sludge to be treated (Yoshida et al., 2013).

Usually, LCA studies on sludge management are focused on comparing different technologies, thus supposing that all the functions to generate the sludge are the same in the studied scenarios. In these cases, the most usual approach is to assume the concept of zero burden, also called the *cut-off approach* (Pradel et al., 2016; Schrijvers et al., 2016). This simplification means that the LCA study is only focused on the sludge treatment, excluding all the functions that generate the sludge. This has implications for the allocation of environmental impacts, since it implies that the sludge is free from environmental burden regarding its production. According to Pradel et al. (2016), this approach could be valid if the sludge is considered as a waste with no further treatment. However, if sludge is considered as waste-to-product (energy and/or materials are recovered), the application of the zero burden concept may be questionable.

Waste management systems can be modeled in LCAs by using different approaches regarding the main functions considered. If only one function is considered (for example, to treat the waste), the system could be modeled as a mono-functional system. However, this is not the approach commonly used, since sludge management systems comprise additional functions (energy/material recovery), and consequently, these must be taken into account, so that the system is modeled as a multioutput process.

According to Pradel et al. (2016), multifunction wastewater sludge management systems are usually solved by modifying the system boundaries by expansion or reduction (adding or excluding functions, for instance, those related to co-products). When studying wastewater sludge management systems by LCA, the most usual objective is to compare different technologies, and therefore, the systems are simplified by excluding functions related to water and/or part of sludge lines. On the other hand, substitution is also applied in LCAs of sludge management systems. This approach consists of giving an environmental credit to the process due to avoided impacts of secondary functions with respect to the product/service replaced (Heimersson et al., 2017). Finally, an allocation approach is also included in ISO 14044:2006 as a possible alternative way to solve multifunction systems. This approach consists of allocating the environmental impacts between the different products obtained in the process. This allocation can be performed in terms of product quantity (mass or volume) or by using economic criteria.

13.3.2 Application of LCA to Anaerobic Digestion of Wastewater Sludge (ADWWS)

LCA has been widely used to evaluate the environmental and economic feasibility of AD applied to sludge management. As reported by Yoshida et al. (2013), a large number of works have focused on studying AD from an environmental perspective using LCA.

ADWWS has been traditionally modelled as an input-output system, but this concept has changed toward more complex systems, in which additional features have been included, mainly due to the possibility of recovering valuable materials from wastewater sludge. Although both approaches are different, the most usual way to define the functional unit is in terms of quantity of dry sludge to be managed (Yoshida et al., 2013).

In establishing the boundaries of the system, most LCAs consider ADWWS separately from WWTP. This leads to considering only the processes related to anaerobic digestion, thus not including upstream processes. Figure 13.3 shows the typical processes considered in LCA studies on ADWWS, including all the operations (from sludge pre-treatment to recovery of energy and nutrients). Most LCAs reported in the literature studied ADWWS at mesophilic temperature (35°C) and treating wastewater sludge with total solid content of 3%–6% (Li et al., 2017). The retention time values usually vary within the range of 12–30 days (Mills et al., 2014).

As can be seen in Figure 13.3, the upstream processes (wastewater treatments) are not included; thus, it is considered that the sludge enters AD free of environmental impacts. This implies that the sludge is considered as waste, although energy and/or nutrients are recovered, and actually, the sludge should be considered as waste-to-product. This assumption, called the *zero burden assumption* as already explained, is usually applied in LCAs on ADWWS regardless of the aim of the study (Pradel et al., 2016).

The modeling of ADWWS depends on the consideration of sludge as waste or as waste-to-product (if energy or materials can be recovered). In the first case, ADWWS function would be to treat the waste and consequently, all the impacts would be allocated to the wastewater treatment. However, as shown in Figure 13.3, the LCA of ADWWS is not restricted to the sludge treatment but also includes the

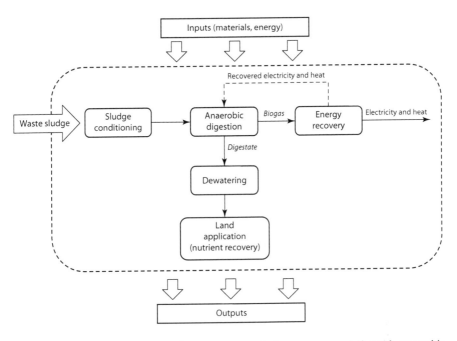

FIGURE 13.3 Scheme process of waste sewage sludge management through anaerobic digestion.

recovery of energy and nutrients. Thus, as reported by Heimersson et al. (2017), the main function of the system would be the sludge treatment and the end-use, whereas the secondary function would be production of biogas and digestate, which in addition can have further functions, such as energy and nutrient recovery. In these cases, the most common approach to solve multifunctionality is by expansion/reduction or by substitution. In the first case, some functions can be included (expansion) or excluded (reduction), as already explained. These approaches are usually applied to make different systems comparable, and consequently, they are most commonly used to compare ADWWS with other technologies (Pradel et al., 2016). In the substitution approach, additional functions are considered (energy and/or nutrient recovery) and the avoidance of impacts related to the replacement of using energy or resources (Heimersson et al., 2017). This approach is the most recommended, according to ISO 14044:2006, when secondary functions (use of biogas and/or digestate) are studied to quantify the avoided emissions related to the substitution of energy and nutrient resources. Finally, the allocation approach is hardly used in LCAs of ADWWS, since the functions of this system (waste treatment, biogas, and organic digestate) cannot be related by a common factor. Regarding economic allocation, great uncertainty exists in determining the profits and cost of the different functions (Heimersson et al., 2017).

Quality of inventory data is essential to ensure the reliability of LCA results. There is a wide variety of ways in which inventory data are obtained for LCAs on ADWW. The most suitable option should take data from real systems (WWTPs) or use experimental data obtained in the laboratory. This is the most common option for work reported in the main reviews (Yoshida et al., 2013). Usually, these inventories are completed with estimates for those data that are difficult to obtain. Data on infrastructure and materials are not always considered within LCI data. Inventories of ADWWS are usually very detailed, and only in very specific studies, such as the report by Hospido et al. (2010), does the lack of reliable data appear as an aspect to be solved.

Although methodologies to assess environmental impacts are under permanent discussion within the LCA community, in the case of LCAs on ADWWS, the most commonly applied is CML methodology in its different versions. Currently, the application of ReCiPe methodology is emerging, as it harmonizes mid- and end-point approaches, and is thus more flexible and uniform than other methodologies (Goedkoop et al., 2009). As mentioned, the variety of functions present in ADWWS systems implies the evaluation of a large number of impact categories. Thus, climate change, acidification potential, eutrophication potential, photochemical oxidation potential, ozone depletion potential, human toxicity, ecotoxicity, and depletion of abiotic resources are commonly evaluated in LCAs on ADWWS.

Table 13.2 summarizes the main LCA methodology aspects of works focused on ADWWS. Herein, only contributions specifically studying LCAs on ADWWS have been considered.

As explained earlier, the usual scheme for studying the feasibility of ADWWS using an LCA approach includes biogas production and use as well as land application of the organic digestate. Although most LCAs present this process scheme, LCA results are not comparable between the different works reported in the literature, mainly due to the different assumptions taken into account.

The first applications of LCA concerning energy production from ADWWS date from the 2000s; they were not specific to this treatment but were included in comparative studies of wastewater sludge management at the WWTP (Bridle and Skrypski-Mantele 2000; Heimersson et al., 2017; Suh and Rousseaux 2002). These studies were not full LCAs, but they were the first to use the life cycle approach to sludge management. These works were basically focused on studying the substitution of the energy required in ADWWS by the energy recovered from the biogas, mainly in terms of electricity and heating. Moreover, these works pointed to the need to recover energy and/or materials from wastewater sludge to provide sustainable sludge management options.

Further LCA studies on ADWWS incorporated more detailed inventory data, thus obtaining more realistic results. The main objective of these works was the comparison between different wastewater sludge treatments by quantifying environmental impacts. In these works, most of the environmental categories mentioned in this chapter were included and quantified. Thus, Yoshida et al. (2013) reviewed LCAs on ADWWS and further use of the organic digestate and compared the obtained results with other sludge treatment schemes (mainly thermal processes). The obtained results showed that AD and further land applications, including energy recovery from biogas and application of organic sludge, reduced GHG emissions relative to other wastewater sludge treatment schemes.

Another issue of interest regarding LCAs on ADWWS is the study of pre-treatment of wastewater sludge before anaerobic digestion. Carballa et al. (2011) pointed

TABLE 13.2
LCA Studies on ADWWS Treatment

System Function	Functional Unit	Inventory Data	LCIA Method	LCIA Categories	Reference
Anaerobic digestion, biogas, and digestate use	1 dry matter ton of mixed sludge	Operational and literature data	Characterization factors reported by different authors	EP, ODP, GWP, AP, POCP, ADP, HTP	Hospido et al. (2005)
Pre-treatment of sludge, anaerobic digestion, digestate use	2 dry tonnes of sludge	Literature and calculated data	Not specified	GWP, HTP, TETP	Peters and Rowley (2009)
Anaerobic digestion, biogas, and digestate use	10 L of mixed sludge	Experimental and calculated data	CML 2 baseline 2000	EP, GWP, HTP, TTP	Hospido et al. (2010)
Pre-treatment of sludge, anaerobic digestion, and digestate use	10 L of mixed sludge	Experimental and calculated data	CML 2 baseline 2000	ADP, EP, GWP, HTP, TTP	Carballa et al. (2011)
Energy recovery from biogas and digested sludge	1 dry matter ton of sludge	Literature and calculated data	CML 2001—Nov. 2010	GWP, POCP, EP, AP, ADP	Mills et al. (2014)
Pre-treatment of sludge, anaerobic digestion, biogas, and digestate use	1 dry matter ton of sludge	Literature and calculated data	ReCiPe v1.08/ Europe	CC, TA, FE, HT, IR	Gourdet et al. (2017)
Anaerobic digestion, biogas, and digestate use	20 tons of thickened sludge	Literature and calculated data	CML 2 baseline 2000	AP, CC, ADP, POP, EP, HTP, ETP	Li et al. (2017)

ADP, abiotic depletion potential; AP, acidification potential; CC, climate change; EP, eutrophication potential; FAETP, freshwater aquatic ecotoxicity potential; FE, freshwater eutrophication; GWP, global warming potential; HT, human toxicity; HTTP, human toxicity potential; IR, ionizing radiation; MAETP, marine aquatic ecotoxicity potential; NRE, non-renewable energy; ODP, ozone layer depletion potential; POCP, photochemical oxidation potential; TA, terrestrial acidification; TETP, terrestrial ecotoxicity potential.

out that these treatments cannot be ignored in LCAs, identifying that mechanical and chemical treatments improve the environmental performance of ADWWS. Gourdet et al. (2016) studied LCAs on ADWWS, including a complete study of all operations related to sludge conditioning (thickening and dewatering as well as the treatment of the resulting liquid waste). The authors pointed out that thickening, dewatering, storage, and agricultural spreading presented the largest environmental impacts. For that reason, the increase of AD efficiency, the reduction of chemicals in sludge treatment, and the improvement of treatments related to waste liquids from sludge thickening were identified as key aspects to improve the environmental feasibility of ADWWS. Along these lines, Li et al. (2017) studied different options to increase the digestion of organic matter present in the wastewater sludge. For that purpose, mesophilic (conventional) and thermophilic AD were compared considering a system similar to that depicted in Figure 13.3. The main conclusion was that thermophilic conditions present environmental advantages, reducing the environmental impacts due to more biogas being obtained, and therefore, the avoided impacts are higher because of the replacement of grid electricity.

As well as the use of biogas, the end-use of organic digestate must be taken into account to evaluate the environmental sustainability of ADWWS using the LCA approach. The digested sludge is a source of nutrients and energy, thus allowing different end-uses. For this purpose, the use of the digested sludge as fertilizer (nutrient recovery) or in a cement kiln (energy recovery) has been studied in the literature (Murray et al., 2008; Peters and Rowley 2009). From both approaches, fertilizer is the most common application in a typical ADWWS system, since the replacement of fertilizers by digested solids avoids impacts (in terms of emissions and energy consumption) to a larger extent than energy applications. Nevertheless, the agricultural application of digested sludge must be undertaken with caution, as wastewater sludge can contain pollutants, whose environmental impacts must be evaluated, as reported by Hospido et al., (2010). This work studied the environmental consequences of applying digested sludge in agriculture by considering the presence of not only nutrients (nitrogen and phosphorus) but also a number of micropollutants (mainly heavy metals, pharmaceuticals, and personal care products) that can be present in the wastewater sludge. The authors identified nutrient-related emissions as the main environmental impacts, heavy metal emissions to the soil being the most important contributors to toxicity.

As described earlier, biogas has been considered as the main product for obtaining energy. However, Mills et al. (2014) proposed the possibility of obtaining energy from both the biogas and the digested sludge. For that purpose, the biogas obtained by ADWWS is burnt in a combined heat and power (CHP) system, whereas the digested sludge is dried and used as fuel, thus replacing coal. Although this scheme is superior to the conventional approach, regulations concerning the combustion of dry sludge could be an important drawback.

REFERENCES

Abbasi, T., S. M. Tauseef, and S. A. Abbasi. 2012. *Biogas Energy* 2: 11–22. doi:10.1007/978-1-4614-1040-9.

Appels, L., J. Baeyens, J. Degrève, and R. Dewil. 2008. Principles and potential of the anaerobic digestion of waste-activated sludge. *Progress in Energy and Combustion Science* 34 (6): 755–81. doi:10.1016/j.pecs.2008.06.002.

Barrera, E. L., E. Rosa, H. Spanjers, O. Romero, S. De Meester, and J. Dewulf. 2015. Modeling the anaerobic digestion of cane-molasses vinasse: Extension of the anaerobic digestion model No. 1 (ADM1) with sulfate reduction for a very high strength and sulfate rich wastewater. *Water Research* 71: 42–54. doi:10.1016/j.watres.2014.12.026.

Batstone, D. J., T. Hülsen, C. M. Mehta, and J. Keller. 2015a. Platforms for energy and nutrient recovery from domestic wastewater: A review. *Chemosphere* 140: 2–11. doi:10.1016/j. chemosphere.2014.10.021.

Batstone, D. J., D. Puyol, X. Flores-Alsina, and J. Rodríguez. 2015b. Mathematical modelling of anaerobic digestion processes: Applications and future needs. *Reviews in Environmental Science and Biotechnology* 14 (4): 595–613. doi:10.1007/s11157-015-9376-4.

Batstone, D. J., and B. Virdis. 2014. The role of anaerobic digestion in the emerging energy economy. *Current Opinion in Biotechnology* 27: 142–9. doi:10.1016/j.copbio.2014.01.013.

Bialek, K., D. Cysneiros, and V. O'Flaherty. 2014. Hydrolysis, acidification and methanogenesis during low-temperature anaerobic digestion of dilute dairy wastewater in an inverted fluidised bioreactor. *Applied Microbiology and Biotechnology* 98 (20): 8737–50. doi:10.1007/s00253-014-5864-7.

Bridle, T. and S. Skrypski-Mantele. 2000. Assessment of sludge reuse options: A life-cycle approach. *Water Science and Technology* 41 (8): 131–5.

Buonocore, E., S. Mellino, G. De Angelis, G. Liu, and S. Ulgiati. 2016. Life cycle assessment indicators of urban wastewater and sewage sludge treatment. *Ecological Indicators*. In press. 12. doi:10.1016/j.ecolind.2016.04.047.

Cao, Y., M. C. M. van Loosdrecht, and G. T. Daigger. 2017. Mainstream partial nitritation–anammox in municipal wastewater treatment: Status, bottlenecks, and further studies. *Applied Microbiology and Biotechnology* 101: 1365–83. doi:10.1007/s00253-016-8058-7.

Carballa, M., C. Duran, and A. Hospido. 2011. Should we pretreat solid waste prior to anaerobic digestion? An assessment of its environmental cost. *Environmental Science & Technology* 45: 10306–14. doi:10.1021/es201866u.

Carlsson, M., A. Lagerkvist, and F. Morgan-Sagastume. 2012. The effects of substrate pretreatment on anaerobic digestion systems: A review. *Waste Management* 32 (9): 1634–50. doi:10.1016/j.wasman.2012.04.016.

Chernicharo, C. A. L., J. B. van Lier, A. Noyola, and T. Bressani Ribeiro. 2015. Anaerobic sewage treatment: State of the art, constraints and challenges. *Reviews in Environmental Science and Biotechnology* 14 (4): 649–79. doi:10.1007/s11157-015-9377-3.

Chitapornpan, S., C. Chiemchaisri, W. Chiemchaisri, R. Honda, and K. Yamamoto. 2013. Organic carbon recovery and photosynthetic bacteria population in an anaerobic membrane photo-bioreactor treating food processing wastewater. *Bioresource Technology* 141: 65–74. doi:10.1016/j.biortech.2013.02.048.

Corominas, Ll., J. Foley, J. S. Guest, A. Hospido, H. F. Larsen, S. Morera, and A. Shaw. 2013. Life cycle assessment applied to wastewater treatment: State of the art. *Water Research* 47 (15): 5480–92. doi:10.1016/j.watres.2013.06.049.

De la Rubia, M. A., V. Riau, F. Raposo, and R. Borja. 2012. Thermophilic anaerobic digestion of sewage sludge: Focus on the influence of the start-up. A review. *Critical Reviews in Biotechnology* 33: 1–13. doi:10.3109/07388551.2012.726962.

Dvořák, L., M. Gómez, J. Dolina, and A. Černín. 2015. Anaerobic membrane bioreactors—a mini review with emphasis on industrial wastewater treatment: Applications, limitations and perspectives. *Desalination and Water Treatment* 3994: 1–15. doi:10.1080/19443994.2015.1100879.

Flores-Alsina, X., C. Kazadi Mbamba, K. Solon, D. Vrecko, S. Tait, D. J. Batstone, U. Jeppsson, and K. V. Gernaey. 2015. A plant-wide aqueous phase chemistry module describing pH variations and ion speciation/pairing in wastewater treatment process models. *Water Research* 85: 255–65. doi:10.1016/j.watres.2015.07.014.

Flores-Alsina, X., K. Solon, C. Kazadi Mbamba, S. Tait, K. V. Gernaey, U. Jeppsson, and D. J. Batstone. 2016. Modelling phosphorus (P), sulfur (S) and iron (Fe) interactions for dynamic simulations of anaerobic digestion processes. *Water Research* 95 (1): 370–82. doi:10.1016/j.watres.2016.03.012.

Foley, J. M., R. A. Rozendal, C. K. Hertle, P. A. Lant, and K. Rabaey. 2010. Life cycle assessment of high-rate anaerobic treatment, microbial fuel cells, and microbial electrolysis cells. *Environmental Science & Technology* 44 (9): 3629–37. doi:10.1021/es100125h.

Fruergaard, T., T. Astrup, and T. Ekvall. 2009. Energy use and recovery in waste management and implications for accounting of greenhouse gases and global warming contributions. *Waste Management & Research : The Journal of the International Solid Wastes and Public Cleansing Association, ISWA* 27 (8): 724–37. doi:10.1177/0734242X09345276.

Georgiopoulou, M., K. Abeliotis, M. Kornaros, and G. Lyberatos. 2008. Selection of the best available technology for industrial wastewater treatment based on environmental evaluation of alternative treatment technologies: The case of milk industry. *Fresenius Environmental Bulletin* 17 (1): 111–21.

Ghosh, S., U. K. Dairkee, R. Chowdhury, and P. Bhattacharya. 2017. Hydrogen from food processing wastes via photofermentation using purple non-sulfur bacteria (PNSB)—A review. *Energy Conversion and Management* 141: 299–314. doi:10.1016/j.enconman.2016.09.001.

Goedkoop, M., R. Heijungs, M. Huijbregts, A. De Schryver, J. Struijs, and R. Van Zelm. 2009. ReCiPe 2008 First Edition Report I: Characterisation.

Gourdet, C., R. Girault, S. Berthault, M. Richard, J. Tosoni, and M. Pradel. 2016. In quest of environmental hotspots of sewage sludge treatment combining anaerobic digestion and mechanical dewatering: A life cycle assessment approach. *Journal of Cleaner Production* 143: 1123–36. doi:10.1016/j.jclepro.2016.12.007.

Gregson, N., M. Crang, S. Fuller, and H. Holmes. 2015. Interrogating the circular economy: The moral economy of resource recovery in the EU. *Economy and Society* 44 (2): 218–43. doi:10.1080/03085147.2015.1013353.

Gu, Y., Y. Li, X. Li, P. Luo, H. Wang, Z. P. Robinson, X. Wang, and J. Wu. 2017. The feasibility and challenges of energy self-sufficient wastewater treatment plants. *Applied Energy* 204: 1463–1475. doi:10.1016/j.apenergy.2017.02.069.

Heimersson, S., M. Svanström, C. Cederberg, and G. Peters. 2017. Improved life cycle modelling of benefits from sewage sludge anaerobic digestion and land application. *Resources, Conservation and Recycling* 122: 126–34. doi:10.1016/j.resconrec.2017.01.016.

Hospido, A., M. Carballa, M. T. Moreira, F. Omil, J. M. Lema, and G. Feijoo. 2010. Environmental assessment of anaerobically digested sludge reuse in agriculture: Potential impacts of emerging micropollutants. *Water Research* 44: 3225–3233. doi: 10.1016/j.watres.2010.03.004.

Hospido, A., M. T. Moreira, and G. Feijoo. 2007. A comparison of municipal wastewater treatment plants for big centres of population in Galicia (Spain). *The International Journal of Life Cycle Assessment* 13 (1): 57–64. doi: http://dx.doi.org/10.1065/lca2007.03.314.

Hospido, A., M. T. Moreira, M. Martín, M. Rigola, and G. Feijoo. 2005. Environmental evaluation of different treatment processes for sludge from urban wastewater treatments: anaerobic digestion versus thermal processes. *The International Journal of Life Cycle Assessment* 10 (5): 336–345. doi: http://dx.doi.org/10.1065/lca2005.05.210.

Hospido, A., I. Sanchez, G. Rodriguez-Garcia, A. Iglesias, D. Buntner, R. Reif, M. T. Moreira, and G. Feijoo. 2012. Are all membrane reactors equal from an environmental point of view? *Desalination* 285: 263–70. doi:10.1016/j.desal.2011.10.011.

Hülsen, T., D. J. Batstone, and J. Keller. 2014. Phototrophic bacteria for nutrient recovery from domestic wastewater. *Water Research* 50 (0): 18–26. doi:10.1016/j.watres.2013.10.051.

Ioannou-Ttofa, L., S. Foteinis, E. Chatzisymeon, and D. Fatta-Kassinos. 2016. The environmental footprint of a membrane bioreactor treatment process through life cycle analysis. *Science of the Total Environment* 568: 306–18. doi:10.1016/j.scitotenv.2016.06.032.

Jaouad, Y., N. Ouazzani, L. Mandi, M. Villain, and B. Marrot. 2016. Biodegradation of olive mill wastewater in a membrane bioreactor: Acclimation of the biomass and constraints. *Desalination and Water Treatment* 57 (18): 8109–18.

Kazadi Mbamba, C., S. Tait, X. Flores-Alsina, and D. J. Batstone. 2015. A systematic study of multiple minerals precipitation modelling in wastewater treatment. *Water Research* 85: 359–70. doi:10.1016/j.watres.2015.08.041.

Krzeminski, P., L. Leverette, S. Malamis, and E. Katsou. 2017. Membrane bioreactors—A review on recent developments in energy reduction, fouling control, novel configurations, LCA and market prospects. *Journal of Membrane Science* 527: 207–227. https://doi.org/10.1016/j.memsci.2016.12.010.

Li, H., C. Jin, Z. Zhang, I. O'Hara, and S. Mundree. 2017. Environmental and economic life cycle assessment of energy recovery from sewage sludge through different anaerobic digestion pathways. *Energy* 126: 649–57. doi:10.1016/j.energy.2017.03.068.

Liao, B.-Q., J. T. Kraemer, and D. M. Bagley. 2006. Anaerobic membrane bioreactors: Applications and research directions. *Critical Reviews in Environmental Science and Technology* 36: 489–530. doi:10.1080/10643380600678146.

McCarty, P. L., J. Bae, and J. Kim. 2011. Domestic wastewater treatment as a net energy producer—Can this be achieved? *Environmental Science & Technology* 45 (17): 7100–7106. doi:10.1021/es2014264.

Martinez-Sosa, D., B. Helmreich, T. Netter, S. Paris, F. Bischof, and H. Horn. 2011. Anaerobic submerged membrane bioreactor (AnSMBR) for municipal wastewater treatment under mesophilic and psychrophilic temperature conditions. *Bioresource Technology* 102 (22): 10377–85. doi:10.1016/j.biortech.2011.09.012.

Massara, T. M, O. T. Komesli, O. Sozudogru, S. Komesli, and E. Katsou. 2016. Benchmarking of low environmental footprint biological processes for the treatment of industrial waste streams. *CYPRUS2016 4th International Conference on Sustainable Solid Waste Management* 44 (0): 1–22.

Masullo, A. 2017. Organic wastes management in a circular economy approach: Rebuilding the link between urban and rural areas. *Ecological Engineering* 101: 84–90. https://doi.org/10.1016/j.ecoleng.2017.01.005.

Matassa, S., N. Boon, and W. Verstraete. 2015. Resource recovery from used water: The manufacturing abilities of hydrogen-oxidizing bacteria. *Water Research* 68: 467–78. doi:10.1016/j.watres.2014.10.028.

Meneses-Jácome, A., R. Diaz-Chavez, H. I. Velásquez-Arredondo, D. L. Cárdenas-Chávez, R. Parra, and A. A. Ruiz-Colorado. 2016. Sustainable energy from agro-industrial wastewaters in Latin-America. *Renewable and Sustainable Energy Reviews* 56: 1249–62. doi:10.1016/j.rser.2015.12.036.

Mills, N., P. Pearce, J. Farrow, R. B. Thorpe, and N. F. Kirkby. 2014. Environmental & economic life cycle assessment of current & future sewage sludge to energy technologies. *Waste Management* 34 (1): 185–95. https://doi.org/10.1016/j.wasman.2013.08.024.

Murray, A., A. Horvath, and K. L. Nelson. 2008. Hybrid life-cycle environmental and cost inventory of sewage sludge treatment and end-use scenarios : A case study from China. Environmental Science & Technology 42 (9): 3163–9. doi:10.1021/es702256w.

Nishio, N., and Y. Nakashimada. 2007. Recent development of anaerobic digestion processes for energy recovery from wastes. *Journal of Bioscience and Bioengineering* 103 (2): 105–12. doi:10.1263/jbb.103.105.

O'Connor, M., G. Garnier, and W. Batchelor. 2014. Life cycle assessment comparison of industrial effluent management strategies. *Journal of Cleaner Production* 79: 168–81. doi:10.1016/j.jclepro.2014.05.066.

Pan, S. Y., M. A. Du, I. T. Huang, I. H. Liu, E. E. Chang, and P. C. Chiang. 2014. Strategies on implementation of waste-to-energy (WTE) supply chain for circular economy system: A review. *Journal of Cleaner Production* 108: 409–21. doi:10.1016/j.jclepro.2015.06.124.

Peters, G. M., and H. V Rowley. 2009. Environmental comparison of biosolids management systems using life cycle assessment. *Environmental Science & Technology* 43· (8): 2674–9. doi:10.1021/es802677t.

Pikaar, I., K. R. Sharma, S. Hu, W. Gernjak, J. Keller, and Z. Yuan. 2014. Reducing sewer corrosion through integrated urban water management. *Science* 345 (6198): 812–14. doi:10.1126/science.1251418.

Pintilie, L., C. M. Torres, C. Teodosiu, and F. Castells. 2016. Urban wastewater reclamation for industrial reuse: An LCA case study. *Journal of Cleaner Production* 139: 1–14. doi:10.1016/j.jclepro.2016.07.209.

Pradel, M., L. Aissani, J. Villot, J. C. Baudez, and V. Laforest. 2016. From waste to added value product: Towards a paradigm shift in life cycle assessment applied to wastewater sludge—a review. *Journal of Cleaner Production* 131: 60–75. doi:10.1016/j.jclepro.2016.05.076.

Pretel, R., A. Robles, M. V. Ruano, A. Seco, and J. Ferrer. 2013. Environmental impact of submerged anaerobic MBR (SAnMBR) technology used to treat urban wastewater at different temperatures. *Bioresource Technology* 149: 532–40. doi:10.1016/j.biortech.2013.09.060.

Pretel, R., A. Robles, M. V. Ruano, A. Seco, and J. Ferrer. 2016. Economic and environmental sustainability of submerged anaerobic MBR-based (AnMBR-based) technology as compared to aerobic-based technologies for moderate-/high-loaded urban wastewater treatment. *Journal of Environmental Management* 166: 45–54. doi:10.1016/j.jenvman.2015.10.004.

Puyol, D., E. Barry, T. Huelsen, and D. J. Batstone. 2017a. A mechanistic model for anaerobic phototrophs in domestic wastewater applications: Photo-anaerobic model (PAnM). *Water Research* 116: 241–53. doi:10.1016/j.watres.2017.03.022.

Puyol, D., D. J. Batstone, T. Hülsen, S. Astals, M. Peces, and O. Jens. 2017b. Resource recovery from wastewater by biological technologies: Opportunities, challenges and prospects. *Frontiers in Microbiology* 7: 1–23. doi:10.3389/fmicb.2016.02106.

Renou, S., J. S. Thomas, E. Aoustin, and M. N. Pons. 2008. Influence of impact assessment methods in wastewater treatment LCA. *Journal of Cleaner Production* 16 (10): 1098–105. doi:10.1016/j.jclepro.2007.06.003.

Schrijvers, D. L., P. Loubet, and G. Sonnemann. 2016. Developing a systematic framework for consistent allocation in LCA. *International Journal of Life Cycle Assessment* 21 (7): 976–93. doi:10.1007/s11367-016-1063-3.

Skouteris, G., D. Hermosilla, P. López, C. Negro, and Á. Blanco. 2012. Anaerobic membrane bioreactors for wastewater treatment: A review. *Chemical Engineering Journal* 198–199: 138–48. doi:10.1016/j.cej.2012.05.070.

Smith, A. L., L. B. Stadler, L. Cao, N. G. Love, L. Raskin, and J. Steven. 2014. Navigating wastewater energy recovery strategies: A life cycle comparison of wastewater energy recovery strategies: Anaerobic membrane bioreactor and high rate activated sludge with anaerobic digestion. *Environmental Science & Technology* 48: 5972–81.

Spirito, C. M., H. Richter, K. Rabaey, A. J. M. Stams, and L. T. Angenent. 2014. Chain elongation in anaerobic reactor microbiomes to recover resources from waste. *Current Opinion in Biotechnology* 27: 115–22. doi:10.1016/j.copbio.2014.01.003.

Suh, Y. J., and P. Rousseaux. 2002. An LCA of alternative wastewater sludge treatment scenarios. *Resources, Conservation and Recycling* 35: 191–200.

Vaneeckhaute, C., V. Lebuf, E. Michels, E. Belia, P. A. Vanrolleghem, F. M. G. Tack, and E. Meers. 2017. Nutrient recovery from digestate: Systematic technology review and product classification. *Waste and Biomass Valorization*, 1–20. doi: 10.1007/s12649-016-9642-x.

van Lier, J. B., F. P. Zee, C. T. M. J. Frijters, and M. E. Ersahin. 2015. Celebrating 40 years anaerobic sludge bed reactors for industrial wastewater treatment. *Reviews in Environmental Science and Bio/Technology* 14 (4): 681–702. doi:10.1007/s11157-015-9375-5.

van Loosdrecht, M. C. M., and D. Brdjanovic. 2014. Anticipating the next century of wastewater treatment. *Science* 344 (6191): 1452–3. doi:10.1126/science.1255183.

Vera, L., W. Sun, M. Iftikhar, and J. Liu. 2015. LCA based comparative study of a microbial oil production starch wastewater treatment plant and its improvements with the combination of CHP system in Shandong, China. *Resources, Conservation and Recycling* 96: 1–10. doi:10.1016/j.resconrec.2014.09.013.

Verstraete, W., P. Van de Caveye, and V. Diamantis. 2009. Maximum use of resources present in domestic "used water." *Bioresource Technology* 100 (23): 5537–45. doi:10.1016/j.biortech.2009.05.047.

Verstraete, W., and S. E. Vlaeminck. 2011. ZeroWasteWater: Short-cycling of wastewater resources for sustainable cities of the future. *International Journal of Sustainable Development & World Ecology* 18 (3): 253–64. doi:10.1080/13504509.2011.570804.

Wang, Q., W. Wei, Y. Gong, Q. Yu, Q. Li, J. Sun, and Z. Yuan. 2017. Technologies for reducing sludge production in wastewater treatment plants: State of the art. *Science of the Total Environment* 587–588: 510–21. doi:10.1016/j.scitotenv.2017.02.203.

Wei, C.-H., M. Harb, G. Amy, P.-Y. Hong, and T. Leiknes. 2014. Sustainable organic loading rate and energy recovery potential of mesophilic anaerobic membrane bioreactor for municipal wastewater treatment. *Bioresource Technology* 166 (0): 326–34. doi:10.1016/j.biortech.2014.05.053.

Weiland, P. 2010. Biogas production: Current state and perspectives. *Applied Microbiology and Biotechnology* 85 (4): 849–60. doi:10.1007/s00253-009-2246-7.

Winkler, M. K. H., R. Kleerebezem, and M. C. M. Van Loosdrecht. 2012. Integration of anammox into the aerobic granular sludge process for main stream wastewater treatment at ambient temperatures. *Water Research* 46 (1): 136–44. doi:10.1016/j.watres.2011.10.034.

Wu, J.-G., X.-Y. Meng, X.-M. Liu, X.-W. Liu, Z.-X. Zheng, D.-Q. Xu, G.-P. Sheng, and H.-Q. Yu. 2010. Life cycle assessment of a wastewater treatment plant focused on material and energy flows. *Environmental Management* 46: 610–17. doi:10.1007/s00267-010-9497-z.

Yoshida, H., T. H. Christensen, and C. Scheutz. 2013. Life cycle assessment of sewage sludge management: A review. *Waste Management & Research* 31 (11): 1083–1101.

Zeeman, G., and G. Lettinga. 1999. The role of anaerobic digestion of domestic sewage in closing the water and nutrient cycle at community level. *Water Science and Technology* 39 (5): 187–94. doi:10.1016/S0273-1223(99)00101-8.

Index

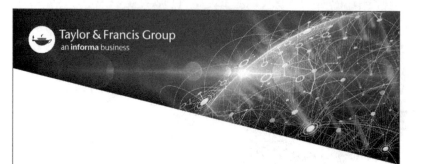